Starting from Scratch

This book compares the beginning of symbolic thought in human infancy with that of our close primate relatives, the chimpanzees. The author investigates the precursors of symbolism by studying the actions and interactions of a small group of these intelligent, non-human primates who live in Singapore Zoo.

Drawing upon his years of detailed observations, Matthews offers an in-depth analysis and interpretation of chimp behaviour to present an unprecedented account of the beginnings of symbolic thought. The book shows that the actions the chimpanzees perform have structural and semantic similarities with the actions of emergent expression and representation we find in human infancy. Of great importance is the finding that chimpanzee mark-making activity is not an artefact of human interference, but part of chimpanzee culture. Young chimpanzees seem to be introduced to acts of pretence and imagination by older and more experienced ones and *taught* the rudiments of expression, representation and symbolism.

The implications for our understanding of symbolism, language, art and education are enormous, as are those about our origins and our place within nature. The book is written in an accessible style for both specialist and non-specialist readers, and illustrated with the author's drawings and photographs.

John Matthews is an artist and educator. He has taught all age groups from nursery children to University level, at institutions including Goldsmiths Art School and Goldsmiths Faculty of Education at the University of London, and the National Institute of Education, Nanyang Technological University, Singapore. His main research area has been early symbolism in very young children; more recently he has become involved in studying the precursors of symbolic thought in non-human primates.

Starting from Scratch

The origin and development of expression, representation and symbolism in human and non-human primates

John Matthews

Psychology Press
Taylor & Francis Group
LONDON AND NEW YORK

Published in 2011
by Psychology Press
27 Church Road, Hove, East Sussex BN3 2FA

Simultaneously published in the USA and Canada
by Psychology Press
711 Third Avenue, New York, NY 10017

First issued in paperback 2015

Psychology Press is an imprint of the Taylor & Francis Group, an informa business

Typeset in Times by RefineCatch Limited, Bungay, Suffolk

Cover design by Design Deluxe Ltd

This publication has been produced with paper manufactured to strict environmental standards and with pulp derived from sustainable forests.

British Library Cataloguing in Publication Data
A catalogue record for this book is available
from the British Library

Library of Congress Cataloging-in-Publication Data
Matthews, John, 1945 Nov. 11-
Starting from scratch: the origin and development of expression, representation, and symbolisation in human and non-human primates/ John Matthews.
p. cm.
ISBN 978–1–84169–689–8 (hb)
1. Mental representation. 2. Perception. 3. Symbolism. 4. Primates. I. Title.
BF316.6.M38 2010
156'.33—dc22

2010015925

ISBN13: 978-1-138-87761-0 (pbk)
ISBN13: 978-1-84169-689-8 (hbk)

To the humans

In the beginning is the gestural act (inherently grammatical).
Roger Fouts, 2007

I wonder what a chimpanzee would say to this.
Charles Darwin

Contents

List of figures

Foreword

The title of the book, *Starting from Scratch*, has multiple meanings. It is about a 3-year study of a group of chimpanzees in Singapore Zoo. It is about the precursor actions which form the beginnings of expressive, representational and symbolic thought. It is a comparison between the development of symbolisation in both human infancy and in chimpanzees. It suggests one of the ways in which art, writing and counting might have started in evolution, from patterns of movement performed by our pre-human ancestors, perhaps 6 million years ago or even earlier.

It is also a metaphor which refers to this author's own re-evaluation of a lifetime's research into the origin and development of semiotic thinking in modern, human children. In this earlier work I had focused upon the example of children's drawing.

After studying this for about 35 years, I thought I was entitled to have a rest from it. But something happened. Suddenly, I rediscovered the drawings and paintings of the Great Apes. A friend lent me a copy of Thierry Lenain's book *Monkey Painting* (Lenain, 1997). The first thing that struck me was how similar the apes' drawings were to those of very young human infants.

Now, I want to be very clear about this: I don't just mean that they *reminded* me of the drawings of children, I mean *they had the same structure and organisation as those of the earliest drawings made in human infancy and early childhood.* It looked to me as if these apes were concerned with the same issues as those which confronted human infants in their drawing.

Of course I knew of the artwork of apes from the 1960s when I was a young art student but (I am ashamed to admit) I had not taken it seriously. I won't try to excuse myself for this mistake but perhaps I had neglected this work because, after a brief flurry of media interest in it, it was soon scorned by many critics. Lack of good theory on children's drawing and the disastrous error of likening ape art to modernist, abstract art, compounded by the demeaning media circus act animal painting soon became did not help serious scholars.

A very similar fate befell ape language studies; no sooner had some us become interested in these, than critics dismissed the evidence for various reasons (which we will consider later).

Actually, in his book, Professor Lenain is also rather sceptical about ape art. He does not feel it has anything to tell us about human art. However, his fascinating analysis and painstaking observations of ape behaviours gripped my attention. It made me carefully re-read the work of Desmond Morris and look at the drawings and paintings of the chimpanzee Congo (Morris, 1962). From Dr Morris's work, I started to read avidly everything I could about primatology, including studies of non-human primates' emergent semiotic capacities and about the apes' use of signing. I read the arguments about whether or not apes had the capacity for language. I read up everything I could about ape 'art'. I soon realised that I had to re-think all my work on the human child's development of expression, representation and symbolisation.

Introduction

The relationship between emergent representation in human and non-human primates

Studies about the origin and development of visual representation in human infancy and childhood (including my own) seriously underestimate the importance of the development of drawing and painting in non-human primates. There is an important relationship, barely investigated, between the development of symbolisation in very young children and that of the ape.

This writer is indebted to many pioneer thinkers in primatology including Schiller (1951), Kohler (1925), Khots (1935), and Lenain (1997). The work of Roger Fouts (1997) has been of vital importance, as has that of Sue Savage-Rumbaugh (Savage-Rumbaugh, Shanker, & Taylor, 1998). Betsy Herrelko's new work on chimpanzees and their use and understanding of videographic technology is of great importance too (Herrelko, 2010).

Desmond Morris's work on the art of the chimpanzee, Congo (Morris, 1962), has especially been a source of inspiration. A re-reading of his work in the light of new studies about the origin and development of representational thought in human infancy reveals that there are deep similarities between the developmental trajectory in human infancy of emergent expressive and representational structure (including that found in drawing, painting and play) and that of non-human primates, especially chimpanzees.

At the time when Dr Morris was writing, there was a scarcity of research work on children's development, and that which existed was often seriously flawed. Though some of the writers were very influential pioneers and advocates of children's art (for example, Lowenfeld, 1955), they usually couched development in terms of a hierarchical stage theory, moving from supposed inferior to successively superior forms of representation. The beginnings of drawing in early mark-making received scant attention – a failure which has had continuous repercussions on the way we understand and provide for the growth of semiotic thought.

Perhaps it was the paucity of research on the development of children's art which obliged Morris to refer to Rhoda Kellogg's famous (1955) study. Unfortunately, Kellogg's work does not help illuminate or support Morris's ideas in any useful way. This is because Kellogg's study is not derived from direct observation of the drawing processes over time but reliant on the analysis of the end products alone – the finished drawings and paintings.

Kellogg projects her own ideas into these finished artefacts and in doing so forces children's drawing into misleading categories, turning it into a static vocabulary which has nothing to tell us about development. Kellogg's thesis is not really a *developmental* theory at all.

However, this flaw should in no way detract from the importance of Morris's revolutionary thesis. His 1962 book is one of the major works of the twentieth century. Recent work on the development of children's drawing reveals that the similarity between the development of anthropoid paintings and drawings and those of very young children is not a superficial one. Relatively new work on the origin and development of children's drawing reveals semantic and organisational characteristics from the very first markings made in infancy (Athey, 1990/2007; Matthews, 1999, 2003, 2004, 2010; Maurer & Riboni, 2010; Stamatopoulou, 2009; Wolf, 1983; Wolf & Fucigna, 1983).

This recent work shows that children's earliest drawing is not meaningless, motoric marking but is an expression of emergent concepts and rich in representational, logicomathematical and linguistic thought. This means that when very young children are engaged in early mark-making they are simultaneously generating concepts and expressive ideas. These emergent concepts are echoed and reiterated in behaviours mostly disregarded by scientists as extraneous to drawing proper. Yet such actions turn out to be an important part of emergent semiotic thinking.

Studies of non-human primate drawing and painting have to be coupled with other evidence from primatologists, who, after years of hard work, showed that chimpanzees are capable of acquiring American Sign Language or are able to use other systems for using and linking signs together in a meaningful and grammatical way (Fouts, 1997; Gardner & Gardner, 1966; Savage-Rumbaugh & Lewin, 1994; Savage-Rumbaugh et al., 1998). This work too has come under harsh criticism yet it seems to me to offer evidence that cannot be ignored; that is, chimpanzees and bonobos are on the borders of symbolisation. If this is so, then the exciting possibility arises that non-human primate art also has conceptual layers of meaning. If chimpanzees and bonobos have the potential to converse linguistically (albeit in signs), and if, as some studies suggest (Matthews, 1999, 2010; Maurer & Riboni, 2010; Stamatopoulou, 2009; Wolf, 1983), pre-linguistic human children are producing semantically organised drawings (not really 'scribbles', as was once thought), then does this not lead to the intriguing possibility that non-human primate drawing and painting (similar in visual and dynamic structure as that of human infants) also has meaning? If human children are engaged in emergent representation before pictorialisation (Wolf & Fucigna, 1983), could not apes also be so engaged? Perhaps, like human children, their drawing is not merely motoric and meaningless after all, but is the embodiment of emergent concepts.

Close encounters: Observations of chimpanzees and human infants

Using detailed, naturalistic observations of both human children's development and that of chimpanzees, the present book describes deep patterns of action and meaning suggestive of structural and semantic similarities across some primate species.

If we put to one side some traditional but wholly inadequate models of children's development in drawing and adopt a new one which illuminates the infrastructural, the investigative, the representational, the expressive and the symbolic aspects of actions which occur during the sensorimotoric phase (activities often disregarded as mere 'messing about'), then the possibility emerges of a different approach to the construction of meaning and one which simultaneously undermines the traditional dichotomy between man and beast.

Using new, naturalistic data, the book will compare the structure of 'action-assemblies' (Thelen & Smith, 1994) in the activities of human and non-human primates and how these form deep patterns of meaning which underlie representational, expressive and symbolic thought. The book will present evidence of a cross-species dimension to this symbolic substructure.

It is hoped that this book will shed light on the nature and significance of representation, expression and symbolisation in human and non-human primates and hopefully offer insights into the responsibilities which come with a new realisation of our position within nature.

Expression, representation and symbolism in human infancy

Emergent representation in human infancy has been concealed for nearly a century by definitional problems about what constitutes a meaningful representation (exceptions including Atkinson, 2002; Matthews, 1984, 1999, 2003, 2010; Maurer & Riboni, 2010; Stamatopoulou, 2009; Wolf, 1983; Wolf & Fucigna, 1983). In numerous accounts, both old and new, the beginning of representational and expressive thought has gone unnoticed. This is because the dawning of semiotic thought is embodied in a cluster of exploratory and play behaviours formed within early infancy which, if observed only superficially, appear trivial and chaotic. Closer scrutiny and systematic observation of these behaviours, made in detailed, naturalistic, longitudinal observations, reveal their semantic and structural organisation.

If the early emergence of semiotic systems in human infancy has remained undiscovered until recently, it should come as no surprise that the beginnings of symbolisation in non-human primates has been misconstrued, neglected or just plain denied altogether. A prejudice in western science dating from Plato and consolidated in the work of Descartes has ensured that many humans (or, at least, 'technologically advanced' humans) consider themselves unique in their cognition, imagination and especially in their linguistic skills. This tradition has ensured that we tend to think of ourselves as quite distinct from, and superior to, the Great Apes. These beings have, for the most part, remained classified as mindless brutes.

There are notable exceptions to this position, from at least the time when Samuel Pepys, on the 24th August 1661, surmised that the great, man-like beast he saw shipped into London (perhaps a chimpanzee) might be capable of learning language. More recently, the painstaking studies of chimpanzees' social and jungle skills by Jane Goodall (2000); the intriguing work in their signed language by Alan and Beatrix Gardner (1966) and Roger Fouts (1997); the study of apes' symbolic and linguistic skills by Sue Savage-Rumbaugh (Savage-Rumbaugh et al., 1998); and also studies in chimpanzee drawing and painting by great pioneers in this field, including Paul Schiller (1951) and Desmond Morris (1962), have helped to challenge the old, negative image of these extraordinary primates and has made some of us at least question our treatment of them. Rather than being crude, ferocious beasts, research reveals our close relatives as sensitive beings who have much to teach us about our origins and, perhaps, our future.

1 Expression, representation and symbolism

Children's drawings

In my earlier work on children's drawing, I discovered that the earliest mark-making and the later 'scribbling' was important in the formation of symbolic thought. I found that so-called 'scribbling' had structure, organisation and meaning. Soon I found that even scribbling was not the beginning of the story but had itself a developmental pre-history which stretched even further back in time. What people thought of as 'scribbling' turned out to be a rather sophisticated type of drawing, emerging relatively late on in developmental time.

So, scribbling has predecessors. Now I wanted to go even further, back into deep time, to the origins of representational, expressive and symbolic thinking. In my child studies, I located the act of drawing within intellectual and emotional development as a whole. I realised that the role of *play* was crucial to development in symbolisation. I focused especial attention on the example of drawing but I located drawing within a background of trivial-seeming actions which had received little attention. These chaotic looking actions set the context for drawing. Note carefully that we are not here discussing drawings as *depiction*. Most other work (exceptions include Athey, 1990/2007; Bruce, 1987, 1991; Costall, 1993, 1995; Maurer & Riboni, 2010; Stamatopoulou, 2009; Wolf, 1983) habitually equates *drawing* with *picturing* and in doing so misses an opportunity to study a family of activities which accompany drawing and which are critically connected with it. These actions are usually considered extraneous and irrelevant to the act of drawing. On the contrary, I found that, to fully appreciate the meaning and significance of the acts of drawing and writing, it was vital to study precursor actions which led toward symbolic thought. I wanted to show that expressive and representational thought was built gradually, incrementally, in a process involving evolution, development and learning. My work with children taught me that symbolisation was built slowly upon behaviours which were not initially symbolic in themselves. This approach to the dawning of semiotic thought is in contrast to a prevailing view at present that symbolic thought has no developmental or learned aspect. Currently, many psychologists think that representation is the province of human thought alone and that it appears 'just like that', as Alan Costall (2010, personal communication,

Tate Modern Art Gallery) aptly writes – the product of special 'modules' unique to the human brain (Leslie, 1987).

The ideas contained within the present book do not fit into this paradigm at all. Nor do my studies seem to fit in with the idea that human learning is so vastly different from that of the chimpanzee. Tomasello, Kruger and Horn Ratner (1993) think that only humans consciously 'teach' their young but my observations challenge this, revealing some rather active, sensitive, and structured teaching offered to the young chimpanzee by those more experienced.

In order to counter the prevailing notion that expression, representation and symbolic thought are unique to humans and just appeared 'out of the blue' I knew I had to look very carefully and in an unbiased way at both the child's and the chimpanzee's actions and interactions with the environment in a range of behaviours which had received scant attention. These were actions which, on the face of it, seemed trivial, chaotic and meaningless.

The book focuses on a group of chimpanzees in a zoo but we immediately see in their world our own human society. We can see an eerie reflection of the human family and how the child grows up within it and how his or her natural development and learning is supported. We can see our own, human condition; our moments of despair and transcendence.

I thought that my 35 years or so of study of children's drawing and play was complete. After learning about the chimpanzees I realised I was wrong. I had to start again. I had to start from scratch.

Expression, representation and symbolism in non-human primates

This book then, compares the development of expressive, representational and symbolic thought in human infancy with that of one of our closest primate relatives, the chimpanzees (*pan troglodytes*). In the pages which follow I investigate the precursors of expression, representation and symbolism in the actions and interactions of a small group of these intelligent, non-human primates who live out their days in Singapore Zoo.

By the term 'expression' I mean when a sound, a movement or an image is used, individually or in combination with others to form a single unit, to communicate or – more fundamentally – *expel* an emotion in relation to another. This 'other' may be a real-life, physical other, or it may be a virtual other, existing in one's own mind. By 'representation' I mean when a sound, movement or image is used individually, or together with others as a unit, to 'stand for' something other than itself, whilst, paradoxically, still remaining just itself. This means that any object, action, or combination of actions, might be a representation; it just has to temporarily take the place of something else.

For a representation to become a *symbol*, however, requires it to form part of a symbol *system*, in which each part or element in a work (a sentence, say, or an artwork) contributes to the meaning of the whole piece or work, according to rules which organise the symbol system. Language is of course a good example of a symbol system, but an artwork also relies on the organisation of elements within it. It is in this sense then, that I use the term 'symbolism' to differentiate it from 'representation'. Additionally, a symbol (as used here) can stand for a generic type or class of object, whereas a representation cannot: A representation stands for a specific object or, in its more advanced form, encodes the specific features of a unique object.

Expression, representation and symbolism in humans

In my earlier work, I was fascinated to discover that all of these possibilities are present during the investigation and play of very young *human* infants. When I looked at very young children playing and investigating things, and especially when I saw them starting to record and express experiences in mark-making, I realised that I was witnessing the genesis of semiotic thought. I went to the literature to find help in understanding what I was seeing but I found a paucity of any helpful research.

This was in the 1970s. I found two main schools of thought about children's art. One certainly appreciated its beauty but had little to say about how it came about, preferring to describe it as the creative products of a golden age of innocence. This approach however was on the wane and being replaced by a long tradition in science in which children's development was described in terms of its supposed deficits; that is, in terms of the child slowly overcoming severe obstacles or limitations in his or her thinking until he or she represented reality in a fashion deemed 'correct' by the establishment. This prejudice was obvious in accounts of the child's development in the visual arts, especially in drawing. Here, the child was thought to slowly progress from thorough incompetence through a series of stages until he or she could draw objects and scenes *visually realistically*.

In contrast to this general model, I found the beginnings of drawing started with the first markings; that these were not merely haphazard but exhibited semantic and organisational characteristics right from their outset. Traditionally, such mark-making activities were relegated to a so-called 'scribbling stage' and were considered to be without meaning or control. From here (it was thought), the child might have a lucky break when he or she makes a *fortuitous* representation – that is, if and when the child happens, by accident, to draw something recognisable to an adult. From here, the myth continues with the child learning to repeat purposely those configurations which were initially the product of accident.

In fact, the accidental and the fortuitous do play an important role in the development of drawing and, as I will show in this book, in the development of semiotic thought overall. This was first suggested by George Luquet, a great pioneer in the analysis of children's drawing but many subsequent interpretations of his theory have distorted what is actually a very powerful idea (Costall 2001; Luquet 1927/2001).

From here (the story goes) the child is launched along the correct developmental trajectory. Usually, the endpoint of this development is assumed to be the production of *visually realistic* pictures. With a few notable exceptions, the 'visually realistic' was never defined by the majority of researchers who apparently found the term unproblematic. The assumption seemed to be that the term 'drawing' was synonymous with a *perspectival* representation of the object. To compound the problem, very few of the investigators seemed to have any clear idea of what 'perspective' meant either. Nor did they much care. Everyone knew what a 'good' drawing was, didn't they?

The concept of play

Before we go any further, I need to introduce the crucial concept of *play*. An understanding of play is central to what follows. Without the capacity for play, the symbolic act would not be possible.

Play will be a key theme in this book because of its role in the development of expressive, representational and symbolic thought. There are many kinds of play and in this book I am concerned only with *spontaneous play*. In both human childhood and in the chimpanzee's development, this is their self-initiated, self-motivated, self-sustained and unsupervised play. This type of play is still rarely studied properly with regard to its structure and meaning. Both chimpanzees and children fare badly in this respect. In educational research recently there has been tendency to study only those aspects of children's play which can be shown to measurably increase cognition. Some kinds of play are barely studied at all and discouraged or even forbidden in many educational settings. Proper outdoor play is nowadays rare. So is rough-and-tumble play. Kathy Sylva, Roy, and Painter (1980) seem to think that it offers nothing of worth to children in terms of their cognitive development (Jarvis, 2007). And of course we must not forget the dreadful effects of pretend 'gun-play'. For decades now wrongly associated with machismo 'male' violence by the female-dominated, early-years profession, contemporary children can no longer experience the important thoughts and feelings which take place in pretend fighting – not in school anyway.

Shortly I will be describing to you the facilities supplied for the play of a group of chimpanzees. These animals are held in captivity in a zoo in Singapore yet a brief look at the supposed 'free' play of children in the playgrounds of our cities offers an eerie, dreary comparison to the zoo enclosure. Indeed, after watching children playing on the tamed and domesticated plastic and metal junk supplied them one might be forgiven for coming to the conclusion that the apes are better off in *their* prison.

From my work with young children in early childhood development I have learnt about forms of play not generally considered remotely important in education. These are the frowned-upon, anarchic messings about which remain largely unanalysed by scholars; behaviours which seem at first glance chaotic, without rules and certainly without 'aims and objectives'. These are the behaviours never associated with 'high-achieving schools' and which never get printed into what Gunther Kress (1997) succinctly describes as today's 'profoundly unimaginative' national curricular. Yet these forms of play turn out to be ludic experiments of enormous emotional and intellectual worth and probably the most important types of activities in the formation of the human semiotic thinking. They also appear to serve the same function for one of our last remaining primate relatives, the chimpanzee, whom we seem bent on exterminating.

Complex mammals including the whales, dolphins and seals and many land-dwelling ones including cats, dogs, smaller mammals, plus ourselves, as well as our close cousins the Great Apes, have access to two important and interpenetrating modes of consciousness. It may even be that we cannot restrict intelligent ludic play to mammals. For example, the manta ray also displays play behaviour and the sentient curiosity which accompanies it (Marshall, 2009).

We are all capable of two quite different types of orientation to the world. On the one hand, the actions we perform have evolved to help us survive. Involvement with the everyday world requires the development of efficient, adaptational actions in order to accommodate to the external environment and its demands. Included within this class of action are tool and object mastery. The manipulation and use of objects and the refinement of the skilled actions we can perform upon them or with them have assisted our survival. The making and using of tools has been especially important.

However, we also need access to quite another type of orientation to action and objects. Our learning could never be complete were it not for a quite different type of interaction with objects and with other persons. We have access to an arena of activity which is released from adaptational constraints and in which the components of task demands and object mastery are set free from immediate concerns about survival. This arena of activity is called *play*.

Consider the handling and use of an object. By relinquishing the object from any function and liberating the components of the actions involved in handling it from their usual adaptational contexts and performing them instead in a highly subjective mode, with no utilitarian object in mind, paradoxically, objectifiable properties are revealed, properties which would remain undiscovered if we simply had a set of fixed and routinised responses to the object (Lorenz, 1944–1948).

The imaginary object

Batman hides from Buffalo Bill (in a coffee-grinder)

08 August 1977, 35 mm colour-slide film-recorded child-observation and hand-written transcription

Joel (3 years 3 months) has hold of a small, plastic 'Batman' figure and is moving him around and around and up and over and down the sides of an old fashioned coffee-grinder. The wooden coffee-grinder forms a near cube with a metal lid on top which can be slid shut or open for pouring in the coffee beans. A central, metal rod penetrates the metal top of the grinder and this serves as an axis around which a metal handle can be manually rotated – to grind the beans. At the base of the coffee-grinder is a small, wooden drawer into which the ground coffee falls.

Joel has never made a cup of coffee in his life so far, yet he knows a great deal about the mechanism and uses of this object as well as what I think of as its *shape*, *form* and *structure*. I will define what these terms mean shortly. Joel has often seen the coffee-grinder being used to grind coffee beans for making a grown-up's drink and some of his understanding comes from simply watching this process. A far deeper understanding however derives from his transformation of this machine in his imagination. This transformation causes the actions he performs upon it to be organised in a completely different way yet at the same time still remaining based on the essential mechanism and function of the machine. The object – 'coffee-grinder' – has become in his mind part of the architecture of imagined world peopled by imaginary agents imagined to have their own thoughts, feelings, intentions and motivations. Batman is being pursued by 'Buffalo Bill', another small plastic figure, held in Joel's other hand. Batman travels around the four sides of the coffee-grinder, making use of the corners behind which to conceal himself from his pursuer. As well as chasing around and around the wooden cube, the toy figures, flown in Joel's fingers, also fly up and over and down the other side of the box. A final trajectory is pursued which takes, first Batman, then Buffalo Bill, down, into and through the coffee-grinder to emerge from the little wooden drawer at the bottom. First of all, the metal door at the top is slid open and the figures are inserted into the well inside; last of all they emerge from the little drawer at the bottom.

This is like a story. Of especial interest to us is that the central part of the story and the main point on which it all hinges – the figures' passage down to the drawer below – can only be carried in Joel's imagination, for the toy figures cannot be physically milled into plastic dust! This being so necessitates his physical transportation of the toys to the drawer below, whilst simultaneously 'editing out' this act of 'stage management'. The passage of the characters through the interior chamber of the coffee-grinder is sustained in his imagination alone. Finally, after physically placing the toy figures in the wooden drawer below, he has to 'inter-cut' (as it were) their re-emergence into the world.

This is an imaginary, subjective scenario, yet Joel actually utilises and exploits the real, objective properties of the object, describing in effect, its main axes. He describes a roughly circular axis around the device, a *continuous-rotation*, utilising the corners around which Batman hides. In this way, Joel shows his understanding of line-of-sight and the occluding nature of the 'corner'. The other main axes utilised are *going-over* the coffee-grinder and finally *going-through* it. These are also the axes along which his narrative runs yet they remain tied to the physical mechanism and function of the coffee-grinder.

In this book, these are also the main spatial axes we will see the chimpanzees exploring and investigating. Joel's use of these directions of movement in relation to the shape, form and structure of the objects support an event which is partly, at least, imaginative. Do the chimpanzees also use objects and the actions they can perform upon them in this way? Does chimpanzee investigation and play have this added dimension of imagination? Well, we shall see.

When adaptive constraints involved in task and object mastery are abandoned and processes and strategies set free to be explored as entities in themselves, a fluid state of consciousness is created. Such a state may enable creativity and learning. Ordinary actions, events and objects are set free from a utilitarian dimension and become the currency of free-flow thinking (Bruce, 1991).

The interaction between play-mode and adaptational action is a dialectical relationship. This is to be seen in the merest whim of an action to the most complex tasks. This operation works not just with objects but with actions and people too. Although it can be said that when children play they distort the world, they are simultaneously obliged to accommodate to real objects in a real world and, if they are playing with others, to their playmates too; collaborating with them and creating situations which would not have been possible on their own.

Play has been termed a 'simulative modality' (Bruner, 1972; Reynolds, 1976) because in order to play, the primate is required to interiorise a mental model of his or her own body and those of others. Such internal representations are essential to the social organisation of primates and their interaction together. Like humans, in play chimpanzees discover important (yet not always obvious) properties of objects and possibilities of space. In play they also form relationships with others which are complex, creative and extraordinarily subtle. This is because, in play, their interactions with each other are no longer rooted in utilitarian matters.

Naturalistic observations

From the observations I was making of children, I realised that deficit approaches to their development in visual representation were based upon measuring children's development against a hypothetical endpoint conjured up in the minds of adults. These models completely masked the true structure and meaning of children's emergent symbolisation. I wanted to capture symbolic aspects of children's natural play, especially that which occurred in their spontaneous drawing and painting. I wanted to describe the structure, organisation and meaning of drawings that children wanted to make, not what some adult wanted them to make; not the atypical drawings produced in artificial tasks given them by psychologists at the time, nor yet artwork from the school art curriculum which was being put through its initial stages of domestication. [This was the beginning of the period during the 1980s – the Thatcher–Reagan era – in which, first the USA and next Britain and then a great deal of the rest of the world forced a generation of educators to abandon the progressive approaches that, contrary to the mythology systematically propagated by successive capitalistic governments, had from the 1940s been so beneficial for children's intellectual and emotional development (Berliner & Biddle, 1995; Kelly, 1990; Simon, 1988).]

I realised that special methodologies were required if I was to capture the kinds of actions which formed the precursors of expressive and representational thought and its development. I developed a naturalistic, non-invasive approach in which, rather than testing their abilities in copying two-dimensional shapes or in drawing objects set before them (which most of the studies did at that time), I followed and carefully recorded the child's own *spontaneous* drawing. Instead of testing children with artificial tasks and noting what children didn't or couldn't do, I made careful, detailed, longitudinal studies of what they *were* doing and what they were *interested* in doing by themselves. This approach was going to serve me equally well, when, many years later, I studied the chimpanzees.

The pre-history of meaning

One thing I soon discovered in my child studies was that all early acts of symbolisation made by the child, including drawing actions, had a pre-history in earlier actions which were not in themselves representational or symbolic. Young children would perform actions which, on the face of it, seemed to have no relationship to drawing, writing or thinking and had no meaning. Persistent, regular and detailed observations revealed a cyclic reiteration of activities emerging and then apparently disappearing only to reappear transformed, as though each time revised and saturated with successive layers of meaning. Later, I read that Jean Piaget called this process 'vertical decalage' (Piaget, 1951). I was to find the same process in chimpanzee development.

What sort of actions are we talking about here? Well, they are to be found in commonplace actions including all sorts of self-movement and interaction with others and with objects. They are the elemental units of movement which make up ordinary acts of running, jumping, singing, waving hands, moving fingers, kicking, rolling, banging, stamping, shoving, shouting, spinning, scribbling, tapping and touching. Borrowing Esther Thelen's and Linda Smith's term from their dynamic-systems theory I called them 'action-assemblies' or play-assemblies (Thelen & Smith, 1994). These units of action are initially performed singly but soon are combined with others to form more complex orchestrations. They might be performed on a large scale, with the whole body or they might be performed on a small or even tiny scale with coordinated movements of hand, fingers and eyes.

Development seemed like an evolutionary process in that actions originally designed for adaptation and functional purposes or for mastering objects or tools, were adopted and adapted for purposes of expression, representation and symbolisation. Initially sans meaning and function, trivial-looking actions formed the substrate for semiotic and expressive thought. I realised that I had to make a careful study of the actions which preceded representational thought yet which signalled its dawning.

The chimpanzees

The ability to form representational, expressive and symbolic thought is considered, by some people, the last remaining ability unique to humans. Up until the latter half of the twentieth century, tool-making and tool-use had for a long time been thought restricted to humans and this, supposedly, defined their superiority over other animals. When, in the 1960s, in her revolutionary study of wild chimpanzees, Jane Goodall discovered that humans were not alone in making and using tools, the claim for the privileged status of humans within the animal kingdom was seriously undermined. The only remaining claim to human uniqueness is our expressive, representational and symbolic thought and this too is being eroded.

One way in which the supposed uniqueness of human semiotic thinking is being challenged is in ape language research. Earlier, I mentioned a prejudice dating from Descartes and before in which it is assumed that only human beings have linguistic capabilities. In his time, Descartes offered a radically new approach to defining humans and animals in which language, or more specifically, speech, played the key role. He said that animals were really like machines, and that only humans had true consciousness and that the evidence for this was in our speech. This dualistic philosophy, with humans totally disconnected from animals, has persisted to the present day. Perhaps the most notable recent example of this philosophy and possibly its pinnacle is in the work of Noam Chomsky who claims that language is unique to human beings. He and his colleagues have always rejected the notion that apes could use language, arguing that humans – and only humans – have the fundamental rules of language hard-wired into their brains (Chomsky, 1966).

Apes can make a range of sounds but it is true that they cannot be said to speak. In a 1940s study by Keith and Cathy Hayes of the chimpanzee Vicki, a lengthy attempt was made to get her to pronounce a few English words (Hayes, 1951). The chimpanzee found it extremely difficult to utter more than a few breathy words. The usual explanation for this has been differences between the chimpanzee and human vocal apparatus. Recently, some scientists have said that the difference between human and chimpanzee vocal tract has been exaggerated and that what really prevents chimpanzees from speaking is a difference in their brains (Fitch, 2009; Fitch, 2009). This idea has been quickly taken up by those who believe that language is unique to humans. However, it might be that the ways in which Vicki was trained to speak were rather old-fashioned. Later on (page 93), I will mention work by Marina Davila Ross which shows apes laughing when they are physically tickled. Perhaps tickling apes in order to make them laugh is a better direction to take toward understanding primate speech (Davila Ross, Owren, & Zimmermann, 2009 and later in this book).

The work of Cathy Price (2009) and others shows just how complicated is the production of speech. Many different parts of the brain have to cooperate together when we talk. These include parts which control the meticulous movements required to manipulate our mouths and tongues. It may be that such fine motor control of the lower part of the face is beyond the capacity of the chimpanzee though, from my observation, they appear to have rather mobile mouths and labile lips. There are several occasions in which chimpanzees seem to be trying to imitate the speech actions I make to them in strange mouthing movements of their own. The controversy of whether apes have the capacity for language rages on.

Apes and human scientists

Sue Savage-Rumbaugh is amongst a group of modern researchers who realise that the *linguistic* need not be tied to the *verbal*. Alan and Beatrice Gardner and Roger Fouts are amongst others who have started from this new premise. Another important thing these scientists have in common is that they worked closely with their non-human primate subjects in what has come to be called cross-fostering. In this type of study, the ape is brought up by humans in a human environment. The Gardners were the first to try to find out if a chimpanzee, named 'Washoe', could acquire American Sign Language. They brought her up in an environment in which only sign language – no speech – was used. Roger Fouts worked with them and did some extremely important work in cross-species communication (Fouts, 1997).

Apes as subjects in human experiments

In the work on the origins of both language and art, non-human primates have been studied within close relationship to human investigators. Some of this work has been ground breaking in our understandings of the origin of semiotic thought and language particularly. In research by Alan and Beatrice Gardner, Roger Fouts and Sue Savage-Rumbaugh especially, the apes are treated with the kindness one would expect to be accorded to beings so like us. These studies raise our awareness of the sensitivity and intelligence of these animals. In their research, humans and chimpanzees interact freely with each other. The chimpanzees are not forced to perform artificial tasks.

Fouts, Savage-Rumbaugh and the Gardners all realised the importance of the chimpanzees' natural play and exploration. They also realised that language development required that there be a constant human companion to interact linguistically with the subject. Both these aspects, the availability of play and intellectual and emotional companionship are vital to human development in language. Yet these conditions have not always been met in other ape research on language.

Nim Chimpsky

In Herb Terrace's famous (1979) study of signing in a chimpanzee, the subject, 'Nim Chimpsky' (named after Noam Chomsky!), is drilled into learning signs (Hess, 2008). He is prevented (as far as is possible with a young, lively chimp) from engaging in his natural play and investigative behaviours. When, despite severe constraints placed upon him, such behaviours burst out anyway, they are disregarded in much the same way as important evidence about children's spontaneous play and investigation have been traditionally ignored in studies of children's drawing and other forms of meaning-making.

In 'Project Nim', the young chimpanzee has no constant companion. Terrace hires one teacher after another to train the chimp in sign language. All researchers in childhood development know that the availability of continuous companionship with another language user is essential if the child is to properly develop language and other semiotic systems.

During one phase of the experiment, one 'helper' forces the chimpanzee to sit at a school desk for his language 'lessons'. He even has to hang his coat up on a hanger before he starts work! (There's a rather sad irony here, because, as we will later see, the 'Hanger' is one of the chimpanzee's very own inventions.) Worst of all, he is locked up in a trunk if he misbehaves. Eventually, Terrace announces that his experiment shows that chimpanzees cannot, after all, learn language.

Surprise, surprise. As many have pointed out, under such circumstances it would be incredible if a child could learn to speak.

That some sensitive members of Terrace's team are discouraged from recording naturalistic observations of Nim's play activities and of his natural investigation is a great pity. Careful recording and analysis of true emergent symbolisation in Nim's ordinary activities may well have provided a far more meaningful study.

In all these experiments, good and bad, these non-human primates are reared in human contexts, in experiments devised by humans and using materials and technologies which the apes would not encounter in the wild. The question remains as to how to interpret the outcomes of these studies. What are the studies actually telling us? Are the human experimenters teasing out potential cognitive processes which remain unused by the animals when in the wild? If so, this does not fit well with our ideas about how evolution works. Biologists tend not to like the idea of animals having superfluous abilities – evolution seems to be very thrifty in the abilities it gives living beings. How then should we regard these unusual acts of cognition, made by apes, in human settings? Are we witnessing potentialities within a non-human primate which could possibly unfold during evolution? Are the experiments telling us something about how we, as humans, might have evolved into symbol-using primates? Or are the extraordinary cognitive abilities of the animals the product of human contrivance?

It is important to note at this point that it would be a fundamental mistake to think of chimpanzees as a sort of underevolved, pre-human. They are not. They are a perfectly adapted species different from us humans. Chimpanzees and people do however share a common (and, as yet, unknown), ancestor from whom we parted company, evolutionarily speaking, about 6 million years ago. Chimpanzees are genetically extremely close to humans (sharing about 99 per cent DNA with us) and so it is very rewarding for scientists to study them to help imagine what might have been the beginning of semiotic thought.

The painting and drawing of apes

The same sorts of questions about chimpanzees' linguistic abilities have been raised about captive apes' drawing and painting abilities too. Are we right to view these results as the genuine precursors of an aesthetic dimension of Great Ape thinking? Or are they, as some writers insist, the product of situations wholly artificial to the animal, which certainly reflect the cleverness of the captive animal but only in opportunistically exploiting his or her human captors for food or other rewards, or perhaps simply for care and attention. It is undeniable that in some of these studies, the chimpanzee (or other non-human primate) has discerned the possibilities and use of an art medium – but to what purpose? Do such studies have anything of value to tell us about the genesis of artistic expression, representation and symbol-making? For example, in captivity, and with the supportive attention of a human, some Great Apes produce startlingly powerful paintings – and not for reward, either. Those painted by the chimpanzee artist Congo whilst in the care of Desmond Morris are good examples (see Figure 1.1).

But some authors argue that ape drawing is merely the product of human contrivance and cannot occur in the wild (Janson, 1961). Even Jared Diamond, whose work on the art of the bower bird is of extreme importance (not least because it extends the possibility of art even beyond the mammalian world), writes: '. . . ape painting is just an unnatural activity of captive animals. One might insist that, since it is not a natural behaviour, it could not illuminate the animal origins of art' (Diamond, 2002, p. 156).

In my book, I will offer evidence which seriously challenges this verdict. I am going to argue that, far from being unnatural activities, basic drawing and painting actions are perfectly characteristic of the primate condition.

Figure 1.1 A painting by the chimpanzee Congo made with a brush and in black, water-based paint on cream paper, No. 11, 19 August 1958.

Methodology: The approach

Methodology means more than how you gather the data in the sense of the tests and the tools you use and your ways of measuring and evaluating. It also includes your underlying conceptions and approach from which your research tools and methods arise and upon which they are based. As long as one uses an inappropriate methodology, emergent expression, representation and symbolisation will remain hidden. In my chimpanzee studies, I use the methodology similar to the one adopted in my earlier studies of the children. It is a naturalistic approach; I simply look at and listen to the spontaneous behaviours of my subjects. My video-recorded and written observations are interpreted according to appropriate theories of development which illuminate the raw data and suggest their organisation. No tests are involved. One difference from my child studies is that, in the present study, neither I (nor any other human) have contact with the subjects – the chimpanzees.

This is a big difference but in every other way the research approach is much the same. I do not interfere with what the chimpanzees are doing. I do not reward (or punish) them or train them in any way. My method is to simply stand and watch and wait, hour after hour, day after day, week after week, month after month, year after year, in the blazing equatorial sun or the pouring, monsoon rain, in Singapore Zoo.

Originally, in addition to my studies of the chimpanzees, I start off observing Singapore's famous orang-utans (*pongo pygmaeus*) too. I do this for about 18 months before exhaustion obliges me to concentrate on the chimpanzees alone. However, the work I did with the orang-utans was extremely useful and I will refer to it occasionally in the present book.

Concentrating on the chimpanzees then, I video-record what they are interested in doing, how they do it, and if and when and how what they are doing changes over time. As in my child studies, changes in the way the chimpanzees approach things seems to involve development, practice and culture. Yes, the chimpanzees' learning also takes place within what I can only term education and culture.

Coupled with my readings of primatology, I use the theory I developed in my child studies to guide and illuminate the observations I am making. This theory identifies the modes of representation and expression children move through as they grow older. My approach owes a great deal to many people. Jean Piaget (1951) sets the scene for it with his work on sensorimotor intelligence and Jerome Bruner's (1964) work about the development from *enactive*, to *iconic* to *symbolic* thinking is also central to it. Another part of the central core of this book is Vygotsky's (1966) theory about play and the *pivotal object*. Thelen and Smith's (1994) work on a dynamic-systems approach to learning and development has been most helpful. John Bowlby's (1953) work on the mother and infant dyad has been most useful as has Donald Winnicott's (1971) ideas about the *transitional object* and *transitional phenomena*. The work of John Willats (1997, 2005) on drawing and other forms of visual representation is crucial. Of great importance too has been the ecological psychology of Alan Costall (1995), as has Francis Pratt's (1983) work on intellectual realism in children's drawing. Chris Athey's (1990/2007) and Tina Bruce's (1987, 1991, 2001, personal communication) work has helped me understand something of how children move through a sequence of emergent concepts about spatiotemporal relations and structure within a social and interpersonal context. Of the utmost importance has been Athey's work on the child's systematic and orderly progression through a series of dynamic and figurative *schemes*.

I have mentioned some of the theoretical influences from the fields of art research and early childhood. Other influences are more recent and are from the field of primatology and ethology. The work of Paul Schiller (1951), Desmond Morris (1962, 1967, 7 December 2009, personal communication) and Roger Fouts (1997) has been of great importance to me. I have been very influenced by the work of Jane Goodall (2000), Birute Galdikas (2005) and Frans de Waal (2000, 2005) and, following their examples, I use an ethological approach and naturalistic methods. Frans de Waal's thoughts about the importance of naturalistic observations have been encouraging as have his ideas about the value of making observations in zoos.

A third influence is my long experience as a diver. I started snorkelling and free-diving at 11 years of age back in the 1950s around the coast of Devon, England. At that time, very few people had any idea of what I was doing. I was asked interesting questions like: 'How much oxygen does that breathing-tube (snorkel) hold?' and 'Is there anything to see down there?' I have made over a thousand SCUBA dives all around the world. My encounters with marine animals taught me that we are not alone. A critical experience for me was diving with some captive dolphins and an orca, the latter whose teeth, tongue and gums he allowed me to caress with my hands. (Dolphins and whales have thick, tough skin like perfect wet-suits and so the most sensitive parts of their bodies are their mouths.) I have no idea what possessed me to put my hand into a killer-whale's mouth. Down there in the blue depths I somehow knew that he would not hurt me.

The fourth influence is my work as an artist and art educator. I have been at art school since the age of 14. I learnt a great deal about drawing from the then young artist, Terence New. Also, from that time (starting in the 1950s) I produced and edited films, using the beautiful 8 mm Kodachrome of the time and, later, 16 mm film, with my friends, photographer Martin Thompson and designer Rick Miller. Art educators John Hart, Tony Collinge and Tom Osbourne, at Goldsmiths Art School, helped open my eyes too.

At art school they let me do what I liked. (Well, almost.) This is now a forbidden approach to education. At art school I learnt about art praxis with traditional methods, including perspective drawing, and painting from life and observation, to modern and post-modern methodologies and technologies including electronic media. In the 1960s I was one of the first performance artists. I have exhibited artwork in London and Singapore.

These experiences have heightened my sensitivity to *shape*, *form* and *structure*. By 'shape' I mean the optical projection of a single view of an object as seen from a single, stationary point in space (or *station-point*) and by 'form' I intend the overall volumetric configuration of the object irrespective of any particular station-point.

By 'structure' I intend another subtle differentiation made in perception. This term refers to the main axes and coordinates which link shape and form together and which create the kinematic and kinaesthetic configuration of the object as both an object and event in time and space. My life as an artist has helped me understand these distinctions.

I am initially concerned that working in a zoo is not going to be good enough but since going out into the wilds of Africa is out of the question I have no option. As a professor in Visual and Performing Arts at a University I have a full-time job to do, part of which is concerned with arts education; part with educational research.

I do not have a grant from the university but I do have access to semi-professional videographic and computing equipment. Additionally, Singapore Zoo kindly allows me free access to the zoo and I am very grateful to the director and the staff for their help. It turns out too that rather than being a serious shortcoming, there are important advantages in studying captive chimpanzees. I note above that Frans de Waal makes the same point. In the wild, it would be virtually impossible to capture the patterns of behaviour so crucial to my kind of investigation.

Fieldwork

I visit the chimpanzees once a week for 3 years 7 months. Each visit is of about 3 hours. I arrive early on a Friday morning, before the zoo is opened and before the visitors arrive and just before the chimpanzees emerge from their night quarters or 'The Den', as the Keepers call it. I then spend about 3 hours video-recording them. The chimpanzees all get to know me. I cannot claim that I am very important in their lives but they all recognise me, listen to me sometimes and, if they are in the mood, will wave to me, greet me, smile and nod at me. After a long time has passed, some of them show me what they are doing. The adult females even bring their babies to show them to me – an action I take to indicate their trust in me, since female chimpanzees are often highly secretive about their infants.

Waving 'goodbye' seems more difficult for them – I rarely saw a chimpanzee indicate an unequivocal 'goodbye' in any form – but I do once see the glimmering of one when Grumpy Ganga spots me through the trees going by on the zoo-tram when one day I am leaving. She recognises me immediately and gives me a side-to-side 'no' head-shake. Of course, she could be just saying 'hello' again! (By the way, she never gives a clear, up and down 'yes'-type nod but always appears to be 'saying' 'no' to everything!) Now that the study is over, I miss the chimpanzees very much and hope that they are managing okay. I had no idea that I would feel this way when I started.

Initially, I have naive ideas about being allowed in the enclosure with the chimpanzees. I am told straightaway that this will not happen! Later I am to learn that captive chimpanzees are more dangerous than wild ones, since the latter at least have no idea how puny humans are.

I am told that the chimpanzees have never had any human contact. I am only to be allowed to view the chimpanzees from the same pathway as the other zoo visitors. I can video-record them over a moat and an electric wire but I can go no nearer.

After my initial disappointment, I soon realise that this situation is going to be a big advantage. That the chimpanzees have no human contact means that their behaviour is going to be fairly natural. There is no interference from humans. Even though the chimpanzees are captive animals they still live out many of the same forms of behaviour they would in the wild. Of course, there are some important differences and as Crickette Sanz and David Morgan (2007) note, human presence of any kind, even the most delicate, will alter their natural behaviour (see also Foer & Nichols, 2010). But for my purposes, I soon realise that, if I simply stand and watch for long enough, from over the fence, where the other ordinary zoo visitors stand, I will be able to make valuable observations about how chimpanzees might form the precursors of symbolic thought.

Methodology: The tools

Naturalistic observations, video-recording, analysis and interpretation

Using semi-professional equipment, including a digital video-camera supported on a tripod, I shoot about 470 hours of videotape. All of the raw footage is reviewed. All my movie material is analysed for content and meaning on my computer and I write detailed transcripts of about 80 per cent of this recorded material. The written transcripts of the observations amount to about 400,000 words. I use a small, selected sample of these observations for this book. I edit the raw footage in an analytical way, constructing a series of movies each of which reveals the aspect of behaviour I have discerned and the point I want to make about it.

Eventually a Digital Versatile Disc (DVD) is made of each theme or topic or aspect of chimpanzee behaviour or development I discern. From the 470 hours of raw footage about 290 edited DVDs are made. Most of the edited movies are around 30–40 mins in length, though a few are an hour long and a few others just 5–10 min. An estimated 250 hours of movie are edited, analysed and transcribed. Many of the DVDs are different versions of just the same few minutes or seconds of material, each DVD edited to show what I come to think of as a different layer of time. Each DVD bears a date, a title and which 'Edit', or 'Version' or 'Part', it is. There may be as many as 14 versions of the same few minutes of raw footage. If the DVD material is unedited, then the disk bears the legend, 'raw footage'. Unless otherwise stated, each observation is recorded on a dated and titled DVD. Other observations derive from drawings, hand-written notebooks, cine-film and videotapes and this information is also cited in this book at the beginning of the relevant observation. I decided to include this information in order to suggest that important phases of development, and even evolution, happen at specific times, rather than at vaguely defined 'stages' or 'periods'. For the same reason, I also show, where possible, the approximate minute and second when an event starts by citing the number on the footage counter of the video before each written entry. Such detailed, temporal notation is useful because it shows how some key moments in development (and, perhaps, evolution) take place in incredibly small intervals of time – just seconds or fractions of seconds. I am trying to suggest that developmental time and evolutionary or *deep time* coincide with clock time and that this time is ticking away.

I have shown a little of the footage to about 50 colleagues, both academic staff and students at Nanyang Technological University, Singapore and none of these people has seriously contested my interpretations. Colleagues are invited to double check for themselves the veracity of my descriptions. The way in which I have annotated the movie material means that people can match the text to the exact footage or even rearrange the movies according to their own interpretations of the events, rather than just be stuck with my version. (The DVDs are arranged chronologically in cardboard boxes in my house in London. To obtain the address, make contact through the email address at the end of this book. I will leave the key under the mat – no – no, only joking.)

On my computer I edit the footage to tease out main themes I discern. I use slow-motion techniques and frame-by-frame analysis. It is an extremely labour-intensive, painstaking task, sometimes gruelling, entailing my going over and over the same few seconds of movie for hours, days or weeks at a time in front of the computer. (I can become violent when I hear the term 'anecdotal'.)

But the work is extremely exciting. I difficulty tearing myself away from the computer. I get so much practise using the simple editing software that, on a good day, I can estimate a 25th of a second (the temporal resolution of the movie material) completely intuitively and with great speed, without needing to count individual frames or 'scans' on the editing suite's display panel.

The editing of the movie is the first stage of analysis. By going over and over the same footage one sees more and more. All of the observations have to be transcribed into writing and the intertwined processes of viewing, editing and writing together heighten my perception of the significance and meaning of the chimpanzees' actions. The act of writing and rewriting is itself a major research tool. Struggling to find words to describe the actions of beings who have no words is a Herculean endeavour but one which gradually uncovers a wealth of undreamt-of structure. I believe Charles Darwin uses the process of writing in this way.

Standing on a ladder (or 'heightened perception')

In my first observations I find I cannot always see the chimpanzees' activities when they are on the upper levels of the wooden construction provided for them (see information about the enclosure on page 23). To obtain shots from a higher viewpoint or 'station-point', I take an aluminium step-ladder with me. This necessitates making the tripod taller too, so I ask my friend, Chinese artist, Frederick Tay, to make three long wooden poles which may be attached to the tripod legs with cable ties. Standing on the top of my step-ladder, with my camera on its taller tripod, I have a higher viewpoint and am able to capture sufficient videographic information about the chimpanzees for my purposes. (Of course, having no research student I have to lug all this stuff back and forth through the zoo every week in the baking sun or pouring rain.)

Even with a second, taller ladder and especially lengthened tripod, I still cannot see some activities the chimpanzees perform in the really high parts of their environment. This is a major weakness in my research. Sometimes I can see the actions involved in drawing but not always the drawing itself, in terms of visible marks on a surface. I eventually obtain good shots of some of the chimpanzees' drawing but in many cases, although I am sure that the actions I saw would leave traces, the fact remains that I did not always see them. Even so, I resist the temptation of obtaining an even taller ladder since I sense that this will be too obtrusive and will disturb the chimpanzees. I feel they need some privacy. Frans de Waal (2005) notes that chimpanzees do not like to be constantly looked at. I remember his advice throughout the work. The first priority in my research with children too, and in my underwater photography, is to avoid disturbing the subject. I prefer to miss good shots rather than infringe this rule.

Despite this obvious weakness, I feel that I have video-documented a sufficient amount of the drawings themselves appearing from the hands of the chimpanzee; enough to suggest that further research (of a highly sensitive, friendly and unobtrusive kind) might confirm my findings and help move us to a new level of understanding about them and us.

Apes at the frontiers of symbolisation

The chimpanzees

In the wild, chimpanzees usually live in social groups of 2 to about 80 but they often split up into sub-groups which change in size from time to time. They live for about 40 years in the wild but may reach the age of 50 or 60 in captivity.

They come from a number of regions of equatorial Africa. About a century ago there were an estimated one or two million of them but now their numbers have dwindled to roughly 200,000. You don't need to be told why.

They are genetically very close to humans, sharing with us about 99 per cent the same DNA. The males are about 1.70 m tall and the females about 1.30 m. Their arms are longer than their legs and proportionately longer in relation to their bodies than those of humans. Like us, they have nails, not claws. A chimpanzee's body is not covered in fur but black hair. The human body has the same number of hairs growing on it but they are usually so tiny that we are thought of as 'the naked ape' (Morris, 1967). As I will describe later, chimpanzees, like us, favour their hands (rather than their feet) for handling things in order to look at them. (For example, those chimpanzees who paint and draw do so using their hands.)

However, it is important to note that the foot-hands often play an extremely important supportive role in any task which involves skilful manipulation accompanied by careful visual monitoring.

Because chimpanzees are easily able to use both feet and hands very skilfully, to grip, carry and manipulate things, I will refer to their feet and hands both as 'hands', dividing them between a pair of 'front-hands' and a pair of 'foot-hands'. (I will, therefore, when necessary refer to the soles of the feet as 'palm-soles'.)

Chimpanzees have very good manipulative skills with all four hands and can do almost anything with their front-hands. They have a very good finger and thumb opposition and pincer-grip, though not quite as precise as that of a human. To make up for this, they have a good pincer-grip in their foot-hands. Their lives are divided between the trees and the ground. Chimpanzees move in the trees with great agility, their arboreal locomotion surpassed only in spectacle by the brachiation of the graceful gibbon.

Another noteworthy aspect of their front-hands and foot-hands is that the palm and sole of each is sufficiently long to allow the gripping (when necessary) of a branch by palm or sole alone. This leaves the fingers free for finer manipulation coordinated with acts of looking whilst investigating and playing. That this is so with these and other primates (for example, the orang-utan) may have had portentous evolutionary consequences for the emergence of thinking.

Chimpanzees also come down to the ground, where they can walk around on all fours, their foot-hands palm-sole to the floor; their front-hands on their knuckles. They are sometimes called 'knuckle-walkers' for this reason but they can also stand and walk upright, or *bipedal*, on their foot-hands. Their visual system is similar to ours, with front-facing eyes providing binocular, stereoscopic depth perception. Their visual acuity is excellent. They can see into the distance in fine focus but accommodate to close-up vision too, fine focusing within a few centimetres of their eyes. Their auditory, olfactory and gustatory faculties are also excellent.

The leader of the group is usually the dominant male, or 'Alpha' male and he has a harem of females. He also has first choice of any new female. There is often a high-ranking female too. Chimpanzees mature by age 14 and the females can start giving birth at this time. The females give birth to babies usually singly and have a gestation period of 8 months. The female may have a baby every 5–6 years. A typical mother looks after her baby with extreme care and affection for at least 4 years. Even after this period, the mother and infant pair have an emotional bond which often lasts for life. The mother performs most of the care of the infant, though the father does some too and also protects the young ones and everyone else in the whole group. In terms of time spent, although the old Alpha male in my group, 'Togo', seems to do less than the mothers, this study shows him helping the youngsters with the learning of key skills and also offering them companionship in certain, important activities. The females share responsibility for looking after the babies including those infants who are not their own. Both juvenile females and males learn how to look after their infant siblings too and will take turns in carrying them around, playing with them and introducing them to a range of play-modes and skills.

Of central importance to the present book is that chimpanzees are highly intelligent. There is evidence that some chimpanzees and other primates have visual memory for number (Cantlon & Brannon, 2007; Matsuzawa, 2007). Indeed, skills of spatiotemporal calculation and measurement may not be limited to primates. Even the humble rat, for example, seems able to measure time intervals (Davis & Memmott, 1983). How far such skills are 'natural' or the product of the laboratory setting is controversial, though one study of wild rhesus macaque monkeys (Flombaum, Junge, & Hauser, 2005) seems to suggest natural numeracy.

But it is not just because they are smart which makes the higher primates attractive to humans. After all, many other animals show remarkable intelligent traits. Recent work on bears and dogs show them to be sensitive and intelligent creatures. The dog especially has an emotional empathy with humans which in certain respects exceeds that of the chimpanzee (Kaminski in Child, 2010). Large marine mammals, including seals, dolphins and orcas show that they have thoughts and feelings which they occasionally share with humans. Scientists have recently recommended that dolphins be thought of as 'non-human persons' (Reiss, 2010; Reiss & Marino, 2001). Even the big-brained fish, the manta ray, shows extraordinary levels of empathy with experienced and sensitive aqua-nautical humans. So it is not just because chimpanzees are smart that makes them and other 'higher' primates interesting to humans. It is because they are like us in more humble ways. It is because they have bodies and minds quite like our own; that they experience the same kinds of pain and pleasure. We feel for them not only because they share the same kind of intelligence with us but also because they share the same tragic limitations. Really, when it gets down to the deepest level of description, the human condition is the primate condition.

Although whether or not chimpanzee use of signs qualifies as true language remains in dispute, their capacity to understand when spoken too, to solve complex problems and to grasp social situations is undeniable.

Their consciousness is a fruitful balance between caution and curiosity which enables them to investigate and explore. Like humans, they are not rigidly locked into routinised responses but change their behaviour in the light of experience or if they become interested in something new. Although the social grouping of the chimpanzees is hierarchical, it is not rigidly so as is the case with some other primates (for example, the baboons) and there is room for some social flexibility. This is important because it allows for individual and interpersonal creativity to take place. Discoveries or inventions can be made and shared by the group.

Like humans, chimpanzees spend many years learning. Aside from suckling, very little else of their behaviour is instinctive. Even clinging to the hair of one's mother has to be learned. Chimpanzee society has a potentially rich combination of familial and social stability coupled with sufficient flexibility to allow the young to devote many years to learn, play and make individual discoveries in safety. Any new discoveries or inventions a chimpanzee makes may be added to and change the culture of the group. I have not the space here to write more fully about their complex, personal, social and emotional lives. Their real lives, in the wild or in captivity, are as complex and individual as our own.

For more about the life of chimpanzees and other primates please read Jane Goodall (2000), Frans de Waal (2000, 2005) and Desmond Morris and Steve Parker (2009). For more about how many chimpanzees have endured tragic lives in captivity at the hands of humans, read Roger Fouts (1997).

The chimpanzees in Singapore Zoo

These basic characteristics are true of the group of captive chimpanzees I studied in Singapore Zoo.

When, on 8th September 2005, I make the first of my weekly observations, there are nine chimpanzees – four males and five females. Because of problems of overcrowding, in the third year of my study (in February 2008) two young brothers, Makhulu (about 7 years of age at the time) and Manzu (about 8 years of age), are flown off to a zoo in South Africa. (The family are clearly saddened by this, especially the remaining brother, Nkosi.)

Toward the close of the study (my last observation is made on 13th May 2009) there are 12 chimpanzees including three babies; two born in August 2007 (one on the 13th and one on the 29th) and a third baby on the 16th February 2008. Here are the names and particulars (where known) of all the chimpanzees in the group I studied, including those of the two juvenile males transported to South Africa.

Chimpanzee particulars

1. Name: Togo
Sex: male
Date of birth: estimated 1972
Weight: 58 kg
From: Togo, The Congo, Africa
Status: dominant male of the group (Alpha)

2. Name: Ida
Sex: female
Date of birth: 14/08/1973
Weight: 47 kg
From: Hanover Zoo
Status: mother of Rah and Nkosi

3. Name: Suzie
Sex: female
Date of birth: estimated 1973
From: Ruhe, Africa
Status: mother of Makhulu and Ainka

4. Name: Ainka
Sex: female

Date of birth: 29/08/2007
From: Singapore Zoo
Status: father – Togo; mother – Suzie

5. Name: Bianca
Sex: female
Date of birth: 09/11/1979
From: Sydney Zoo
Status: mother of Manzu

6. Name: Zanta (named by John Matthews)
Sex: female
Date of birth: 16/02/2008
From: Singapore Zoo
Status: father – Togo; mother – Bianca

7. Name: Ganga
Sex: female
Date of birth: 08/02/1993
Weight: 47 kg

8. Name: Kimoni
Sex: male
Date of birth: 13/08/2007
Status: father – Togo; mother – Ganga

9. Name: Manzi (or Manzu)
Sex: male
Date of birth: 1999
From: Singapore Zoo
Weight: 15 kg
Status: father – Togo; mother – Bianca
(Taken to South African zoo in February 2008)

10. Name: Nkosi
Sex: male
Date of birth: 27/06/2000
From: Singapore Zoo
Status: father – Togo; mother – Ida

11. Name: Makhulu
Sex: male
From: Singapore Zoo
Date of birth: 07/05/2000
Status: father – Togo; mother – Suzie
(Taken to South African zoo in February 2008)

12. Name: Rah (or Rara)
Sex: female
Date of birth: 07/12/2004
From: Singapore Zoo
Status: father – Togo; mother – Ida.

The chimpanzees' zoo environment

The chimpanzees' enclosure (see Figure 1.2) is a grassy area of undulating knolls and gullies of 638 m^2. It has a thick hedge around approximately half of it and behind this there is a steel fence and then a high, concrete wall (some of which includes the outside wall of the chimpanzees' indoor, night quarters, called by the Keepers, 'The Den'). The remaining half of the land is surrounded by a moat of about 5 m wide and perhaps 2 m deep. The moat contains goldfish. The land area slopes steeply down to the water's edge in some places; more gradually in others. At the water's edge is an electric fence. On the other side of the moat, beyond the reach of the chimpanzees, a strip of land, full of plants, slopes up to a small, low, wire behind which is the zoo-visitors' path that winds around the moat and then off into the rest of the zoo.

Singapore is in an equatorial region and originally rainforest. There remain a few chunks of rainforest and the zoo is situated amongst some of these. Some

Figure 1.2 The chimpanzee enclosure in Singapore Zoo.

of the remaining indigenous creatures roam the zoo freely, including reptiles, amphibians, mammals, insects and many kinds of birds. The chimpanzees encounter these animals from time to time.

The chimpanzees' grassy land area has a few living trees and a number of still upright, dead trees plus some fallen logs. The living trees are encircled by electric wires at a height of about 15–20 m. There are also numerous granite boulders of varying sizes; some quite large, of about 3 m × 3 m × 3 m; some of only 1 m × 1 m × 1 m. Some of the smaller, rounded rocks are embedded in the ground with their slightly domed surfaces showing. I refer to these rocks as 'Platform Rocks'. The chimpanzees use them for many purposes including seats, as well as for playing and as working surfaces. When they get to know me, some of the chimpanzees sit or stand on the Platform Rocks as podia from which to greet me.

The nearest position a chimpanzee can get to be with a human though is on top of a steep cliff above the moat. Some of this cliff is composed of real granite whilst some of it looks like rock but turns out to be artificial, made from resin. This fabricated rock looks quite convincing until a chimpanzee drums upon it, betraying its hollow interior, or when one occasionally manages to break a hole into it. I come to call this area 'Hollow Rock'.

An artificial stream runs from high up in the land area, dividing into two channels, one which runs down a narrow, rocky fissure of about a metre's width to the moat by Hollow Rock; the other running through a jumble of big boulders ('Waterfall Rock') to form a shallow pool ('The Pool') before tumbling over a little waterfall ('The Waterfall') down about 2 m into a gully where it forms a wider, shallow stream before joining the moat. This gully and its stream are traversed by a long, old log, horizontally placed to form a bridge between the two sides of the gully. I call this 'Horizontal Bridge Log' or sometimes just 'Bridge Log' and it is over 1 m in width and about 9 m long. There is a drop from this log of a couple of metres down to the shallow stream. Each end of Bridge Log is marked by an upright, dead tree. To make the spatial relationships reasonably clear I will privilege my own station-point as if facing the enclosure from the Visitors' Path and refer to the positions of features and landmarks in relation to my own viewpoint. Hence, if I face Horizontal Bridge Log, there is, at its 'left' end (my left) a broad, old, dead, twisted tree, bone-white in colour ('Twisted Tree'), whilst at its 'right' end (my right) is a taller, straighter tree. Because this tree at the right end of Bridge Log is the most distant from me when I arrive at the enclosure by Hollow Rock, I call it 'Faraway Tree'.

In roughly the centre of the enclosure is a very large wooden structure, about 6 m high by about 8 m along one side and covering perhaps an area of about 40 sq. m. In the book I refer to it as the 'Construction'. It is an irregular edifice, strongly built and added to and modified over the years in an organic and intuitive fashion by the zoo Keepers and builders. It is made from stout wooden poles (like telegraph poles; circular in section and of about 30 cm in width), discarded timber and wooden platforms, all built around some living and dead trees left vertical, still rooted.

The Construction is divided roughly equally into three Levels by a series of wooden platforms supported by vertical wooden columns ('Corner Posts'), circular in section, like old telegraph poles, and traversed by 'Horizontal Beams', more telegraph poles aligned roughly horizontally. The lowest platform I designate Level 1, being about 1.80 to 2 m from the ground; Level 2 which is very roughly 3.5 m in height; and Level 3, the uppermost level, roughly 6 m high. The chimpanzees can climb, play and rest on and in the 'Construction'. Until the 21st March 2008, when some changes are made to the Construction, Level 3 is made from one of the zoo's original, old wooden signposts and on its underside a human can read the embossed words (in English) 'Entrance to the Great Rift Valley' carved in relief in block capitals into the timber. The chimpanzees sometimes finger the raised, wooden letters. I call this platform, for a time the highest one, 'Word Platform'.

I call the lowest level, the ground level, the 'Void Deck' after the custom of Singaporeans who so refer to the open, ground floor in their apartment blocks. Parts of the floor of the Void Deck are old and worn concrete, the fragmented remains of earlier paths, whilst most of it is bare earth. In some parts of the Void Deck there is space enough to allow bipedal walking. Until 9th November 2007 the Construction and some other parts of the enclosure is traversed by old fire-hoses, flattened by age and use into battered, brown straps, upon which the chimpanzees can climb and travel. Some fire-hoses hang vertically whilst others form links between certain trees and the Construction and between other landmarks. Fire-hose linkways which run roughly horizontal in long, sagging lengths between posts and trees I term 'Slack Straps'.

Another important feature is the 'Rotten Stump'. This is the ancient remains of a sawn-off tree stump, about 118 cm high and 50 cm wide – a decaying, crumbling ruin of rotting wood, partly overgrown with moss and weeds, which from time to time a chimpanzee will try to uproot, but which despite its decrepitude stubbornly refuses to be budged. Rotten Stump is located a couple of metres from the bank which slopes down to the moat.

This remains the basic structure of the enclosure up until the changes are made. Even then, the overall arrangement remains more or less the same but an additional, very high bridge ('Lateral Log') about 10 m high, is formed between two trees, which (again privileging my own station-point from the Visitors' Path) I term 'Left Tree' and 'Right Tree', or sometimes 'Tall Tree' because its sawn-off top has now replaced Level 3 as the highest place a chimpanzee might sit or stand upon. An important change to the environment is the replacing of the old fire-hoses with artificial vines or lianas. These are made from plastic with an inner skeleton of nylon rope and wire. To complete their realistic appearance their outer skin is covered with a substance referred to by the Keepers as 'zoo-poo'.

Further changes are made to the Construction in the second week of March 2008 when the rectangular platforms, including the old 'Word Platform', are replaced with triangular ones. Another new feature dating from this time is an additional level between Levels 2 and 3, which is a sloping, rectangular platform I call the 'Mezzanine Level'.

Occasionally, to make my descriptions as precise as possible, it will be necessary for me to designate the different sides of platforms with a numeral; three numbers for triangular ones, four for rectangular ones. To make it simple, and using the example of a rectangular platform, I again favour my own viewpoint in numbering the sides, so that Side 1 is the side directly in front of me, Side 2 is the one to my left, Side 3 is out of sight, round the back (opposite Side 1) and Side 4 is the remaining side to my right.

There are less permanent features which are occasionally installed, including two versions of a 'Swinging Platform', a small platform hanging from steel cables high up in the Construction which can be set in motion by the chimpanzees; a variety of hanging, swinging, large plastic barrels and dustbins; a 'Seesaw Log', which, as its name implies, may be seesawed up and down; a 'Lever-Log' which can be used to lever another log, atop it, up and down; and a 'Tremble Post', which can be violently shaken so that it vibrates. From time to time, the Keepers also install various versions of a 'Hanging Branch'; a bushy branch, sometimes leafy, sometimes bare, hung upside-down from a Level 3 Horizontal Beam by a chain and a padlock ('Padlock Point'), upon which the chimpanzees can climb and swing. The thick chain which secures Hanging Branch encircles the middle of the Horizontal Beam, running through a deep groove cut into the timber.

Every day in the wild, adult chimpanzees build a fresh nest, high in the forest canopy, in which they sleep at night. They build these from leaves, twigs and branches. The Singapore Zoo chimpanzees sleep indoors at night but they still like to build nests and the Keepers often supply nest-building materials for them. These materials consist very often of palm branches and other long stems and branches. The chimpanzees' skills in weaving these nests are prodigious and worthy of a study in themselves. Additionally, and of significance in the present work, their skills in weaving are frequently applied to other matters. Also important to note is that sometimes, within these bundles of nest-building materials, other objects from the human world are inadvertently introduced into the chimpanzee world.

It remains for me to mention the places where the chimpanzees can feed. There are two concrete, fake, termite towers which I term 'Tower 1' and 'Tower 2'. They are a couple of metres apart and are imitations of the genuine termite towers Jane Goodall saw in the African jungle. Dr Goodall was the first to observe and record wild chimpanzees using slender sticks as *tools* which they inserted into the holes in the towers and up which the termites would run, to be caught and eaten by the chimpanzees. This practice is sometimes referred to as 'fishing'. The imitation concrete towers in the zoo are not filled with termites but with milk and honey. The chimpanzees design, make and use specialist tools to get this food.

There are also two Food-Boxes; one small metal one attached to a vertical post which, because of its colour, I call 'Grey Food-Box' and into which a chimpanzee may insert his or her fingers to extract food, and another, longer 'Puzzle-Box' as it is called by the Keepers. This is a translucent plastic box with a metal lid on it. The box is usually installed so that it rests obliquely on a branch or log so that the food inside it falls to its lower end, away from an opening in the box at the higher end. The metal lid cannot be opened by the chimpanzees but it has a variety of different shaped slots made in it. The chimpanzees learn to obtain the food inside by use of a stick-tool which they insert into one of the slots in the lid. The food inside can be obtained by either spearing it or flicking it up to the entrance, depending on the type of food it is. The chimpanzees become extremely skilful in doing this. They use a variety of strategies but an interesting one is when a chimp flicks a piece of food from the bottom end of the box with the tool held in one hand and retrieves the food with the other hand which is held in a waiting position at the opening at the other, higher end of the box. Because of the colour of the box, a translucent yellow plastic, I call it 'Yellow Puzzle-Box' though, by the end of my study, the yellow colour has been all but bleached away by the ferocious Singapore sun.

2 Genesis

In my studies of children's drawing, I found that children move through successive generations of visual and dynamic structure. At the very beginning of the infant's life these take the form of physical actions but it soon became apparent that these actions are linked to emergent concepts the infant is forming. Dennie Wolf (1983) has described these as 'waves' of different approaches to emergent symbolisation. According to my observations, there seem to be three successive generations of visual and dynamic structure which sweep through the child's interaction with a range of objects and media like waves.

The First Generation of structure is largely topological and dynamic in character. It is to do with patterns of movement and the shape of trajectories in space and time. This generation of structure manifests itself in actions performed by children's entire bodies in time and space and my observations show that this is so of the chimpanzees. In children's paintings and drawings this structure is translated onto a two-dimensional surface by three basic actions of the drawing or painting hand. These actions issue from the natural oscillations of the infant's muscular and skeletal frame (Smith, 1983) and comprise a side-to-side fanning motion of the drawing hand and arm, issuing initially from the shoulder (*horizontal arc*); an up-and-down stabbing gesture of the arm and hand (*vertical arc*); and a pushing and pulling movement to-and-from the self (*push–pull*). These basic drawing actions will be described in full later but they are present in the painting by the chimpanzee Congo on page 12.

The Second Generation or 'wave' of structure redefines the first and elaborates upon it, manifesting itself in concerns about *proximal and distal relationships, adjacency, closure, inside and outside relations*, and the *beginnings and ends* of lines in space and time. Again, these are expressed through the child's whole body in space-time as they are by the chimpanzee. Like the child, the chimpanzee can also encode these relationships onto the drawing surface and, again, can be seen in Congo's painting. Here, just like a human child, Congo has as it were 'opened-up', push–pulling and horizontal arcing movements so that loops and closures are described. In other paintings by Congo, the *beginnings and ends* of lines are clearly demarcated.

The Third Generation of structure is about emergent concepts *going-through, connectivity and attachment* and also *moment-of-turn* and *angular-variation*. It also brings in structural variations which make possible ensembles capable of the expression of feelings and ideas. In Congo's painting there are some examples of rapid direction changes of his brush resulting in sharply angled lines but other aspects of Third Generation Structure, and the concepts which, in children's painting, emerge along with them, are not clearly represented.

Additionally, the child's use and organisation of Third Generation Structure is accompanied by a new ability to use visual and dynamic structure in reference to further layers of meaning and metaphor. Whether chimpanzees are also capable of imbuing physical and visual structure with such meaning remains a mystery. During this level of representational thinking the child may also encode *projective geometrical* understandings of *depth, three-dimensionality* and *line-of-sight* or *viewpoint.* Again, I have not detected anything resembling projective relationships represented in chimpanzees' drawings and paintings.

Over the 3-year period of my research in Singapore Zoo, I see all the first two generations of dynamic and visual structure in the actions of the chimpanzees and, with the exception of projective geometry, glimpse the beginning of the Third Generation too, in terms of an interest in *connectivity* between forms, *going-through* hollow or tubular forms and *angular-variation* or *moment-of-turn* (for example, in the way in which a branch might be bent through points of the compass). All of the emergent structural principles and concepts which underlie human children's drawing during the First and Second Generations of visual and dynamic structure are present in the exploratory, investigative and ludic actions of the small group of chimpanzees who live in Singapore Zoo.

Toward the latter stages of my research I also see dynamic and visual structure briefly encoded on the drawing surface by the chimpanzees just as it is in human childhood. Whether this has, as it has in children's drawing, additional levels of representation or symbolism within it, I do not know. Still, it would seem strange if there was not even a flickering of this possibility within the drawing actions of the chimpanzee. After all, we once thought that children's early marking was devoid of meaning. An interesting possibility is then, if, during the sensory-motor phase of children's drawing there are significant levels of expressive and representational thought, then perhaps there are for the chimpanzees too.

The chimpanzees' deployment of drawing actions is certainly very similar to those I observed in human infancy. They are clearly not random, anymore than those of human infants are, but sensitive, intelligent, intentioned, exploratory and investigative. They are sometimes performed quite slowly and carefully – not at all like the stereotype of the wild, human, abstract-expressionist artist. Frans de Waal (1999) also makes this important point and, although Thierry Lenain disagrees that what the apes paint or draw is 'art', he too notes the characteristic care with which chimpanzees paint and draw (Lenain, 1997).

As in human infancy, these are *dynamic-structures*, in the sense of events having internal principles which hold them together as forms. They emerge in a certain developmental sequence, became interwoven together and are mutually reciprocal in the way they nourish each other and in the way in which they cause structural variation.

Actions and field-of-view

There exist important physical actions shared by both human and non-human primates from which all representational, expressive modes and systems have their origin. It seems to me that the actions which underlie the emergence of expressive, representational and symbolic thought are of two kinds: *whole-body action* and *close-up field-of-view-and-action.*

Whole-body action

One kind of action involves the whole body moving through space and time. In this book I will term this *whole-body action* (and, depending on the exact context, sometimes *whole-body play*, or *whole-body expression* or *whole-body representation*).

Humans use it in their everyday lives in a huge range of bodily actions made in three dimensions of space plus the dimension of time. So do the chimpanzees and many other primates. Unlike the chimpanzees however, when we do it we never leave the ground, or at least, not for very long. Yet some of our greatest efforts are made when we attempt to do so. Many of our pastimes, sports, entertainments and art forms give us a taste of what it would be like to escape the tyranny of gravity. This happens for short durations of time on the stage, sports field or street. Even activities based on the earth, for example, gymnastics, football, dance, skate-boarding, half-pipe snow-boarding, free-running or parkour and other forms of street gymnastics and dance, all give us brief moments of movement in all three dimensions plus the dimension of time. Longer sessions away from the pull of gravity are sought underwater in SCUBA diving where we clumsily imitate the experts, the marine creatures, the fish, dolphins, seals and whales. Aviators copy the birds of the air and astronauts experience the magic of perfect weightlessness.

We like watching whole-body play too. Used expertly in theatre, in films and on the sports field it can reach high levels of artistry and stimulate our emotions.

Much of the content of our dreams and desires has to do with temporary departure from a 1 G gravity field. This whole-body expression is central to what I have come to think of as chimpanzees' precursor symbolisation.

Close-up field-of-view-and-action

Another kind of action also contributes to expressive, representational and symbolic thought. This kind of action involves a more specialised use of arms and hands moving within the visual field of the human and non-human primate; that is, within a small, spatial sphere extending to about arm's length from the face before the eyes and into which the conscious agent can insert his or her hands. Humans have developed this kind of action to a high degree in their formation of expressive and representational thought, but my study shows that chimpanzees too use this kind of action. In this book, I term this *close-up field-of-view-and-action.*

In the present theory, both kinds of action provide the basic structure for primate and human-primate acts of expression, representation and symbolism. As in human infancy, the chimpanzee is quick to exploit the expressive characteristics of movements of his or her body actions (Trevarthen, 1995). Chimp and child are launched on a developmental trajectory along which both learn to differentiate and combine actions, first of all empty-handedly; soon with handheld objects, for what will become communicative and expressive purposes. In the child's case, these actions will be put to representational and symbolic purposes and we may see the beginning of this in the chimpanzee.

We will start by considering an example of whole-body play as used by a very young chimpanzee.

Whole-body play: The example of the pendulum-swing

11 August 2006 (Parts 4 & 4a) Video-recorded observation (Edited DVD) Singapore

Rah (1 year 7 months) and her mother Ida (33 years) are sitting on the Yellow Puzzle-Box. Rah hangs by her left hand, on an old and worn fireman's hose, flattened with age into a rubbery strap, and provided by the Keepers for the chimpanzees to climb and play on. Holding onto the fire-hose, Rah swings outward from and back inward toward the steel and plastic food-box.

In this book, I term this type of swinging motion, a *pendulum-swing*, a two-way swing made from a central, overhead fulcrum and describing a simple curve which may repeat itself over and over again along the same course. The pendulum-swing is an example of one of a family of dynamic-structures the chimpanzee spontaneously generates and purposely puts together. Because these are consciously constructed by the chimpanzee, I think of them as special dynamic-structures which I call *action-assemblies* (Thelen & Smith, 1994). This is our first example of a *whole-body action-assembly*. For a chimpanzee, it is a relatively simple movement in three dimensions, yet this action serves as the basis of a virtually infinite series of expressive variations. This starts to happen in our present observation of Rah.

Rah starts to add some extra features to her basic curving line in space. First, she completes the apogee of her inward flight with an overarm slap of her right hand against the box, tentatively at first, then with gradually increasing strength.

I call this type of action-assembly, *trajectory-and-moment-of-impact*. I will explain later how the chimpanzee uses this to demarcate the finishing point of a linear trajectory in space.

A further variation is made when she translates her finishing slap into a securing hook of her fingers over the lip of the box. This action-assembly undergoes, in turn, a further transformation. Sometimes, when she flies inward toward the box, she touches her mother Ida for reassurance, using two, gentle, overarm arcs; *vertical arcs*, as I will term this type of arm movement throughout.

Perhaps these contacts against Ida's body are tests of her mother's tolerance, for Rah then starts to develop her swinging; each incoming flight now ending with a vertically arcing swipe at Ida's arm – something Rah would not risk if Ida was in a bad mood.

This is an example of a *whole-body* action-assembly. Though relatively simple in structure, this pendulum-swing is the basis of important precursor expressive and possibly representational forms which I will describe and discuss later.

We too, as humans, also start off in life using our whole body in acts of expression in which arms and legs, hands and feet are used in histrionic synergy. For most of us however our movement through three dimensions is soon restricted to the ground in acts of walking and running. Stuck on earth, all of us nevertheless use very complex whole-body action-assemblies. Even ordinary walkers and talkers do complex things in three dimensions of space.

As well as doing things with our whole bodies, both of us, chimpanzee and human, are good at moving from these larger-scale actions to smaller-scale ones made with our hands before our own eyes and often on flat, working surfaces. Some of us spend a large part of our lives making quite fine actions on flat surfaces including sheets of paper and computer screens. Many of us become specialists in using flat surfaces in conjunction with specialised movement of arm and hand. The work of the carpenter, the builder, the dressmaker, the electrician and the plumber are all extraordinary orchestrations of *close-up field-of-view-and-action.*

Like humans, chimpanzees also have skills in specialist activities situated just centimetres from the forward-facing eye.

Close-up field-of-view-and-action: Tool-making and tool-use

The spear

14 October 2005 (Part 5) Video-recorded observation (Edited DVD) Singapore

00:00:57 Grey-haired Ida (32 years 2 months) knuckle-walks unhurriedly toward Yellow Puzzle-Box. The Puzzle-Box has fruit inside. Ida knows about boxes. She knows about their inside and outside relations and she knows that in order to get the food which is inside, outside, she needs a special tool that she will have to make. After finding a suitably slender stick, she strips the leaves from it except for a bunch at one of its ends. Using her powerful teeth, she sharpens its other end into a spike whilst leaving a small section of bark which seems to serve as the textured grip of a handle which she grasps in a front-hand. She pokes the sharpened stick through one of the slots in the box's metal lid and spears fruit at the bottom of the box.

The chimpanzees are not restricted to one type of tool but can make other types to fit the requirements of a particular job. Another tool my chimpanzees make and use is the *mop*.

The mop

08 September 2005 Video-recorded observation (Edited DVD) Singapore

00:19:00 Ganga (11 years 9 months) is poking a thin stick into a hole in a concrete, fake termite tower (Tower 1) to obtain milk and honey placed there by the Keepers. She makes skilful hand-over-hand penetration and withdrawal of the stick. The end of the stick which she inserts into the hole does not have a sharp spike but a whitish, bushy mop of frayed bark and pale, shredded stem.

Ganga, a female adult, is adept at designing, making and using tools for inserting into the holes in the concrete tower. Like many other chimpanzees she can design and fashion different tools for different purposes. The specialist tool she is using now is of the 'mop' type. She made this by chewing away the bark and the innards of the stem at one end of the stick so that, instead of forming a sharp spike, this end is frayed into a mop-like or sponge-like head. The first few trial runs of the mop-tool will further roughen its working end and so improve its mop-like properties. This tool is ideal for soaking up the milk and honey within the tower which then can be withdrawn from the imitation termite tower. Sanz, Schoning and Morgan (2009) have observed wild chimpanzees also making mop-like tools though they use them to capture crowds of a special type of ant they like to eat.

Chimpanzee tool-making and tool-use was first observed by Jane Goodall in the Gombe forest of Africa. The use of tools is not uniform across the chimpanzee specie. Each group of chimpanzees has its own specialist tools, fashioned to suit specific needs. The revelation that chimpanzees fashion and use tools is one of the first indicators that each group of chimpanzees is unique in other ways too; in the interpersonal relationships of its members, in the way in which its society is ordered and also in the way it educates its young. This is one of the reasons why some primatologists say that chimpanzees have culture.

In the chimpanzee culture of Singapore Zoo, the beings sometimes carry a few different tools to a site, holding spares in one hand or in their teeth, whilst using, with the other hand, usually a front-hand, one special or main tool, designed for the task in hand (or one favoured for reasons harder to define). A tool-kit might consist of a long, slender stick with a sharp, spiked end; a thicker stick, sometimes with a blunt end, more like a probe than a spear; and a mop-tool. Some tools have defined 'handles' and, like the one used by Ida, some have leaves which have not been stripped off but retained at the non-working end of the tool. My observations of the young chimpanzees learning tool-use suggests that this leafy end is left there for a reason. Sometimes, a young chimpanzee will accidentally drop a stick and lose it down a hole in a concrete termite tower. It is easier still to lose a stick-tool down one of the many drilled holes in the wooden, 6 m high Construction. The young chimpanzees often practise poking a tool down into these holes. Accidentally dropping a tool is a nuisance since it means a descent to the ground to retrieve it. Retaining the leaves on the handle-end of the tool means that, if it is dropped accidentally, these leaves prevent the entire tool from slipping down through the hole and being lost. Perhaps the tassel-ended stick-tool is unique to my chimpanzees because it is designed for the special features of their environment. It might be that it is designed by experienced chimpanzees for their young as a sort of 'starter' tool.

Multiple-tool use has also been observed in wild chimpanzees in the Congo (Sanz et al., 2009). As in my observations these authors also show that chimpanzees will re-use a tool abandoned by another. A chimpanzee may just pick up a tool because he or she has just seen it discarded by another. Or they may smell the scent of another upon the stick. But it might also mean that they recognise the stick's function from its form. (One of the reasons they are interested in human artefacts is, I believe, that they know these are 'made' objects and are intrigued by the mystery of their manufacture and use.)

These examples of tool-making and tool-use are not ludic but adaptational; behaviours which have evolved for survival. Originally, these useful tools may well have been discovered by just messing around with sticks. The chimpanzee alters sticks into utilitarian objects but, within the same close-up zone of action, the same tools are often released from the constraints of tool mastery to become objects of creative play.

In this book, body actions are seen as the basis of semiotic development. Differences in lifestyle between human and chimpanzee mean that the ways in which actions are combined and used will differ between the species. Chimpanzees move in four-dimensional space-time; they are quite happy to interact together navigating through three dimensions in their arboreal environment, even dangling upside-down. I have noted already that they possess essentially four hands and although they favour their 'front-hands' for special tasks leaving their 'foot-hands' to play a subsidiary role, the skilful use of all four hands makes a powerful difference to the way in which expressive and representational gesture will develop.

Therefore, in contrast to my studies of human babies, when it seemed natural to start with their manipulation of objects (and later on, drawing materials) within a close-up, visual field and with the use of their 'front-hands' alone, it did not take long for me to realise it was not there that I should start my study of the chimpanzees. My interpretation of the chimpanzees' emergent acts of expression and representation would start by watching them high up in the trees, under the sky.

Breakfast timing

21 September 2005 Video-recorded observation (Edited DVD) Singapore

00:13:27 Nkosi (5 years 2 months 24 days) swings, pendulum-like, holding on with two hands to one of the old and flattened fire-hoses. This is a non-vigorous, relaxed pendulum-swing which he develops into a faster swing, accentuating it by little thrusts of his body at its two apogees.

Like Rah's swing this is no mere mechanical pendulum. It has rhythmical aspects. Nkosi seems to imbue this simple back-and-forth sway with a rhythmic accentuations and emphasis. He is clearly enjoying the movements he can produce with his own body in conjunction with this simple device. His happiness does not take an overtly excited form; rather both his face and body adopt a peaceful, even serene demeanour.

00:13:27 The last in his series of pendulum-swings takes him to a horizontal beam at Level 2 where he alights at Grey Food-Box which is fixed to a vertical beam at Corner 3. He stops at this box and starts to feed from it, using his fingers which he inserts into an aperture at the bottom of the box. Inside, his fingers can coax food to fall from higher up inside the box onto a tray below from which he can retrieve it.

What happens next in this observation is extremely interesting for it reveals a special state of mind without which semiotic and expressive actions could not take place. This special state of mind is *play*. We will need to understand play if we are to properly comprehend emergent expression, representation and symbolisation. This observation is a good place to start because it demonstrates how any action, even such a basic one as eating, may be released from its constraints in survival strategies; liberated momentarily from its adaptational and subsistence orientations and transformed to acquire quite different meanings in *play*.

21 September 2005 (observation continued)

When Nkosi starts to dip his finger into Food-Box he immediately follows this action with a sideways turn of his head to check on the whereabouts of the others. This is an important survival strategy. Nkosi is low down in the chimpanzee hierarchy so he has to keep a watch out for elders who might supplant him at the Food-Box. Togo, the Alpha male of the group (who is also his father), or Ganga, a high-ranking and somewhat bad-tempered female, might well approach and push him out of the way so that they can get to eat the best stuff first.

Nkosi's head-turning strategy is repeated several times as he checks on everyone's whereabouts. Each head-turn is interspersed with an interval in which he concentrates on eating. After he has repeated his head-turning he seems to realise that this reiterated action can be translated for its own sake into a rhythmical sequence. This rhythmic sequence is composed of the following elements in this order:

Finger-dipping
Head-turning
Finger-dipping
Head-turning
Finger-dipping

. . . and so on; a steady rhythm maintained throughout, so transforming his feeding episode into a rhythmical *play-assembly*. Note that nothing is added to the actions themselves. These remain essentially unchanged; they are simply subordinated to an emergent timing and rhythm.

00:14:07 Now he adds to this ensemble a rhythmical swinging of his right hand in synchrony with his head-turning, so that his whole body starts to oscillate from side-to-side, whilst the finger-dipping passage now becomes 'sharpened', as it were, having to take place within a single beat of time in order that the rhythm is maintained. Thus emphasised, this touch of the fingertips seems to punctuate each point-of-arrival of both his turning face and his swinging arm at the Food-Box. None of this interrupts the passage of food to his mouth which occurs easily during the time in which his body sways around again in an arc. This arc originates from first his pelvis, then his shoulders and finally his head, where it ends in preparation for the next swing, back to the Food-Box.

Then he adds something new. In primates' symmetrically arranged bodies, with arms and legs in pairs, one on each side of the body, it is natural that when one arm is swung, the other arm will also swing in opposition to the first arm. This is something I have learned in the Chinese martial art, Tai Chi Chuan. Now, Nkosi consciously and purposely starts to build up what was initially the natural and incidental swinging of his other arm and hand – the arm and hand not involved in the gathering of food and passing it to his mouth.

Based in the natural oscillations of his body, his act of eating is turned into play. Even though there is the ever-present danger that a higher-ranking chimpanzee may usurp his place at the Food-Box he still manages to enjoy the moment by turning it into play. As he continues this dance of body, arms and hands, he increases the amplitude and extension of swing of his right, non-feeding arm and hand – their involvement not one of subsistence but of play, their inclusion made solely to fulfil the requirements of its dynamic-structure.

Hand to mouth

In this observation, what started off as adaptational actions involving skills of hand-and-eye coordination and object mastery, are smoothly set free from their constraints in survival strategies and transformed into a rhythmic play-pattern. No major changes are required in the actions themselves; rather the change occurs largely in Nkosi's psychological orientation to actions which remain essentially unchanged. The play-assembly starts with a pendulum-swing which we may think of as *whole-body* play but quickly becomes what I am thinking of as *close-up field-of-view*.

Rhythmic patterns

The chimpanzees seem very sensitive to rhythmic patterns intrinsic to their own body actions. Nkosi happens to repeat a sequence of two actions: (1) dig for food and (2) look around. Making these actions twice running is enough for a rhythmic phrase to suggest itself to him and for it to quickly become a thing of interest in itself. Knowing himself to be safe for the moment, he can develop the action-pattern, even incorporating other body actions which echo the rhythm and structure of the initial process but which are not in themselves linked to the original acts of eating or reconnaissance.

Some have suggested that chimpanzees cannot be taught a simple rhythm but this is misleading. The task of copying someone else's rhythm requires a different skill from that required to generate a rhythm oneself. Exactly the same is true of very little children's music (Young, 2000). Because very young children may not be able to follow another's rhythm does not mean that they cannot spontaneously create rhythmic music and songs – they do. Something very similar happens in children's drawing of shape, too. Very young children may be able to produce many more shapes spontaneously than they are able to when asked to copy them. Being asked to copy a pre-drawn shape requires quite a different set of skills from those needed to produce your own, original ones (Matthews 1999, 2003, 2010; Piaget & Inhelder, 1956).

In the above observation of Nkosi, we can see the two main types of actions I introduced earlier, starting with *whole-body* locomotion and play, in which the primate's entire person is moved through large-scale, three-dimensional space plus the dimension of time, to a smaller, more finely focused, *close-up field-of-view-and-action*, most of this latter mode conducted with the front-hands in synchrony with eye focus and visual attention.

The former aspect of play, the large-scale, full-body play is seen in Nkosi's use of the pendulum-swing. Although this can be used for locomotion it is not so used here. He uses it as a form of enjoyment and expression. It is rhythmic and he varies the instances of apogee by little kicks of his legs, spins of his body or added stresses of his limbs. These instances seem to act as markers or punctuations at each end of his swooping arcs. Even though this is a relatively simple action-assembly it exhibits organisational characteristics far beyond adaptational or routinised response.

The second type of play, which begins as soon as he comes to rest by the Food-Box, is structured on a smaller scale, within arm's reach and before his eyes, forming a transitional arena in which specialised skills of manipulation are not only practised but studied carefully too, as entities of interest in themselves.

Mark-making, drawing and writing

In humans, close-up field-of-view play is very soon expanded and amplified by the graphic act and by the drawing and writing technologies available within our societies. These range from media as basic as food (as in the observation of Nkosi), or even as milk regurgitated from our stomachs, to media only slightly less rudimentary, like milk spilt from cups, to the more sophisticated pencil and paper and electronic technologies, including the digital camera and the computer. For some human children, the use of a flat plane, including many different types of drawing surface and screen, forms an analogue space (Wolf, 1983) which has a unique feedback effect into full-scale play and investigation, fine-tuning their interactions with the world. There is a mutually reciprocal relationship between their use of drawing and writing media and their behaviour in the world at large.

Full-scale or miniaturisation

In my studies of children I found that they were moving back and forth and between acts of representation which occurred in one of two kinds of spatial theatre; one, a kind of 'miniaturised' space (drawing being a very powerful one) and the other, a whole-body expressive and representational use of action. In human childhood, experiences in three dimensions plus the dimension of time eventually find their way back onto the drawing surface where they are encoded into lines, shapes and patches of pigment. Reciprocally, in dialogic fashion, play-episodes beyond the drawing or painting medium are influenced and guided by the analogue spatiotemporal theatre unfolding upon the drawing surface.

At first glance, it would seem that little, if any of this dialectic is involved in chimpanzee play or in their expressive actions. However, this might not be quite true. The studies suggest a glimmering of an interactive cycle in which actions performed in whole-body space are echoed and analogued by the chimpanzee too in the miniaturised space of their close-up field-of-view-and-action. Additionally, although the chimpanzees cannot transform their experiences of events into speech, they can and do reiterate rhythmic muscular experiences across different sensory communicative channels.

As I will later describe, a few of the chimpanzees in Singapore Zoo do produce some simple drawing. Of course, how far they develop this, as do children, as a tool for thinking is debatable. But perhaps we should not discount them too quickly. Just like human children, the chimpanzees oscillate between a field-of-view play-space and whole-body play-space. In human infancy, this process signals the beginning of visual and kinaesthetic mapping in which one set of actions may become an analogue for another set of actions. This cycle seems to be one of the indicators of expressive and representational thought.

A note of caution is warranted however. Evolution does not repeat itself and the ideas suggested in this book should not be taken to mean that, given sufficient amounts of evolutionary time then perhaps the chimpanzees would develop semiotic behaviours in much the way that humans have. As I noted above, chimpanzees are already a perfectly adapted species, they are not primitive humans and even if I were able to hang around the zoo for 6 million years (instead of just the three) there is no guarantee that the chimps would develop a culture of drawing and writing! In the present book it is useful to imagine the chimpanzee as representing an incomplete teleology but, though everything in the book derives from detailed observation, in the end, an act of imagination is all that it can be.

3 Spontaneously generated expressive structure

As I will later describe, my observations eventually capture evidence of the chimpanzees making mark-making actions. This finding is a challenge to those who, whilst conceding that chimpanzees living with humans may become interested in painting and drawing, insist that those in the wild will never do so. There seems to be sufficient evidence in the present study to contest this assumption. Basic mark-making actions may well occur naturally in the wild.

In human childhood, early mark-making seems to move through successive levels of complexity, resolution and differentiation. Elsewhere, I have grouped these marking actions (along with other, related but non-marking actions) into *First*, *Second* and *Third Generation Structure* (Matthews, 1999, 2003, 2004, 2010). The first drawing actions used by the younger chimpanzees seem to be of the same class as those produced by human children. I term these *First Generation Marking Action* (or *First Generation Structure*).

First Generation Structure comprises three basic actions. These three each has its own journey through an epigenetic landscape and, for the purposes of analysis, each may be treated as a separate entity. In real life, however, the developmental pathways of one criss-cross those of the others; the development of one action is linked with the development of those of the other two.

Three actions: Vertical arc, horizontal arc and push–pull

First Generation Structure derives from important physical actions shared by both human and non-human primates. These actions play an important role in the initiation and development of expressive and representational thought. There are three categories of action in particular which require description. These are *vertical arc, horizontal arc* and *push–pull*. Their developmental trajectories are entwined together and ultimately have to be understood in this way.

The actions originate from the natural oscillations, swayings, wavings, fannings and rockings of the muscular and skeletal frame (Smith, 1983). Though there are important differences between the ways in which humans and chimpanzees produce and hone these actions – due to muscular and skeletal differences and the different lifestyles in which these are embedded – the actions are quite similar in both humans and certain non-human primates including the orang-utan and the chimpanzee.

As we have noted above, in studies of human children it has been assumed for hundreds of years that these early mark-making actions have no representational, expressive or aesthetic value. Most people still think it to be haphazard 'scribbling'. Relatively recently though, several authors challenge this (Athey, 1990, 2007; Stamatopoulou, 2009; Matthews, 1984, 1999, 2003, 2010; Maurer & Riboni, 2010; Wolf & Fucigna, 1983), finding scribbling exhibits structure and meaning.

In human infancy, these are not merely motoric actions but signal the start of expressive and representational thought. This book shows that chimpanzees too move through the beginning of this sequence of development in their drawings and paintings and also in actions they make in – if not the wild – then in the kind of captivity in which they have no physical contact with human beings and in conditions which are as near to the chimpanzees' natural ones as is possible in a zoo.

Since the earliest markings made by children were thought until recently to possess no meaning and now we know that they do; and since chimpanzees move through the same sequence of mark-making activities, then the intriguing question is whether their markings, like those of infant humans, also have hitherto, unexpected meaning behind them.

Vertical arc

Vertical arc with empty hand: Human infant

9 July 2002 Video-recorded child-observation (Edited DVD) London

Poppy (13 months) strikes a yellow, plastic, toy house four times with the palm of her left hand, increasing the force of each succeeding blow until she finally succeeds in hurting her hand. She holds the palm of her hurting hand before her eyes, looks at it, and then turns her head to regard the offending part of the toy house which hurt it. Then she looks up toward me, wincing, wanting advice and comfort.

We see similar, empty-handed vertical arcs used by the baby chimpanzee.

Vertical arc with empty hand: Chimpanzee infant

15 September 2006 Video-recorded observation (Edited DVD) Singapore
(Version 2, Interpretation 2)

00:29:35 Baby Rah (1 year 9 months 8 days) arranges nest-building material, thumping and pressing it down with two front-hands in symmetrical vertical arcs. She seems to enjoy making these actions and she does not hurt herself. Big brother Nkosi (6 years 18 days) comes along as if to oversee, bending a stem over for the baby.

Next we see Rah making a powerful series of six, very quick, vertical arcs with her right hand on the nest-building material before she wanders off a few steps to inspect Auntie Ganga's bottom.

As with Poppy the human child, Rah's use of the vertical arc it is made to test materials plus her own strength in relationship to them. An important similarity between the two observations is that they both happen within companionable learning situations; Nkosi comes to have a look at what his baby sister is doing and even bends a stem for her, just as I keep an eye on Poppy and she looks to me for assistance or advice.

One obvious difference between Poppy and Rah is that Rah makes some of her vertical arcs using two hands in unison but human babies will sometimes do this too. A less obvious difference but one which will become increasingly significant as time goes by is one of control, skill and power. Rah, in the variations she makes in her vertical arcs, is already showing signs that she accommodates to the particular materials she is using. In addition to switching from two hands to one hand, she also modifies her downward strikes into a pressing action upon the pliable nest-building materials. It is noticeable that Poppy does not do this; she does not modify the type or force of her action until she hurts herself.

Both human baby and chimpanzee learn to modify their basic, First Generation expressive and investigate actions according to the kind of situation and the properties of the materials they encounter. The difference between them is that the chimpanzee does this more quickly. Consider the following observations.

07 March 2008 Video-recorded observation (Edited DVD) Singapore
(Tape 1, raw footage)

Kimoni (5 months 24 days) is atop the Yellow Puzzle-Box with his mother, Ganga (14 years 29 days) close by. Ganga will not allow him to go far.

Kimoni gently slaps the metal, top-end of the box with the palm of each of his front-hands several times. These actions are made in response to the cries of his brothers who have been kept in the Den today.

Both the episodes recorded above, that of Poppy (human) and that of Kimoni (chimpanzee), show the vertical arc offering up information about objects, materials and surfaces though Kimoni's actions are less wild and are used in response to his brothers' cries Additionally, the particular qualities of the box, including the resonance of the metal and plastic when struck, may be more consciously sought out and exploited by the chimpanzee. Although Poppy uses salient features of the toy house as targets (mainly its chimney), these are forcibly struck with an unadjusted palm in an action which inevitably causes the baby pain. In contrast, Kimoni's vertical arcs are more controlled and, whilst also directed toward local landmarks, the configuration of his palm is adjusted in mid-flight to the specific form of the target. For example, a couple of his vertical arcs are adjusted into a good grasp of a corner of the box. In this way, Kimoni's vertical arc is already being modified to become a grasp rather than a hit.

But both Poppy's and Kimoni's vertical arcs have communicative and semiotic aspects. Consider the continuation of the same observation made of Kimoni.

07 March 2008 (observation continues)

Later on in the same episode, this time sitting up against his mother's body, he makes another vertically arcing action but this time whilst he is looking at me.

This time, the contact of his palm against the box is again only slight, and in this respect is different from the powerful slaps used by Poppy. His vertical arc is already becoming a *signal* rather than a physical blow.

Techniques of representation: Deceleration

Because of the controlled *deceleration* he applies to the same class of movement, the vertical arc, plus the fact that he is looking at me, I take it that his action is made in relation to me and my gaze. Kimoni has slowed down his action for me. It is like a *pretend* slap. It is an act of communication rather than the testing of an object. Rather than the somewhat aggressive impact made a moment ago against the resistant surface of the box, this slowly executed vertical arc is at the beginning of a developmental journey toward becoming a *sign.* Rather than a strike, Kimoni makes a forerunner 'wave'. His purposeful slowing down of a downward strike, an action which is made usually at fast speed, is significant here. *Deceleration* is one of a cluster of techniques with which the chimpanzee and the human infant transform a utilitarian action into a sign. As we move through the book, we will encounter other techniques used by the chimpanzees to transform natural actions into vehicles of expression, and possibly, of representation and symbolisation. It is also important to note that this is an instance of cross-species communication. Kimoni has made a special effort to communicate with a human.

We have considered two observations of the use of the vertical arc: the first, by a human baby; the second by a baby chimpanzee. In the first example (performed by human infant), the vertical arc is used as a basic strategy for the testing of surfaces and objects. The other First Generation Actions are used in this way too, as we will shortly discover.

During this part of her development, Poppy uses the vertical arc to test any new object or surface she encounters and in doing so is rewarded by a range of sensory-cognitive feedback. This is also true of Kimoni's vertical arc. Additionally, I suspect that both Poppy and Kimoni are testing their own strength and power against highly resistant materials and objects. Whenever the two infants encounter a challenging and adamant material, they will strike it with the vertical arc action. They need to find the extent of their own strength and they need to assert their own power.

In Poppy's case, note that after she has hurt her hand she holds it before her eyes, looks at its stinging surface, then looks directly at the part of the toy house against which she hurt it and finally she looks at me. This *hand-regard* is indicative of a dawning realisation of the 'Self' as conscious agency in the world. Later, I will describe another example of hand-regard in a non-human primate.

These acts are also *expressive* in that they are driven by emotion. These are emotional acts performed upon inanimate objects. They mark the beginning of a journey toward acts of imagination. Both Poppy and Kimoni vent their feelings upon objects which have no feelings. This suggests to me that the objects are precursors of symbols with which actions of love and hate may be performed upon an imaginary other.

My feeling is that there exists a gradual continuum from acts which do not yet have an unequivocal imaginative dimension to those which do. There is no clear line before which the human child does not possess representational capability and after which he or she does. Perhaps this will also turn out to be true for the chimpanzee. Acts of representation and imagination are built upon precursors which have no representational or imaginative content in themselves (Mitchell, 2002a) but which, because of their structure and character, and because of the relationship between actor, act and the object acted upon, serve as the substructure for later, more fully developed acts of representation and symbolisation. With the proviso that we should avoid invoking a deterministic narrative, we might imagine that with the human child we see the entire developmental trajectory; with the chimpanzee we see only its beginning – the rest of it we can only imagine as a hypothetical, evolutionary trajectory.

This empty-handed action is later amplified with tool-use. As Poppy discovers, bashing an object with the hand may not always be satisfactory. Using some kind of stick or beater is a good way of causing percussive effects without hurting one's hand. Hitting this against a specially designed musical instrument (a xylophone, say), may be even better (Young, 2000). In such cases, the musical aspect is enhanced. With other media, the visual effects may be enhanced too:

04 December 1983 PAL VHS videotape-recorded child-observation (Hannah) (Slomo 2) plus hand-written transcription, London

Hannah (2 years 2 months 28 days), standing on a chair at a solid, wooden table, repeatedly stabs a brush, loaded with blue paint, down upon a sheet of paper which rests upon the table. Each impact of the brush makes a heavy bang and as she raises the brush after her final impact she sighs as she sees the blobs of a blue she has created.

March 2001 Video-recorded child-observation (Edited DVD) Singapore

In Singapore, a 2-year-old Chinese boy stabs his pencil down over and over again on his piece of paper which lies on the table at which he sits. He looks with apparent interest at the cluster of dots he has made.

In these cases, the use of a tool, brush and pencil, respectively, 'amplifies' the action (Bruner, 1964) because leaving traces of normally invisible actions has opened up the possibility of recording and representation; propelling the thoughtful act into a new sphere of influence. Later we will see chimpanzees cause trace-making effects. Like us, they tend to use their front-hands, with or without a handheld tool, to make them. Again, the big question is whether they have an interest in them comparable to that of the human child.

One further point before we leave these observations. Note also that these actions contain the germ of *rhythm*. The possibility that the chimpanzees produce and take note of rhythm becomes more certain in further observations.

14 October 2005 Video-recorded observation (Edited DVD) Singapore

00:18:37 Nkosi (5 years 3 months 17 days) has found a metal pipe or tube of about 33 cm long. He takes the pipe onto a Platform Rock – a flat boulder of about 1 m in diameter – and investigates it. This is one of the rocks the chimpanzee use as work-surfaces (upon which to crack open hard shells, for example), and to investigate novel objects, just as Nkosi is doing now.

He holds the pipe palmer-fashion in his left (?) hand by one end and stabs it downward with great force so that its other end impacts squarely against the flat, rock surface, producing a sharp, sonorous clank.

Where the pipe came from I do not know. It may be part of a previous construction which Nkosi has unearthed or it could have been prised out of the bed of the stream, as other objects are to be. Sometimes, a simple human-made artefact is accidently introduced into the enclosure along with the provision of bundles of nest-building materials.

The metal pipe is attractive to the chimpanzees I think because of its unusual straightness and hardness. It is also of a weight and size which fits comfortably into the hands of a young, strong chimpanzee. I myself could imagine being happy with its satisfying heft in my own hand, just as Nkosi seems to be. The metal pipe seems to be of an ergonomic scale and relationship to the chimpanzee body. I am to see many manifestations of what I come to think of as an aesthetic in the relative size of a favoured object to the chimpanzee physical stature. There is a special relationship between size, form and weight of the object and the primate physique which we find satisfying. The people who advertise and sell us things know all about this (Miller, 2000).

Non-accidental structure

The metal pipe is hard; iron and steel being the hardest materials the chimps discover and which they clearly value. It is human-made, of course, as are many of the found-objects the chimpanzees admire (including this metal pipe, a wooden rod, a metal handle of a bucket, even a slab of concrete). These objects possess *non-accidental* structure and perhaps this is what interests them. They know that these have been 'made'. They know (in effect) that these objects are artefacts. Perhaps Neanderthal people's relationship to the Cro-Magnon's and their artefacts was a bit like this; they found these beings and their things an interesting but insoluble mystery.

Nkosi hammers the pipe downward in a vertical arc, holding it from its far end just as any good workman would. He does not 'strangle' the downward force of the pipe by gripping it near its 'hammer' end. [Thus, the chimpanzees discern both the natural 'handle' and 'hammer' ends of objects and take advantage of these in a workman-like way.]

Nkosi, seemingly stimulated by the repetition of sharp sounds he makes, starts to create a rhythmic succession of blows with his metal pipe. There takes place a transformation from simple duplication and replication of a sound which pleases him to an authentic rhythm.

The synchrony between dynamic, kinaesthetic and auditory feedback is significant here. Human expression starts from analogies made between events transmitted through different sensory communication channels. Nkosi and Manzu are at the beginning of this journey.

The index and the icon

14 October 2005 (observation continued)

00:20:18 Nkosi (5 years 3 months 17 days) and Manzu (6 years 3 months 14 days) play-fight, Nkosi defending his right to use the metal pipe. He stands in the Stream just above the Waterfall, raises the tube in the air over his head and holds it there for a moment as a gesture of threat. In fact, he would never hit his brother with such a heavy, dangerous object. The threat of a blow is composed of a vertical arc intentionally frozen into a dramatic image. The stilled and upraised rod and the arm and hand in which it is held are configured into a unit to form an *iconic* signal.

As we go along through the book we will encounter emergent signs of two different types. One type is *iconic*, the other *indexical* (Armstrong, Stokoe, & Wilcox, 1995). The iconic sign presents a figurative reference to an event or object. It is a bit like a picture. Nkosi's raised hand and metal pipe, frozen in time for an instant, presents a dramatic picture.

The indexical, on the other hand, is not intended to look like anything but is numerical and abstract, usually indicating and marking a point in time, for example, the start or end of an event. Some emergent signs possess both these aspects and this may be the case here. Nkosi's emergent signing is *iconic* because, by pretending to be about to strike Manzu, he creates for his brother a visual and kinaesthetic image of that imagined action – a *threat*. The semiotic process does not end there however because he also wants let his brother know that this is a mock threat. He does not want to really scare Manzu because this will spoil the intended play. In the sense that it signals the beginning of a desired play-episode Nkosi's action has an *indexical* aspect too.

Techniques of representation: Curtailment and abbreviation

In this example, the act of *curtailment* is the principal technique used by Nkosi to make the semiotic transformations possible. Because he can exercise restraint means that he can compose an image. *Curtailment* is another member of a group of techniques used by the chimpanzees for altering actions in such a way that they gradually acquire a semiotic significance. In this example, curtailing an action is at the start of a developmental route toward its *abbreviation*. These and other techniques yet to be mentioned all overlap as strategies for transforming actions into signs.

The vertical arc is one of the first actions chimpanzees and humans deploy to investigate a novel object. It is used as both a first and last resort to gain an effect. Even human Information Technology geniuses have been known to give a recalcitrant computer a little whack to get it started. It is one of a battery of tests run on the object. Nkosi and Makhulu additionally investigate the rod in other ways too.

14 October 2005 (Part 4) Video-recorded observation (Edited DVD) Singapore

Underwater

00:20:35 Moving the pipe through water seems to them to be an irresistible test to run. Nkosi holds the metal pipe in the current of the stream, near the Waterfall, perhaps testing the strength of the current by interrupting the flow of water. He and the other younger chimpanzees make similar tests by running their hands through the water, or, using a hand, interrupting the flow of water poured out from the occasional, illicit, plastic water bottle. Nkosi may be simultaneously testing the strength of the pipe, or even (in effect) 'measuring' the depth of the stream, as other chimpanzees appear to do.

He holds the metal pipe in the current at the edge of the Waterfall, perhaps feeling the force of the current upon it, but preventing it from being washed over the Waterfall. He holds it in his right hand by one end and bites the other end of it. Makhulu watches with great interest, throughout.

00:20:47 Makhulu splashes water at Nkosi, using an underarm scooping action. Nkosi holds the pipe under the water and studies it through the surface. The water is sparkling clear and both chimps bend down to study the pipe's behaviour beneath the surface and perhaps to keep an eye on its whereabouts. Children too like to look at how things look when they are underwater.

00:21:10 Makhulu scampers off and Nkosi, crouching in the stream, watches his retreat and flings water at him with a right-handed, underarm scoop.

This action is a derivative of a First Generation Action now modified in relation to water.

03 November 2006 Video-recorded observation (Edited DVD) Singapore

It is over a year later. Holding a human-made, stout, straight, hardwood stick he has found, Manzu (7 years 7 months) makes powerful vertical arcs against the hard surface of Platform Rock. I am commentating on his actions into the built-in microphone of the video-camera and he seems to be listening to my quiet conversation with the machine. I notice early on in the study the chimpanzees' interest in human speech, especially extended conversation between humans. They are also curious about my monologues to the video-camera, which is, after all an inert object.

The canon and the canonical

There are several other important points to note here. Manzu's vertical arc is a perfect, one might say, *canonical* example of this action. Manzu is total master of this action and he can now form a perfect exemplar. The perfect example of something (a type or class of object or action), the *canon*, is itself recognised by the primate as unusual, not typical, not accidental. Once it is recognised as such by the primate, it can be used in many situations as a sign. This is another way in which a sign or a symbol might be created.

Manzu knows what this action is good for and how to adapt it to any number of situations. It is here chosen as an action to test the strength of the tool against the stone surface. With this action the chimpanzee simultaneously tests his own strength too. The sensory feedback gained from performing the action gives him valuable information about the stick, its relationship to other materials, to his body and to his thinking. His action makes a loud sound too, not as loud as that made with the metal pipe the chimpanzees found a year ago, but a sharp percussion nevertheless.

A musical dimension is present. The sound of the stick cracking down on the rock, once repeated, suggests to Manzu a rhythm. He does not pursue this musical possibility as Nkosi did a year ago with the metal pipe but it is present nevertheless. We will encounter this musical dimension again in other observations.

He has been attracted to the stick, as he was to the metal pipe, by its novel straightness, strength, hardness and density. He might know that this rod, like the metal pipe, is a human-made structure.

With both the metal pipe and the wooden rod, a chimpanzee exploits an object for its functional and expressive attributes. In both cases the human observer, myself, contributes and participates in the chimpanzees' enjoyment. I even sense that Manzu and Nkosi are aware of and appreciate my interest. Additionally, I am starting to think that they understand *why* it is I am interested. It is as though we share common interests.

Chimpanzees support young ones' learning

Some writers claim that young chimpanzees learn just by watching the other, more skilled and knowledgeable chimpanzees performing tasks, and that older, more experienced chimpanzees rarely support or 'scaffold' learning experiences for their young (Mitchell, 2002b). 'Scaffolding' is a term deriving from Lev Vygotsky's (1966) work and refers to the subtle support an adult companion might offer a child in his or her learning. The idea is that the 'more experienced learner' (Blenkin, 1990) anticipates to a certain extent what the infant is trying to understand or do and provides steps for the child to achieve this understanding. The following observation is one of many which challenge the notion that the more experienced chimpanzees do not support their young ones' learning. It highlights the provision of learning opportunities by the more senior members to their young. The following is another observation about the functional application of the vertical arc.

05 May 2006 Video-recorded observation (Edited DVD) Singapore

Breaking open a coconut

00:00:26 Ida holds a coconut in her hand and repeatedly strikes it in vertical arcs against a log. The others gather around in a circle to watch. Nkosi and Manzu make downward head inclinations – a sort of exaggerated 'nodding' – in time with each of Ida's downward strikes of the coconut.

Watching the observation carefully reveals how this old female allows both young males to have a good view of her act and how she performs it. Her support of the young males entails only an adjustment of her spatial positioning relative to the young males but she nevertheless stage-manages her actions so as to present them with a good view.

This synchrony between Ida's vigorous downward blows of the coconut against the log and the young males' vigorous, downward-going head movements shows that Nkosi and Manzu have an empathic response to her actions. In human infancy such empathic and synergistic responses to another (or another's) action come to analogue the model event and form a prelude for representation proper.

Here is another example of the vertical arc used to analogue another action.

13 May 2009 Video-recorded observation (Edited DVD) Singapore

Rah (4 years 5 months 4 days) pummels Ida with a vertical arc motion to induce her mother to play. When this does not work she switches to a repeated, downward, head-thrusting action which seems to reiterate her previous series of vertical arcs but in another form. I suggest that in cases like this, the chimpanzee is creating an analogue in which one action stands for another one. Despite being made with a different body-part (the downwardly nodding head) this action remains essentially within the same class of movement as that which it emulates [the downwardly pummelling fist, or the downward thrust of the bashed coconut in the previous observation]. When this sort of thing happens in human childhood I consider it to be one of the forerunners of representation and symbolisation.

The *nod* itself as a communicative sign may have its roots in such empathic downward jaw thrustings.

07 September 2007 Video-recorded observation (Edited DVD) Singapore

Suzie looks up into my direction when she hears a brief and friendly exchange between me and the Zoo Gardener. He hands me my umbrella which I have absent-mindedly left at another part of the Visitors' Path by the chimpanzee enclosure. He says to me: 'I found this, John.' 'Oh – thank you very much.' I answer. Suzie listens. She looks over at me. I nod at her and she returns my nod.

The 'nod' she gives me is not imitated from human communication but is a natural sign for the chimpanzee. They use it amongst themselves and (more rarely) give it to humans. Primatologists sometimes refer to this chimpanzee 'nod' as a 'head-bob'. Using principles of patterning and contrast, the chimpanzee moves rhythmically a part of his or her body (the head, in this instance) to make a non-accidental structure which is recognised by another as the product of conscious intention. The principle of *contrast* is being employed here. Suzie comes to use it as a greeting to me. According to my observations, Suzie's nod is usually affirmative in tenor though its specific meaning seems to change a little according to context. Suzie, Ida and Nkosi all give me head-nod greetings occasionally. On rare occasions I receive one from Alpha male, Togo, too.

All except for Ganga, their nods are of the up-and-down type, which in some – though not all – human cultures is thought of as an affirmative 'yes'. Grumpy Ganga, on the other hand, when on one of those rare occasions she deigns to give me a head shake, it is a side-to-side one, like she is perpetually saying 'no'! In one of my very last observations it appears that she is making a special effort to give me an up-and-down nod but she just cannot get her head and neck to perform this movement!

The vertical arc, like the other First Generation Actions, undergoes modifications. Here is an example in which a young chimpanzee uses modified vertical arcs to make transformations, or changes of state, take place in malleable material.

Rara and the clay lump

05 October 2007 (Part 2) Video-recorded observation (Edited DVD) Singapore

(Sculpture)

Rah (2 years 9 months 26 days) is on a Platform Rock, using it as a work surface upon which to pummel a football-sized lump of clay with her left hand, knuckles-down. By tugging at the grass growing upon it, she has pulled up this large, roundish, clod of clay-like earth. Rah pummels the lump with the knuckles of her left hand, an action no longer really like the vertical arc from which it is derived, but more an accurate, vertically aligned and repeated downward punch – a strategy used, in fact, by ceramicists. She pounds the malleable, clay-like soil five times in this way. Round about the third stroke she holds the clay in place with the back of her other hand, her right hand, as she pummels it. She looks down at her work for a moment and breaks the clod of earth into two main pieces, the lump to her right being the largest. She chooses the larger of the lumps and breaks a piece off from, what is to her, its left side, and then picks this up and chews it. She puts a small piece of earth into her mouth and as she does so, turns her head to face me. It is as if she is asking for my advice rather like my son Joel does at 13 months, when he touches a paint-covered brush tip to his mouth. Like him, Rah knows she is sometimes restrained by an elder when she puts certain materials into her mouth. It is interesting that in this instance, she looks toward me, a human, for such 'advice'. The reason this relationship is possible between us is because I have known her since she was a baby. Then she rolls the larger clod of earth and grass not quite into her lap but in the little play-and-work surface formed between her knees. This 'play-space' is also formed by very young children, when, sitting on the floor, they play with toys just in front of them. Rah has turned the lump through 360° so that it comes to rest soil-side up with grass-side down. She looks down upon it and hammers it again. She uses her left hand, knuckles down, five times in quick succession. This time, as she pounds it, she pulls her right hand up and out of her field-of-view-and-action. Now she lifts up the lump with both hands and looks closely at the effect of her actions upon the soft clay. She has made the clay smoother, less textured and concave in form, like a simple vessel. The imprints of her knuckles show in the clay.

She holds the form close to her face and then makes a fine adjustment to the orientation of this shallow, vessel-shaped form, so obtaining, for a moment, a perpendicular, direct line-of-sight to the inside hollow of the shallow vessel. She tilts it this way and that to inspect the interior of this concave form, from its deepest point to its rim.

Visual field

She places the lump of earth back on the ground, with the grassy side downward and the concave, clay surface facing upward. She raises her left hand poised just above her forehead, on the left side, 'waiting in the wings' as it were, just out of her theatre of action and vision.

This practice, that of withdrawing extraneous and distracting limbs or other objects from the field-of-view, is also typical of human children when surveying a close-up field-of-view. Children's act of painting is a good example of this organisation of the field-of-view-and-action when they require a clear view of the scene before them, unsullied by irrelevant objects (see Matthews, 1999, 2003, 2010 and also the observation of Hannah on page 236 and the Chinese boy on page 230).

Horizontal arc

The second of the First Generation Actions also involves a movement of the hand and arm in relation to a surface but this time, unlike the brief collision provided by the vertical arc, this action makes a prolonged contact with a surface in the form of a brushing or fanning movement. This is the *horizontal arc*.

In human infancy, the horizontal arc is a strategy in which the hand and arm is fanned from side-to-side, laterally before the face and body, along a smooth surface.

23 June 1975, 35 mm colour-slide film-recorded child-observation and hand-written transcription, (Joel) London

Joel (1 year 1 month 15 days) leans down to put his hands into a pool of milk he has accidentally spilled from his cup onto the floor. Using two hands in synchrony, he makes horizontal arcing actions of his arms and hands; actions which issue mainly from shoulders so that the palms of both hands brush from side-to-side in the pool of milk and make two, symmetrical, arc-shaped patches in it. He may be interested in the changes in shape, texture and colour he is causing to appear. Or, perhaps the episode originates from a desire to wipe-up (clean-up) the milk and the properties of colour, texture and shape are by-products of his action.

Like the other First Generation Actions, the horizontal arc has both functional and expressive aspects and it is sometimes difficult to separate the two. Joel's 'wiping' action (if that is what is) may be such an example. Other ways in which the horizontal arc may be used include that of a strategy for gathering objects to the self, or, the opposite, as an action for discarding them. Sometimes, a human baby, sitting at a table, or in baby's chair with a tray in front of him or her, will gather up a nearby object on the surface using a horizontally arcing sweep, which captures the object and carries it towards his or her midline. A variation of this horizontal arc can be used by the child to discard an unwanted object too, by brushing it away into space. I have also made observations of chimpanzees sweeping a surface to gather up an object; the retrieval of a stick lost in the grass is an example. I have made plenty of observations too in which chimpanzees will use a similar horizontal arc to discard things too, the sweeping away of nest-building materials from a high platform being a good example.

The same action, like the other First Generation Actions, may also be released from any functional use and enjoyed, by both child and chimpanzee, just as an action within itself.

22 September 2006 Video-recorded observation (Edited DVD) Singapore

Drawing in leaves (1)

Rah (1 year 9 months 13 days) makes a horizontal arc in the leaves with a front-hand. There seems to be no utilitarian dimension to this action; it is performed solely because it is interesting as an action in itself. An interest taken in an action for its own sake signals a movement toward an expressive use of gesture.

The chimpanzee is also interested in the effects this action has upon different materials. Note Rah's choice of material. The chimpanzee makes a match between selected action (horizontal arc) and the material in which to make it (strewn leaves). The chimpanzees are sensitive to materials; the way these might be used and the actions of the body which might be performed upon them. I observe several occasions in which a chimpanzee uses strewn leaves in which to make a prolonged brushing action yet minimises the same action upon a hard, rough surface, like a rock.

Here is another example of drawing in strewn leaves from an early observation of Nkosi, made when he is 5 years 7 months 27 days old.

27 January 2006 Video-recorded observation (Edited DVD) Singapore

Drawing in leaves (2)

00:01:36 As soon as I arrive, Nkosi knuckle-runs down to the water's edge to meet me. He looks me straight in the eyes and though this is a relatively benign greeting his interest is tinged with a little warning. These are early days in our relationship and he has yet to get to know me properly. As soon as I have asked, 'How's it going?' he makes a single horizontal arc with his left hand in the deep layer of dried leaves which cover the ground. His hand trails along on its knuckles from his right side to his left. There is a swish of leaves. This arcing movement originates from his foot-hands, is carried through his spread legs to his hips and shoulders where it is translated into a fanning gesture of the arm with the hand moving along on its knuckles through the leaves on the ground. After the back of his hand passes through the leaves there is left a barely discernible furrow or trace. Then Nkosi runs off.

Nkosi's horizontal arcing action in the leaves is made in just the way I have been taught in Tai Chi to make arm movements; the movement to originate and rise from the feet, be translated by the legs, hips and shoulders in this order and finally to be expressed in the movement made with arms and hands.

As Nkosi makes his horizontal arc he sways his body from side-to-side, a look of confrontation in his eyes. This is partially a greeting but it is also a warning. He recognises me and bounds down the path to acknowledge me but at the same time it is as if he is saying: 'Don't let yourself think we can get too friendly.' One might make the pun that he has 'drawn a line' between us. Keeper Mani agrees with my assessment of the situation. 'Yes, he is coming to greet you but he is also teasing you.'

For a chimpanzee, this action, a swaying from side-to-side with one or both arms and front-hands sweeping along or near the ground, is one way a power display can start; a performance made by a chimpanzee as a warning to another or others or a demonstration to others of strength and dominance (Goodall, 2000). In this respect it is different from horizontal arcs made by children. I have not seen a human child stand up and use his or her whole body to make the arcing movement in this way. Nor have I seen a human child use his or her knuckles to make an horizontal arc as Nkosi does. However, although Nkosi's action may derive from a power display, I feel he has separated it from its source and is using it for another purpose, as a partial greeting and signal of recognition. He does not want to be too aggressive toward me because he has learnt that I am not a threat. So, although this action in some important respects is different from that made by children, nevertheless I recognise it as belonging to a generalised class of movement I call *horizontal arc*.

Along with the *vertical arc* we discussed above and a further action, the *push–pull* to which we will be introduced shortly, this *horizontal arc* and its trace-making effect signals the onset of drawing and writing actions in human development.

Although Nkosi does not look at the visual trail he makes with the back of his hand, surely he must know by his haptic and kinaesthetic senses, that he has made one? The ground is quite deep in leaves so surely his impulse to make the action is guided by knowledge of this surface as a soft, impressionable medium? I mentioned above about the rarity of any similar tracing movements made upon rough, rock surfaces. When they do occur they are quickly curtailed.

Note, incidentally that the two examples of horizontal arc cited above, that of Rah and that of Nkosi, are both restricted to a single stroke. It is like a single statement. To repeat it would be to weaken it.

Here is another example from an observation made when Nkosi is 6 years 6 months 29 days old, which is also restricted to a single arc.

26 January 2007 (Part 7) Video-recorded observation (Edited DVD) Singapore

Getting irritated by some children making silly noises at him he performs a little power display on a big, flat rock, adding a miniscule, horizontal arc with his knuckles, but which is constrained to a faint, abbreviated trace, since the surface is hard stone.

This is another example of the use of *curtailment* which will lead to *abbreviation*. In this case, Nkosi may have abbreviated his horizontal arc because, unlike strewn leaves, the granite rock is hard and rough. Chimpanzees have tougher hands than ours but like us they avoid hurting them!

But perhaps most important of all is that trailing one's empty hand over a rock leaves no trace of its passing. This suggests to me that Nkosi is sensitive to the trace-making potential of different surfaces. Is it too much to suggest that when he passes his knuckles through that deep layer of leaves he knows he will leave a furrow? At the very least, he knows that the thick carpet of leaves will allow him to trail his hand in a way which a rough rock will not.

To a certain extent, such First Generation mark-making actions are haptic and kinaesthetic and it might be argued that visual awareness and visual interest is not required to make them.

Back in the 1960s James Gibson and Patricia Yonas made experiments to find out whether young children's first mark-making really required visual feedback (Gibson & Yonas, 1967, 1968). Would the children be just as satisfied by making the actions involved in trace-making but without seeing any trace? To find out, Gibson gave little children dummy pens which left no lines. According to Gibson, the children quickly abandoned these. More recently however, Alan Costall (2009, personal communication) reviews Gibson's experiment and feels it to be too vague. Perhaps the children thought that they were simply supposed to find out if the pens worked or not. Costall finds that children will sometimes be quite happy to move the markers through the air if such make-believe play is suggested to them. It would seem that there are occasions when the children (or the chimpanzee) requires visual feedback but, equally, there are other times when moving the markers through empty space is enough to sustain pretend-play scenario.

From my studies of human children's drawing kinaesthetic, proprioceptive and haptic information guides drawing (and other) actions but once children fully realise the potential of drawing, visual feedback is not merely sought but actually *drives* the urge to draw. This also seems true of some non-human primates, for example, Paul Schiller's chimpanzee, Alpha (Schiller, 1951) and Congo, in the company of another unusual human being, Desmond Morris (1962). At the close of my study, some of my Singapore chimpanzees did show an interest in drawing lines but whether they would ever develop an urge comparable to that of very young children and set out to prolong and extend drawing (whether for the visual effects or for other reasons) remains an intriguing question.

In the next observation, a chimpanzee's drawing actions start from strong proprioceptive, kinaesthetic and haptic (and even olfactory and gustatory) sources. However, these drawing (or perhaps 'wiping') actions are quickly guided by visual information.

06 November 2008 (Part 2) Video-recorded observation (Edited DVD) Singapore

Ganga and the red rag

I give the chimpanzees a bag of soft, magenta-coloured material measuring 23 cm × 30 cm and with a draw-string with which it can be closed. The chimpanzees soon reduce it to a rag and Nkosi gets it wet. Ganga (14 years

8 months 28 days) appropriates it, takes it to her lair on top of Level 3 Platform upon which she starts to make with it pushing and pulling actions and, later on, horizontal arcing actions.

Her actions recall *wiping* actions. Can it be that she is *cleaning*? Her actions and the resultant shape formed by them recall those made in milk by Joel at 13 months, in the observation above on page 53. As with Ganga, I thought that at least part of Joel's intention might have been to wipe and clean a surface.

Ganga now extends the range of her actions, using her left hand to include wiping along the wooden battening which frames the Platform's edge.

She is lying on her back, looking skyward, so cannot see what she is wiping, but she has haptic and kinaesthetic awareness of her actions sure enough. After wiping for a while, she brings the cloth to her nose and sniffs it (here's the olfactory element) and then holds it aloft over her face with her left hand and looks at it.

Though some of her wiping actions are directed from non-visual sources she does importantly pause to visually regard the effects of her actions. She does not look at the marks on the surface but at the marks she makes on the cloth. This does not greatly surprise me. For Ganga, it is probably important for her to find out what this novel, soft material is 'good for'. Young children, after making a mark with a new marker, will sometimes look at the tip of the marker, to see what has happened to it. In Ganga's case, there is some dirt on her red cloth, wiped off from the wooden surface.

However, the next part of the observation shows a grim determination on Ganga's part to see the visual effects of her actions upon the mark-receiving surface. It is as though, having found out what happens to the cloth, she now wants to see what happened to the surface it wiped.

Ganga reaches right under the platform and starts to wipe its flat, smooth, underside using a very clear horizontal arcing action of her left hand. Because the cloth is damp, the *horizontal arc* is described clearly in water on the underside of the platform. She is leaning over the side of the platform as far as she can, her head upside-down, in order to see this effect.

Here is a very good example of a chimpanzee taking great pains to see the visual effect of her drawing (or wiping) actions. In this observation we have an iconic *horizontal arc*, clearly visible, rubbed with a wet cloth through the grime accumulated on a shiny, painted, wooden surface.

Aside from humans and chimpanzees, other primates also make the trace-making action using the horizontal arc.

16 June 2006 Video-recorded observation (Edited DVD) Singapore

An orang-utan draws in urine

00:02:38 Two orang-utans, Miri (female) and Labu (male) are vying for a position on a rectangular swing, up in the forest canopy. Miri is sitting centrally on the platform and will not allow Labu to join her though he keeps trying. But then Miri urinates and Labu gets out of the way quickly! Miri watches with great interest the stream of urine cascading over the edge of the platform. Then she holds up the palm of her left hand and looks at it from a distance of 45 cm, showing some interest in what she sees.

This seems to be the same kind of *hand-regard* that we noted when Poppy the human baby slapped her toy house. Again, it may indicate conscious awareness on the part of the primate of her own agency.

Then, with her right hand she makes a horizontal arc in the urine from her left to her right and back again in a continuous motion. She seems to be looking straight at me as she makes this arc. When she finishes the movement she turns away from me, turning her back to me. However, she just as quickly returns her gaze toward me as if to see if I am still looking.

Orang-utan Keeper, Kumaren, tells me that Miri likes to do this with water and even collects water in her mouth in order to wet a surface which she then wipes (Kumaren, 2006, personal communication). Such trace-making is highly developed in humans where the potentialities of this action are extended with the use of tools.

03 July 1975, 35 mm colour-slide film-recorded child-observation and hand-written transcription, (Joel) London

A couple of weeks after he made the two horizontal arcs on the floor in the milk (see page 53), Joel (1 year 1 month 25 days) enters my studio and picks up a brush laden with yellow, acrylic paint. After finding out, by touching the bristles against the top of a paint pot, that he can transfer paint with this tool to another surface, he then describes, on the linoleum floor, a horizontal arc.

Whereas, two weeks earlier, this action had been achieved using the palm of his hands alone, in milk, now he can 'amplify' this natural action with the use of a brush in paint (Bruner, 1964; Matthews, 1999, 2003).

In both humans and chimpanzees, these earliest drawing actions do not seem just sensorimotoric but are guided and sustained by vision and interest.

In the following section we will consider the origin and development of the third and final First Generation Action – the *push–pull*.

Push–pull

There are two types of push–pull. The first type is made before the infant can support his or her own weight and be able to sit upright. The action involves pushing and pulling movements of the fingertips alone – essentially a scratching movement.

08 November 1974, 35 mm colour-slide film-recorded child-observation and hand-written transcription, (Joel) London

Finger-scratch push–pull

Joel (6 months) lying prone on his stomach on the carpet, scratches with push–pull movements at a small patch of regurgitated milk on the purple carpet.

In this very early example, the 6-month-old human infant is already producing the trace-making effect and is interested in the visual transformations he can cause to happen. This is not really a developed push–pull but it does show the response to a visual target and the visual effects created by a simple back-and-forth movement of the fingers.

The second type of push–pull, the developed push–pull, involves the whole arm and usually emerges when the infant can support his or her own body and sit upright. He or she is now able to push or pull the empty hands and fingers, or pull and push handheld objects to and from his or herself using an alternating outstretching and retraction of the arms, bending and unbending at the elbows.

04 March 1982 Child-observation, hand-written, (Hannah) London

Whole-arm or developed push–pull

Hannah (5 months 28 days) repeatedly pushes and pulls a toy telephone with wheels to and from herself. When the toy is so moved it emits a squeaking sound.

In the above example, tactile, kinaesthetic and especially auditory feedback encourage the baby to repeat the push–pulling action. In the following observation made 2 months and 24 days later she adds her own, vocalised soundtrack to her action.

28 April 1982 Child-observation, hand-written, (Hannah) London

Whole-arm or developed push–pull

Hannah (7 months 22 days) pushes and pulls a plate of bread to and from herself, accompanying this action with an almost donkey-like, in–out, in–out bray of satisfaction.

A range of sensory feedback is also involved in the following example of a chimpanzee push–pull.

18 January 2008 Video-recorded observation (Edited DVD) Singapore

Kimoni (5 months 5 days) pushes and pulls a strand of leaves to-and-from his own body. The mother's supporting stomach and hands form a chair into which he nestles his back, giving him the necessary support to perform this action. This push–pull is repeated with vigour, his arm fully outstretched on the push; fully retracted on the returning pull; the hand, which grips the strand of leaves, touching his lips when it comes back. On some of the strokes back-and-forth, the leaves sweep against the surface of the Lateral Log, like a brush. Whilst he makes these push–pull actions, his mouth opens and moves in a variety of cooing configurations though, from my location, if he does make any sound I cannot hear it.

It would be of extreme interest to know whether he made any vocalisations whilst making these actions. Jane Goodall (2000) tells us that chimpanzees find it very difficult to make vocalisations in the absence of any emotion but Kimoni may well feel strong emotions when he makes these actions. Birute Galdikas (2005) says this happens but I would have to be nearer to my chimps to hear such soft sounds myself. In human infancy, synchrony between vocalisation and actions performed with hand and arm is indicative of analogies being formed between different sensory communication channels. In children's development such cross-modal associations signal the beginning of representational thought.

28 September 2007 Video-recorded observation (Edited DVD) Singapore

Rah (2 years 9 months 19 days) is on Swinging Platform. The Swinging Platform is a long oblong board, roughly 150 cm × 200 cm, which hangs high up in the Construction from four steel cables, one attached at each corner. Rah makes clear push–pulling actions with both hands simultaneously and all fingers. She performs this twice in a 1–2 rhythm. I cannot discern any traces left on the surface. As she pushes and pulls she also sets the platform in motion on its stout cables.

She is to develop this rocking action in later months, often starting it off with some precursor drawing motions, apparently enjoying the precarious combination of movements she sets in train – *whole-body actions* which rock the platform plus *close-up field-of-view*, drawing-type actions.

One way in which the chimpanzees like to make an activity more interesting is simply to make it more difficult, even dangerous, by giving themselves a sort of 'handicap'. We might think of this as *danger play* or *deep play* and we will encounter it again. This observation is also a good example of a combination or *combine* of play-assemblies.

Variations of the push–pull action are soon developed. The following observation is made of Rah about 5 months later.

07 March 2008 Video-recorded observation (Edited DVD) Singapore

Rah (3 years 2 months 24 days) leaning over the surface of the Swinging Platform makes a series of push–pulls with her entire forearms – not just the hands. The chimpanzee arm is hairy and the platform is shiny, making possible this variation of the push–pull on Swinging Platform. Rah has made her hairy arm a sort of 'brush'.

She becomes adept in using sliding techniques firstly, as in this observation, with a body-part, later with her entire body and later still with a handheld stick along a surface in an action suggestive of drawing.

Slipping and sliding

Rah's *sliding* of her body along a smooth surface is an action unique to her play alone and has a developmental trajectory all of its own. I have seen no other chimpanzee in the family use it as intensely as her. She clearly enjoys the feel of smooth, near-frictionless surfaces.

Each of the three basic First Generation Actions has its own developmental trajectory, interlaced with that of the others. We will look at one such trajectory, this one launched from experiments with push-pulling, the *slide*.

24 March 2006 (Part 1) Video-recorded observation (Edited DVD) Singapore

00:00:50 Holding on by four hands, Rah slides down a Slack Strap fire-hose, which is inclined at approximately 40 degrees, and ends up with her mother, Ida, who is waiting at the bottom.

08 September 2006 Video-recorded observation (Edited DVD) Singapore

00:04:08 Rah (1 year 9 months) walks on all-fours on Horizontal Beam, Level 3, but slips and almost falls. She keeps hold of the beam however and slides around it so that she is hanging from it from all fours with her back to the ground.

About 15 min later she seems to already transform that which was initially an accident into a controlled routine.

00:19:50 Rah hurls herself lengthways onto Horizontal Beam at Level 3 and follows through this movement by allowing her body to slide around the beam at which point she takes hold of a branch and makes a swing upon it.

This action-pattern is all very fluid and graceful but it is surely a development of her earlier accident today in which she slipped around upside-down on the horizontal beam. The fortuitous and the accidental play an important role in the development of human art too (Luquet, 1927/2001; Matthews, 1984, 1999, 2003, 2010).

Though this is a personal creation, like all play-assemblies, Rah's slide has a pre-history which includes the ideas and help of others.

27 October 2006 (Part 4) Video-recorded observation (Edited DVD) Singapore

Rah (1 year 10 months 20 days) and Nkosi (6 years 4 months) are on Level 2. Nkosi bends over, sticks his backside in Rah's face and looks at her through his legs and upside-down eyes.

An act like this is sometimes referred to as a 'script-violation', a purposeful insertion of the incongruent into the congruent. It is important in the development of semiotic thought and I will discuss it in depth later. In the meantime, suffice to say that Nkosi intends this act just to amuse his little sister and is perhaps surprised by her unexpected next action.

Rah now jumps onto his sloping back and slides down it. She spills off the edge of the platform but saves herself at the last moment by grabbing the timbers.

Although unprepared for this, Nkosi appreciates the quality and structure of Rah's creation and now cooperates with his little sister to repeat their double-act but with variations in the way in which she avoids falling several metres to the ground (sometimes she grabs a fire-hose; sometimes one of the poles, for instance).

24 November 2006 (Part 2) Video-recorded observation (Edited DVD) Singapore

(New Swinging Platform)

Nkosi (6 years 4 months 27 days) and Makhulu (6 years 6 months 17 days) are playing on a newly installed Swinging Platform. Using inertial force they slide forward on their behinds. Rah (1 year 11 months 17 days) joins in.

They can slide on their bums easily because these are protected by leathery pads called ischial callosities. These pads, rooted right through the haunches to the bone, allow chimpanzees (and Old World Monkeys) to sit for hours on a branch without hurting themselves. Sliding on one's bum is rare in the young 'naked ape' (Morris, 1967) but not unheard-of. If they wear nappies or diapers some babies find they can move around by sliding, swivelling and bouncing along on their bums – even before they learn to crawl. (This is an interesting take on 'tool-use' by the human baby.)

I rarely see Nkosi and Makhulu again practice this or other forms of sliding but Rah seems to instantly pick it up as an idea and translate it into many forms of slide.

26 January 2007 Video-recorded observation (Edited DVD) Singapore

Using her right forearm as a pillow for her eyes and forehead, Rah (2 years 1 month 19 days) lies prostrate on a wooden platform and, with her foot-hands and legs, propels herself forward along the smooth surface in a sliding action.

In this observation, she simultaneously blindfolds herself and cushions her eyes, eyebrows and forehead against the board along which she slides. The structure of the chimpanzee skull means that her nose and her chin will not need protection since, with her forehead pressed flat down on her forearm, her lower face slopes up and away from the board by about 20°. By blindfolding her eyes whilst sliding forward she also gives herself a handicap, making this an early experiment in *danger play* or *deep play*.

26 January 2007 (Part 5 continued) Video-recorded observation (Edited DVD) Singapore

On the same day, a little later, Rah makes a leap onto a Swinging Platform and instead of stopping at the spot on which she lands, translates her impact into glide or slide of her entire, prostrate body along its smooth surface, setting the platform into motion in doing so.

This is a unique variation on an important play-assembly I term *trajectory and impact* which we will discuss later. An important point to make here is that the Swinging Platform slopes a little and its obliqueness can be increased by setting it swinging, which is what Rah does here. A smooth surface which slopes in a regular fashion (a *slope*) is non-accidental structure noted and exploited by both the human and non-human primate for the actions it affords. Some famous human-primates to make use of a slide down a slope are none other than Charles Darwin's children. Darwin, delighting in his children's 'free-flow' play (Bruce, 1991) allowed them to slide down the stairs in his house. This enlightened parent even had carpenter John Lewis build for the children a sort of wooden, stair-carpet sledge (Keynes, 2001). Like Nkosi the Chimpanzee, Darwin the Human appreciated a younger one's ludic interest in, and use of, the *slide*.

Slide, ski and sled

22 June 2007 (3) Video-recorded observation (Edited DVD) Singapore (Rah) (continued)

On the same day, Rah (2 years 6 months 15 days), like Darwin's kids, develops her slide still further with the use of a tool or device. She has no carpenter to build her one but she makes do with sliding along Horizontal Beam upon a stick held laterally across her chest and against the smooth, well-polished wood, so making a ski or sled. Later on in the same observation, Rah combines the idea of the 'hanger' with that of the 'slide', hanging onto a hessian sack, slung over a sloping Slack Strap, like a sort of pulley or ski-lift, and sliding beautifully down the old rubber hose upon it.

I will describe the chimpanzees' invention of the 'hanger' properly later on.

31 August 2007 Video-recorded observation (raw footage on DVD) Singapore (Tape 1)

Rah (2 years 8 months 23 days) has transformed her slide yet again. This time she runs on all-fours with her lips pressed to the edge of Swinging Platform and so slides her opened mouth along it in a sort of long, slithery kiss! Perhaps her wet lips and tongue leave a long smear on the wood. Later during this same observation session, she slides her palm along the same platform edge, as if reiterating her trailing kiss with a trailing hand.

By the time she is 3 years of age she has perfected her trajectory and slide and can perform it with great panache.

18 April 2008 Video-recorded observation (raw footage on DVD) Singapore (Tape 1–1 Tape only)

00:15:36 Rah (3 years 4 months 11 days) runs full pelt along Level 3 and launches herself off its edge into space, hitting the Mezzanine Level below and instantly sliding along it on her hairy chest. I think she does this partly in relation to some appreciative school children who are watching but she also looks toward me on completion of her slide.

'Well done, Rara. That was a very good slide.' She looks at me and then turns her face away, as if shyly.

The sloping or *oblique* surface is differentiated from vertical and horizontal axes and exploited by both human and chimpanzee. Perhaps a crucial difference between the species is that in human children's drawing, there happens a point in its development when this differentiation is mapped onto the drawing surface and organised within an overall system of Cartesian coordinates, horizontal, vertical and oblique (Piaget & Inhelder, 1956). When this happens in children's drawing it signals the start of Euclidean and, later, projective understandings. I have seen no evidence of the encoding of these three axes in the drawings of chimpanzees. However, in their interactions with this carpentered environment in the zoo, the background experience for such understanding is surely present.

4 Structural principles

Modification, variation and extension of First Generation Actions

The interweaving developmental trajectories of the three basic exploratory and expressive actions – vertical arc, horizontal arc and push–pull – eventually lead to what one might think of as emergent *structural principles.*

As indicated above, once First Generation Actions are differentiated from their initial fusion, they are practised, by both humans and chimpanzees, as individual units of action. Both the young chimpanzee and the young human quickly becomes expert in their use. From here, members of both species start to combine them together or with other actions to form new patterns of movement in time and space. (We saw one of these combines above, when Rah put together push–pulling, drawing-type movements whilst setting rocking her entire drawing surface!) The permutations of these actions, their application and the inventiveness of their use are indications that chimpanzees are at the beginning of the trail toward the purposeful composition of acts of expression, representation and symbolisation.

The first play-ensembles are histrionic performances driven by bursts of emotion in which the different actions are fused together.

15 September 2006 Video-recorded observation (Edited DVD) Singapore

Rah (1 year 9 months 8 days) thrashes a frond of leaves in a mixture of horizontal and vertical arcs against the surface of Level 3.

Wild mixtures of horizontal and vertical arc are also typical of the human infant when he or she makes histrionic gestures of self-expression.

06 July 1982 Child-observation, hand-written, (Hannah) London

Hannah (10 months) rubs violently the surface of the tray of her baby chair with a fluid assembly of horizontal arcs and push–pulls. In her case she can also accompany these excited actions with excited vocalisations.

Additionally, during some of this performance, she violently shakes her head from side-to-side, like an exaggerated aversion strategy; a precursor 'no' signal. Hannah seems to take a delight in both the violence and the humour of this action.

Later on, the use of these First Generation Actions is more controlled and any variations made are intentional.

The gentle art of ear-flapping

08 September 2005 (Part 1) Video-recorded observation (Edited DVD) Singapore (Transcription 1)

Nkosi (5 years 2 months 11 days) on Level 3 thrashes Manzu (5 years 5 months 8 days) below him on Level 2, semi-playfully, with a bunch of leaves, using a variety of arcs. Some of these actions are vertical arcs but he also changes position to make an upward, arcing swing instead of an overarm downward one. Manzu cringes below, but tolerates the rain of thrashes. He is a tough guy too and anyway knows his brother will not really hurt him.

He does however shake his head from side-to-side, making the precursor 'no' gesture. Like Hannah's head-shaking it seems humorous.

One difference between Manzu's head-shaking and Hannah's though, is that he can make his big ears flap around. (It is important to mention at this point that, unlike humans, chimpanzees do retain the vestigial ability to move their ears slightly toward a sound source by use of the *pinna muscles* in their ears – Trivedi & Mohnot, 2002.)

As in human development, seemingly trivial actions are not merely abandoned with maturation but retained and developed for expressive use. Here's some more ear-flapping but this time by his brother Nkosi just over a year later.

15 September 2006 Video-recorded observation (Edited DVD) Singapore (Version 2)

(Interpretation 2) (continued)

00:29:35 Nkosi (6 years 2 months 18 days) seems to be helping Rah (1 year 9 months 8 days) with her nest-building techniques. She has been bashing and pressing the straws and twigs but now her big brother presses down on her nest for her with his front-hands. As he does this, he looks at me and nods up and down to me, making his ears flap. The nodding of his head and the flapping of his ears reiterate and underscore the repeated downward thrustings of his arms and hands. Ganga, nearby, has been producing her side-to-side head shake but now she notices Nkosi bobbing his head and wonders to whom he is doing this. Sure enough, she follows the direction of his gaze and finds me, with my video-camera, on the zoo visitors' path.

Earlier, I remarked on an apparent absence of projective understanding in the chimpanzee but this observation shows that this is by no means total. In some contexts – like this social situation, in which Ganga follows Nkosi's line-of-sight – a chimpanzee displays an unerring sense of optical and projective geometry.

Here is another example of ear-flapping used to emphasise a direction.

Techniques of representation: Exaggeration

19 September 2008 Video-recorded observation (Edited DVD) Singapore

Nkosi (now 8 years 2 months 22 days) holding on to Tremble Post with front-hands, violently shakes it back and forth, using what is essentially a push–pull action but one so powerful that even his ears flap back and forth too.

There is a functionally dependent relation between back-and-forth or sudden side-to-side movement of one's head and the flapping of one's ears. The flapping of Nkosi's ears 'goes with' and possibly can be said to reiterate the vibrations of the Tremble Post. I think that both Manzu and Nkosi purposely cause their ears to flap. I think they both use the technique of *exaggeration*, exaggeration of movement; its intensity and force, to draw especial attention to this action.

To me, his ear-flapping looks comical, but does he think it 'funny'? Does it appear so to another chimpanzee? Do Nkosi and Makhulu recognise that flapping the ear is somehow strange and so its addition to interactive communication may be to mark and share with another this unusual moment – the other in this case being a human? The difficulty of answering this question is indicative of the enigma of the chimpanzee.

The mystery of chimpanzee 'humour' arises again in the following observation made 2 weeks after Nkosi thrashed Manzu with the bunch of leaves. Just for a change, Nkosi bashes him with a plastic bottle.

The plastic water bottle

21 September 2005 Video-recorded observation (Edited DVD) Singapore (2nd movie Edit 2)

00:03:16 Nkosi (5 years 2 months 24 days) has found an empty, plastic mineral-water bottle. It is transparent and has no cap.

00:03:19 He hits Manzu (5 years 5 months 22 days) repeatedly on the head with it. Good-natured as he is, Manzu gamely tolerates his brother's teasing. The way in which Manzu just sits there, taking it, like the stooge in a comedy duo, strikes me as 'funny'. Do they find it funny, too? The empty, lightweight plastic bottle makes a repeated, hollow, plunking sound as it strikes Manzu's skull.

To me, even the sound itself is slightly comical, but again, would the chimpanzees find it so? I think it fair to write that they both find the sound interesting enough to want it repeated. This makes a rhythmic pattern. It may be that both chimpanzees are intrigued by the fact that, because the object is both lightweight and hollow, one can hit another quite hard with it yet without causing pain but instead produce an interesting sound. Humour is sometimes created by unlikely juxtapositions like this.

Humour may come from a transitional zone between fear and pleasure (Morris, 1967). In humans (and sometimes in other primates) this emotion may be expressed in laughter. That the action of 'striking my brother on the head' results in a 'funny' sound is what we think of as the 'joke'. Arthur Koestler (1964) defines humour as the collision between two kinds of logic. The present observation may show an evolutionary early example.

Techniques of representation: Repetition

In the above observation, Nkosi again teases his long-suffering brother with the use of the vertical arc. Even though the bottle is just a joke-weapon and chimpanzees are much tougher than humans, Nkosi nevertheless slows down his overarm blow to cause no pain and at the same time to 'tell' his brother that none is intended. Controlled deceleration of a normally fast action is an important, conscious technique of transformation. It will allow simple actions to be given additional qualities and effects which will further transform them and their meaning. Exaggeration is an important effect and one way of achieving it is by simply slowing things down. Its combination with *repetition* is a further step toward the making of the imaginative act. If the identical or near identical event happens once or twice this could be an accident but if a sequence of them happens then (unless it's a pulsar) it is most likely made by someone who thinks.

5 The meaning of the moving hand

The touch (not forgetting the tap, the press, the prod and the poke)

The initial thrashing and stabbing vertical arc is gradually refined into a very gentle contact which has a vast range of potential communicative content embedded within it. The ability to form a skilful, sensitive touch of the fingertip is a little noticed and little discussed aspect of expression, representation and communication. The gentlest touch can take on portentous significance. Carefully placed, as if inserted into a spatiotemporal interval, a touch can be as startling as a slap.

Hitting the spot

06 October 2005 (Part 2 continued) Video-recorded observation (Edited DVD) Singapore

00:10:25 Without letting go of Ida, baby Rah (1 year 9 months 29 days) now reaches out as far as she can along the horizontal beam. She investigates and pokes a small position on the beam, a spot, a physical 'spot', with the index finger of her left hand. She scratches at this spot or mark on the beam with small pulling movements of the nail of her left index finger.

Like 6-month-old Joel's, this is not a fully developed push–pull but it does embody the same back-and-forth scratching or rubbing of a surface. This is the use of push–pull as a means of testing surfaces. Like other First Generation Actions, these early explorations are made within reach of mother. Variations of the touch include the *tap*, the *press*, the *prod* and the *poke*.

Point

The reaching and touching with an index finger toward a small point or spot may presage a *pointing* gesture – though in my observations any pointing the chimpanzees might make appears to be minimal and made in relation to the self and not to direct the attention of another. This seems mysterious to me, since, as I have noted earlier, the chimpanzees do in many other ways recognise that another being has a different visual point of view. They seem to appreciate another's point of view (both geometrically speaking and at further, empathic levels) yet I have not seen a chimpanzee unequivocally point something out to another in the sense of requesting someone else to look at what he or she is looking at.

Though like us in so many ways, as regards the use and understanding of the *pointing* gesture, in that respect they are different. Recent research by Juliane Kaminski (Kaminski in Child, 2010) shows that dogs quickly come to understand a pointing gesture made by a human but that captive chimpanzees do not. Dogs have associated with humans for thousands of years and have come to learn about our emotional states of mind. They will seek a human's face for subtle facial cues of emotions, even looking to the left side of the face because this is the visual pathway for human emotion into the brain sited in its right hemisphere. Dogs even detect human direction of gaze. Although my chimpanzees use concentrated eye-gaze to direct another's attention, dogs can detect where a human is looking simply by watching the human eye. Human eyes have irises situated against a contrasting field of white sclera which makes their movements very noticeable. Perhaps significantly, chimpanzees do not have this area of white around their eyes.

Does their failure to use and understand the pointing gesture have something to do with a limitation in the chimpanzee's understanding of another's mind (a limitation in the chimpanzee 'theory of mind', as some would have it)? Or do chimpanzees sufficiently convey information to each other by means far more subtle than 'finger-pointing'?

Michael Gumert (2009, personal communication) suggests to me that one issue for some primates is that the act of 'pointing' may be interpreted as an aggressive action. He thinks that perhaps the chimpanzees have not the 'cooperative coordination' necessary to direct another to look at, or go to, another place. Dr Gumert makes the interesting comment that though they seem to understand very well some aspects of another's thinking, curiously, other things going on in another's mind seem difficult for them to understand. He suggests that we now need to know what aspects of another's thinking they grasp, and what aspects they have difficulty grasping. The ability to 'point' is related to this issue.

My chimpanzees seem to direct another's attention to another place by staring until the other notices the direction of the stare and follows the line-of-sight of the one who stares. Some of them will physically tug another in a particular direction. Figures 7.1 and 7.2 (see pages 101–2) show Rah physically manipulating Nkosi's direction of gaze by grasping his head in her hands and turning it to where she wants him to look! This means that a chimpanzee has the requisite understanding of another's point of view (in the strictly, optical-geometrical sense) but does not (for whatever reason) translate this into a convergent axis with his or her own straightened arm and pointing finger. Is it a *perspectival* understanding which is lacking, then? Abbreviations of physical tugging to empty-handed gestures made in empty space suggest that the chimpanzees are on the borders of using the 'pointing' gesture along with other signs. Touching a small 'point' in space might also be a forerunner to pointing something out to another.

Even more crucially there occurs the rare occasion on which chimpanzees will use a pointing gesture to indicate a *direction-of-travel* but only if there happen to be other, supporting contextual cues. The example of ear-flapping cited above might seem a strange cue to us humans but there are other kinds which we ourselves employ. For example, nearby instances of clear co-linearity in saliently long objects occasionally supplement and assist the use and meaning of the pointing finger. I will describe one such occasion later.

Before we leave this section it is very interesting to note that, on occasions, and almost in the spirit of evolutionary recapitulation, before the human infant understands the pointing gesture, the frustrated parent may grasp the infant's shoulders or even the head, and physically turn the baby in the direction of a spectacle the adult feels that the child really should not miss. Early years educator Hannah Matthews reports a 3-year-old doing this to an adult. In her nursery class the girl will sometimes tug the teacher's head in the direction she (the child) wishes her to look (H. Matthews, 2010, personal communication).

In the following observation, made 1 week later, baby Rah gradually adjusts the vertical arc to an exploratory touching movement of the index finger.

14 October 2005 Video-recorded observation (Edited DVD) Singapore

00:21:34 Up in the Construction, Rah uses a vertical arc with left arm and hand to strike Ida's right thigh. She bears a slightly impatient and aggressive play expression which reminds me of her brother, Nkosi. In a following strike however, she significantly decelerates her blow and seems to be paying more visual attention to her target. Then she changes the sway of her arm so that it hovers before a nearby vertical pole. She seems to look at the texture of the wooden timbers and it is at this moment that she transforms her initial slapping movement to a carefully focused touch of an index finger.

This seems to give her a further idea when she attempts, with the same hand and arm, to tentatively reach around the vertical pole as far as she can. She looks toward me as if she senses my interest and approval.

In this example, as in my many observations of human children, the vertical arc is gradually modified in the light of experience. In this case, Rah's idea seems to have arisen from some initially bad-tempered-looking whacks she gives her good-natured mum. Then, a final, single strike is noticeably slowed down. She has noticed an interesting visual target on the tree; a knot-hole perhaps, or some other pattern or distinguishing spot. This seems to be the beginning of her new idea. She turns her attention to the timber pole nearby. For some reason, she does not swipe wildly at it but, using the knowledge she has acquired by touching (at first wildly, then more carefully) the safe target of her mother's body, she adjusts her arm and hand configuration plus their speed of approach so that her index finger arrives gently at the pole's surface.

The visible moment

The moment when Rah has a 'new idea' is clearly visible. I see exactly the same behaviour exhibited by very young children (see observation of Hannah, page 236). As one kind of action gives way to a different kind there occurs a momentary pause during which the child has the new idea of what to do next. This phenomenon is observable to the naked eye; there is really no need to invoke, as some do, a 'theory of mind' allowing one to see the invisible inner workings of another's brain. Careful analysis of the action reveals the intention behind it.

Derived from your basic vertical arc, a single tap can also be used to set up a complex action sequence.

The beat

14 October 2005 (Part 4) Video-recorded observation (Edited DVD) Singapore

00:31:33 Makhulu (5 years 5 months 7 days), wet from the stream, approaches Nkosi (5 years 3 months 17 days) and slaps a beam in order to initiate a play-chase with his brother.

The brothers play-chase each other and end up in the stream, water-splashing each other, using underarm scooping actions to do so.

Chimpanzees can give a simple action many different meanings. In this case, the sudden, sharp, slap of Makhulu's wet hand against the timber is a visual and auditory signal to his brother to start a play-chase. Its meaning does not end here however for, in addition to its function as starting signal, it also forms the first beat of the rhythmical sequence it has itself set up. Sure enough, on cue, Nkosi starts the first step of his response pursuit dead-on-time with the second beat of a rhythmic passage initiated by his brother.

Of course, both the chimpanzees have to understand this pattern in order that the play will start and work. It is illuminating to watch an older chimpanzee try to initiate a younger one into such play when the young one has not yet grasped the use of an action as a signal or a summary of a desired event. I will describe one such event later.

The First Generation Actions are modified in other ways too. In the following observations, certain characteristics of a chimpanzee's action are *amplified.*

Techniques of representation: Amplification

18 August 2006 Video-recorded observation (Edited DVD) Singapore

Makhulu (5 years 8 months 11 days) has dipped a hessian sack (given to the chimpanzees this morning by one of the keepers) into the pool above the Waterfall and now, carrying it in his left hand, moves furtively toward his brother. He makes a Tai Chi cross-step along a fallen log toward his brother and flails at him with a horizontal arc of the sack. Water flies off the sack toward his brother, Nkosi (6 years 1 month 21 days).

Makhulu uses the wet sack as a goad to tease his brother. Inertial force is used here to spray the water from the sack at the end of its horizontal arc in space.

We find the expressive use of inertial and centrifugal force in human infant development too.

03 July 1975, 35 mm slide-film-recorded child-observation and hand-written transcription, (Joel) London

Joel (1 year 1 month 25 days) lets fly droplets of paint from a brush he swings in horizontal arcs.

Chimpanzees cannot swim but some like water (Galdikas, 2005). Of the three brothers, Makhulu is the one who really enjoys getting wet.

15 December 2006 (Part 4) Video-recorded observation (Edited DVD) Singapore

Nearly 4 months later, Makhulu makes a horizontal arc with a brush of twigs. Using an overextended version of a similar horizontal arc, he flings his brush away. He goes to collect it. He takes his bundle of leaves and carries it through the Waterfall with him. He gets himself and the leaves thoroughly soaked. Later on, he makes a horizontal arc again with the now wet leaves. This same class of action now creates a cascade of water which is flung off by centrifugal force in an arc of silver droplets.

In this observation, Makhulu makes use of *centrifugal force* and *angular-momentum* to make twigs, leaves and water fly off through the air. The human infant also learns how to use centrifugal and inertial force.

06 April 1984 35 mm colour-slide film child-observation plus written transcript

Spilt milk

Hannah (2 years 7 months) holds a plastic drinking cup which has a lid and a small spout. Holding the cup with the spout turned away from her she makes vigorous vertical arcs in space. Inertial force causes droplets of milk to cascade from the cup's spout into the air.

An important way in which these First Generation Actions are extended is when they leave a trace of their usually ephemeral passing. This is really the beginning of drawing.

After causing milk to spurt onto the plastic table-covering, Hannah then makes a series of horizontal arcs, push–pulling actions and then a rotational action of her hand in the spilt milk.

The meaning of an action

It might be argued that the ludic action-patterns made by the chimpanzee are not so richly polysemous as those performed by the human child. However, perhaps their amorphous nature might set the stage for such symbolic acts over evolutionary time. Nancy Smith (1983) has pointed out that it might be necessary for the formation of symbolic thought that precursor actions be so amorphous.

In the above observations, we see the horizontal arc moving through a wide range of usage, from experiments in inertial and centrifugal force through to the amplification of these forces with handheld objects – sticks, rods, pipes and so on – and also by the adding of other elements (water, for instance). From here, we have also glimpsed its use as a signal and its trace-making effect. In my observations of human children, the onset of the trace-making effects of actions seems to coincide with their beginnings of a communicative function. The humorous act seems to be one way representation and expression might arise. I believe there to be a humorous dimension to the chimpanzee communicative use of body-action too. Both chimpanzee and human infant learn that routine sequences of action may be purposely disrupted for effect. As I mentioned earlier, in primatology, this is sometimes referred to as 'script-violation' (Mitchell, 2002a).

Simple actions are used as elements within more complex action-assemblies. In the following observation, Rah uses a leg and a foot-hand to make a sweeping arc, demonstrating a difference between human and chimpanzee use of first generation structure.

22 June 2007 (3) Video-recorded observation (Edited DVD) Singapore

Rah (Breakthrough) (First edit)

Stretched out on her stomach on Word Platform, Rah (2 years 6 months 15 days) makes repeated horizontal arcs with right foot-hand; a movement which passes through a wide arc right up to her own head!

Again, the combination of flexibility and the protection provided by her body-hair, allows her to do this. Earlier, I suggested that human infants also start off using all four limbs as expressive vehicles but soon give preference to their 'front-arms'. However, human babies retain their flexibility for several years, and if they know that their limbs are protected (by clothing) will still perform acts more commonly associated with the chimpanzee. Here's an example of this I happened to see in a coffee-shop in London.

10 November 2009 Child-observation, hand-written, London

A 3-year-old girl is sitting next to her mother with other adults at a table in a coffee-shop. The girl is modeling plasticine. When mum starts clearing up the table prior to leaving, her daughter helps.

Suddenly, the girl finishes off cleaning with an agile horizontal arc performed with her right leg which sweeps across the table-top! She is wearing woollen tights which protect her calf and heel whilst reducing friction. She flicks her leg onto the table with ease and speed and makes the motion of wiping in a single instant. In fact, the action is so swift and elegant that none of the adults appear even to raise an eyebrow!

This example shows that one of the last vestigial traces in humans of the use of leg and foot as an arm and hand. As well as the agility and flexibility of the child it also reveals her knowledge that her leg's woollen cladding will smoothly wipe the surface whilst at the same time protecting her skin. This recalls the carefree sliding of the hairy arm or leg of the chimpanzee across a smooth surface.

There is one more aspect of the observation of Rah which needs mention.

'Peepbo!'

Line-of-sight

22 June 2007 (3) Video-recorded observation (Edited DVD) Singapore

Whilst she is sliding her leg about she peeks out at me over the edge of the Platform. She then ducks her head back behind the occluding edge of the platform so she disappears from my sight, before suddenly peeking out again.

This is very like a 'peepbo' play that one might engage in with a human infant. Using her sliding technique she develops an alternating 'peep–hide, peep–hide' sequence by gripping the edge of the platform with front-hands and sliding back and forth on her behind, so sliding her foot-hands in and out of my sight along with her face and eyes. Each time she reappears she is looking directly at me.

In this example, we can see that the chimpanzee, even at her young age, is already incorporating this deceptively simple action into a more complex play and – even more interesting – in relation to a human being. Once again, we might also cautiously entertain the idea that there is here the beginning of humour.

At first, because she is nervous and shy of me, she ducks her head out of sight beneath the edge of the wooden platform. This is an aversion strategy which we also observe in human babies and it signals the beginnings of an understanding of *point-of-view* and *line-of-sight*. In human infancy, these are prerequisite cognitive requirements for the understanding and use of *projective* geometry. The emergence of this concept will occupy us later on.

Rah's is also an example of the horizontal arc being used as a ludic action-assembly but this time in conjunction with a familiar human (myself). Note that all I do – all I can do – is *look*, but careful looking is a form of support (so-called 'scaffolding').

6 Directions and locations

Beginnings and ends

> A film should have a beginning, middle and end – but not necessarily in that order
> (Jean Luc-Godard)

Our first understandings of time and space may start from experiences of proximal and distal relations and our feelings about them. The first contrast we may experience may be the contrast between being in close contact with someone who cares for us – of being attached as it were – and its opposite, of being distant, separate and alone – or, at least, the threat or possibility of being so.

Perhaps this contrast is coupled with other bipolar qualities of warmth or coldness, darkness and light, fullness or emptiness, hardness or softness, roughness and smoothness, bitterness or sweetness and, gradually, the spectrum of textures in-between.

This would mean that the objectifiable properties of the world are saturated with subjective, emotional qualities from the outset which shape the actions and interactions of the primate. Perhaps in this way, spatiotemporal and kinaesthetic experiences may form a substrate for emotional expression.

14 October 2005 (Part 4) Video-recorded observation (Edited DVD) Singapore

Ida (31 years 4 months) is sitting on Horizontal Beam, Level 2, is holding baby Rah (10 months 7 days) in her lap. Her older offspring, Nkosi (5 years 3 months 17 days) comes up close to the pair. He tries to play with his baby sister but he is too rough and angers Ida who bites his foot-hand. The young male yelps and hurries away a few metres along the beam, crying and looking sorry for himself. He sits curled and compressed on the end of beam, a blot at the end of a line. After several minutes have elapsed, Nkosi wanders back to Ida, seeking reconciliation and trying to nestle into the same 'lap-space' as that occupied by Rara.

The *lap-space* is the name I give to a special space for infants in front of and against the seated older or adult companion's chest and between the older one's knees. It is an intimate, safe place where the infant hopes to get warmth, comfort and love. It is an important *transitional* arena in between, on the one hand, vulnerability and dependency and, on the other, independence and confidence in the outside world (Winnicott, 1971). Later, we will consider transitional phenomena more closely.

In the present situation Nkosi has fallen out of favour with his mother Ida and he feels unhappy. The old grey female won't allow Nkosi's re-entry into the comfort zone until he behaves himself. He contents himself with stroking her forearm until, at last, she allows him to come close to her. Mother and son are reconciled.

Here is another, similar observation made about 7 months later.

09 June 2006 (Part 2) Video-recorded observation (Edited DVD) Singapore

There is a sudden rainstorm. The family shelter in a tight group in the Construction. A fierce argument erupts, a squall as unexpected as the rain. Nkosi has upset one of the elders and, shrieking, he scampers off 50 m to huddle in the grass; a soaking-wet, miserable looking, black ball of woe. The drenching downpour heightens the effect of his sadness and desolation.

Is he expressing loneliness or, if not loneliness then, of *being-alone*? (See Mitchell, 2002a.)

After sitting, screwed-up in the rain for a few minutes, he wanders back to the group to seek reconciliation, which, after a suitable delay, he receives.

In both observations, proximal and distal relations, or contact versus separation, express, in formal-spatial terms, profound emotions associated with either being together with someone or alone.

These topological and haptic geometries expressed in contact and non-contact may be followed by the discovery of further properties as *inside*, *outside*, *closure*, *hollowness* and *boundary*. From here, the discoveries of *going-over*, *going-inside*, *going-outside*, *going-through* and *going-along* are made, and then the continuities and discontinuities in surfaces, sounds, light and shade.

Going-straight

The act of dropping defines perhaps the first straight line as well as the quickest route between two points. This may form part of the background experience for the start of Euclidean geometrical understanding, which is gradually built upon the developmentally earlier topological thinking. Emergent predispositions for directions of movement emerge from operating within the natural forces of a 1 G gravity field; going straight-down, straight-up or straight-along.

First, topological and then Euclidean geometry defines empty space. Straight lines grow out of sinusoidal meanderings. For some primates, inside every wiggly stick is a straight one wanting to get out. Straight sticks are found easier to use than bent ones. If nothing better comes to hand, a big old twisted branch found in the forest might conceivably be used as a spear or a staff or a tool but it requires great skills to do so effectively. Such skills are formed through the manipulation of objects in processes which uncover their embedded geometry. In many observations we see Nkosi, Makhulu or Manzi find the central and invisible straight axis of a big, wiggling stick in performances which resemble magically taming a serpent.

Clearly defined passages of movement are made and enjoyed by both young chimpanzee and young human. Although no physical line is always or necessarily drawn, such passages of movement soon become what I consider to be *linear routes*. Earlier, I raised the issue of an imaginative aspect to chimpanzee behaviour. The idea of the imaginative is of course complex and its definition controversial but it starts off simply enough as a possibility embedded in ordinary, physical action. For instance, when a chimpanzee or a human child moves directly from Point A to Point B an *imaginary line* is produced in space and time.

Soon, both human and non-human primate starts to demarcate key positions along linear routes they make. These act as landmarks. The first such landmarks to be noted are the *beginnings and ends*.

Going-up, going-down and going-along

There are no real beginnings and ends in natural space-time and giving it these markers might be the first way in which sentient creatures give it structure. In this chapter we consider the chimpanzees' movements through three dimensions of space plus the dimension of time. Chimpanzees move fluidly through any axis in time and space. The routes they make are very complex. To understand how they are organised it is useful to start off by investigating how they are structured according to the laws of gravity. An *ascending and descending* route (a *vertical axis*) is fundamental. If we allow that chimpanzees also spend some of their time on the ground then we might suggest that a roughly *horizontal axis* is important too. We will start with vertical axis and the ascending and descending trajectory.

Ascent and descent

Descent (and then ascent) is discovered in relation to the infant's own body in a 1 G gravity field. Falling is the first and perhaps only true straight-line course produced by natural movement of the primate body. A true vertical axis arises from the natural ballistics of the body in space.

With their kind of lifestyle, much of the time navigating through three dimensions of space in an arboreal environment, chimpanzees quickly learn about gravity. Their interaction together, similar in so many ways to that of humans, is different in the respect that they interact in a fully three-dimensional medium of space, sometimes high off the ground. The adults have incredible strength so that their movements seem effortless. It is curious – and rather telling – that few human researchers [one notable exception being Willie Smits writing about orang-utans (Schuster, Smits, & Ullal, 2008)] of non-human primate intelligence rarely, if ever, mention the fact that chimpanzees move efficiently and skilfully through three dimensions; nor do they consider the phenomenal amount of mental computation required to perform these feats.

Additionally and importantly, chimpanzee *play* may be structured along any axis in three-dimensional space. The closest similarity I can think of is that of astronauts or experienced scuba divers hanging weightless in space.

Divers, like chimpanzees, have to learn to navigate through three dimensions, remembering not only to look all around them along a horizontal axis but up and down too, along a vertical one. Like sports-divers underwater, chimpanzees cannot use speech. Communication between them recalls the subtle gestural language used by divers to each other, when any movement of the body made through three dimensions gains intensity as its semiotic significance changes in this 'silent world' (Cousteau, Dumas, & Dugan, 1953).

A falling leaf

Learning about the vertical axis

When a chimpanzee baby is born, it knows precious little but to root for its mother's nipple. At first it cannot even cling strongly to its mother's hair. Everything has to be learned. They have no 'instinct' for climbing and have to learn this from their mothers and more experienced siblings. Chimpanzees have a culture; all of their infancy and on into their juvenile years is devoted to learning and they pass on their knowledge to the next generation.

We will consider chimpanzee learning and the support given it, later on. Right now, I want to focus on just one aspect of the chimpanzee's investigation and exploration; that made of the *vertical axis*.

Whilst, as a baby chimpanzee, one might not yet be able to climb up or down on one's own, one may experiment with the dropped object and empathise with its fall.

> **14 December 2007 Video-recorded observation (Edited DVD) Singapore**
>
> 00:06:00 Ganga (13 years 10 months 6 days) with her baby Kimoni, just 4 months old, are up on top of the tallest tree (Tall Tree, we will call it) with a bundle of edible leaves. She accidentally drops one stem and Kimoni, hanging from his left arm vertically over the edge of the sawn-off top of the tree on which he sits in his mother's lap-space, watches it fall down. He touches the stripped trunk of the tree, looking down along his arm toward the falling branch.

This act of looking-along a straight or straightish object (a stick; one's arm) derives from reaching-toward and is the beginning of the strategy of *taking-a-sighting*. This act involves embedded knowledge of *line-of-sight*. Line-of-sight and point-of-view are understandings which form part of a *projective* geometrical knowledge. In humans, these presage a non-egocentric, or allocentric understanding. We will return to this important topic of line-of-sight and point-of-view later.

> He seems to be pulling off leaves in order to drop them and to watch them flutter down. This requires Ganga's cooperation. She has to allow him to take one of her twigs.

The act of 'giving' is generally considered rare in chimpanzee culture but my observations suggest that they do a lot of sharing in subtle and complex ways. Simply allowing another to take something of one's own is one such way but there are also occasions in which they will hand each other things. Later, I will write more about chimpanzee giving and sharing.

> Kimoni lets fall the new stem down the long drop to the ground and watches its descent with interest. The baby continues to look down and he makes a series of gentle and controlled vertical arcs with the same left hand as if to celebrate or express his interest in this phenomenon.

Perhaps this up-and-down action is a *synergistic* response to a falling twig, even a precursor representation of it.

Synergy

In this book, 'synergy' is an empathic response to another, in which one identifies totally with that person (see also Gregory & Jessel, 2004). Such synergy may form the background for communication and, ultimately, language. Some synergistic responses are made automatically, without conscious thought or decision, to the actions of others. When little babies thrash and wriggle their arms and legs about in response to someone talking to them, this is a synergistic response (Condon, 1975). If you reach out suddenly to support a stumbling stranger, this is also synergy. Additionally, highly sentient beings, including primates, dolphins, canids and some bears may make synergistic responses to the actions of inanimate objects which happen to get set into motion. These animals seem to respond to such movement as if it is made by a sentient being. Some of these responses are more voluntary than the developmentally earlier, autonomic synergistic behaviours. Such empathic, synergistic identification may form the background to expression, representation and symbolisation.

14 December 2007 Video-recorded observation (Edited DVD) Singapore (Observation continued)

Ganga now wants to distract her baby, Kimoni, from this activity – she has other things to do. She turns away, pulling the baby into the ventral position and turning her back on me. She may also be tiring of my interest.

00:07:00 Kimoni, in his mother's arms, continues to look down, splaying out his fingers and sometimes touching them to his pink, baby face, not yet blackened with age and rosily lit in the morning sun. Then he makes a further series of vertical arcs in space, this time made more dynamically downward, as he stares down at the long drop.

Going-down

Kimoni, now aged 9 months 5 days, has not lost his interest in falling objects. In fact, instead of just watching the phenomenon of descent when chance causes this to happen, he begins to set up situations in which he purposely demonstrates it. In the following observation he skilfully manages the descent of a leaf.

19 September 2008 (Part 2) Video-recorded observation (Edited DVD) Singapore

Kimoni hangs from his right arm at Level 2. In his left hand he holds a yellow leaf which he studies with care. He rotates this left hand, the leaf within it, whilst simultaneously mirroring its movements with movements of his head through the same number of adjustments. In this way, he views the leaf at each of its slightly different positions. He raises the leaf a few centimetres and then, from a limp wrist abruptly formed, he dangles the leaf again and turns his head at a slant to look at it. Opening his fingers with slow deliberation, he releases the leaf from his hand and watches it fall to the ground.

Rah, as a baby, is also interested in things which fall. Like other infant chimpanzees she practises dropping sticks and leaves and watching them tumble down away from her.

Watching something fall

14 October 2005 Video-recorded observation (Edited DVD) Singapore

00:19:17 Baby Rah (10 months and 7 days) is chewing at something as she looks downward. Then she opens her mouth and allows her morsel to drop. She watches it fall to the ground about 8 metres below.

The interest in watching things fall does not die out as the chimpanzees grow older but reappears, transformed by changes in cognition. All through my 3 years of observations at the zoo, Rah returns again and again to her interest in dropping items and watching them fall, repeatedly adding new inventive twists to this idea. Here she is again but 1 year 3 months and 12 days older.

26 January 2007 Video-recorded observation (Edited DVD) Singapore

Rah (2 years 1 month 19 days) reaches down and watches rain running off her fingers. She spits out pieces of leaves and, leaning over the edge of the platform, watches them fall too.

The human baby also becomes interested in watching things fall. They like to watch over the edge of a pram or push-chair whilst they drop objects, toys or groceries or anything which comes to hand, in just the same way as Rah and Kimoni like to watch leaves and twigs fall from their hands over the edge of a beam or branch.

Jumping and leaping

Both the human infant and chimpanzee learn to jump and leap.

18 February 2009 (Part 4) Video-recorded observation (Edited DVD) Singapore

Ainka learns to leap

Suzie (34 years) and Ainka (1 year 5 months 19 days) are sitting peacefully on the grassy bank by the moat. Ainka toddles just less than a metre away from her mother and then practises leaping toward her. She repeats this several times. Suzie watches her daughter carefully.

Ainka is learning her basic leap from Point A to Point B; a rising and falling arc comprising the elements *going-up*, *going-over* and *going-down* plus *point-of-arrival*. This simple dynamic-structure forms the core of much of chimpanzee and human infant play.

Gradually, they both become especially interested in the point-of-arrival at the end of a fall; the end of the line, the moment-of-impact.

The punctuated fall

The point-of-arrival of an intentional fall, jump or leap is demarcated in a number of ways by both chimpanzee and human infant. A simple way in which the salient end of a line in space is noted is with a strike, a slap or a tap of the hand. This transforms the ordinary fall, or the merely functional jump, into what I think of as a *play-assembly*.

Usually, the structure, organisation and semantic characteristics of play-assemblies grow out of natural origins. First of all, an inevitable outcome of a leap in space is that the hands or foot-hands strike a branch or the ground at the point-of-arrival. Sometimes this makes an audible, distinctive smack. Such sharp sounds automatically, inevitably, register the end of a linear trajectory.

The demarcation of the end of a descending line may derive from another natural action too, automatically made to balance the falling body at its moment-of-impact. When a chimpanzee leaps from one position to another, the sudden arrival is sometimes balanced by a tap or a beat of a free hand against the surface of the arrival location. A little slap or stamp provides instant additional support to the primate body at crucial intervals between different actions, within the precarious transitional moments between stable postures.

Eventually, this slap or beat is exaggerated or even added when not strictly necessary for stability. It becomes used (I would argue) simply for aesthetic pleasure, as if to announce one's arrival whilst 'at the same time' emphasising or underscoring the exact moment of that arrival.

28 May 2007 Video-recorded observation (raw footage on DVD) Singapore

Rah (2 years 5 months 21 days) leaps off a bough to land on another and, as soon as the palm-soles of her foot-hands hit its surface, she slaps it with both palms of her front-hands. The slapping sound is sharp and distinct.

When her palm-soles strike the bough she accentuates this moment of impact by – almost but not quite – simultaneously slapping the same surface.

In Rah's case, the moment of arrival is demarcated or stressed by use of her front-hands which slap the branch on the second beat of a 1–2 rhythm. Beat 1 is the first impact made by the palm-soles of her foot-hands at their natural moment-of-arrival the branch; Beat 2 is added an instant afterward with a spare hand. It is as if she is 'making the most' of the fact of her feet slapping the ground with the use of exaggeration or emphasis. This is exactly as we find in human infancy.

Note that here we are writing of jumping-down as an act of *play*. It is of course perfectly possible for the chimpanzee or human child to land quite ordinarily when not in the state of play, simply in order to get from above to below (for example). In the course of a chimpanzee's day this happens a lot! However, here we are concerned with how this ordinary act is transformed by play.

For both primates, chimpanzee and human, the punctuation of the end of a sequence becomes very important. A line described through space terminates with a slap, a thump or a bang – a *beat*, in fact. Sometimes the natural ending of the line serves as the punctuation, as when the foot-hands simply strike the ground, as they must do, at the end of a leap.

Even though this natural kind of demarcation inevitably occurs in their locomotion, in their play-states the chimpanzee (and human child) makes an effort to maximise the moment of arrival in various ways, say, by ensuring that either or both palm-soles hit the ground simultaneously or, alternatively, the impact of first one palm-sole and then the other is staggered ever-so-slightly, so that the second beat is broken into two half-beats, making a 'di-dit' type of rhythm. Or else, the 'dit' part of the phase is provided by a spare front-hand. This second beat is often placed in time with especial care, following the natural impact of the descent and making a distinct 1–2 rhythm. Musical variations are developed by both children and chimpanzees. The production of rhythmic sequences will become very important in the present work.

I call such demarcated ends of descending lines *punctuated falls*. The entire action-assembly from start to finish is an example of *trajectory-and-moment-of-impact*.

Trajectory-and-moment-of-impact

The pounce

In day-to-day life, the chimpanzee will often incorporate trajectory-and-moment-of-impact in their play, when they pounce upon each other in mock attack.

05 October 2007 (Part 4) Video-recorded observation (Edited DVD) Singapore

Nkosi (7 years 3 months 8 days) holds a sheaf of leaves. Rah (2 years 9 months 26 days) climbs up on a boulder above him and then, after a glance at me, leaps down upon him.

In real life, jumping upon someone else like this is very difficult and requires considerable skill. One has to be daring yet sensitive not only to one's own safety but that of one's partner. Consider the following part of the same observation.

05 October 2007 (Part 4 continued)

Even within the jiffy of time it takes for Rah to fly toward him, Nkosi almost succeeds in getting his own strike in first (with his sheaf of leaves), before she has even landed. [Chimpanzee perception and response time is that quick.] For her part, even at the end of this same, miniscule space of time, the instant after her impact on her brother's back and shoulders, Rah manages to fit in a beautifully executed backward-roll over Nkosi's body, landing next to him in the grass.

Observations show Rah learning to make good head-over-heels roll on another's back by the time she is only 1 year 1 month 20 days (27 January 2007 video-recorded observation, edited DVD, Singapore).

Here is another spectacular performance from an observation of Nkosi and Makhulu made 2 years before.

Interpolated trajectories

06 October 2005 (Part 2) Video-recorded observation (Edited DVD) Singapore

Metres up in the air, Makhulu (5 years 5 months 29 days) leaps into space toward Nkosi (5 years 3 months 9 days) at the same time as Nkosi leaps from the opposite direction toward Makhulu. They collide at a midpoint in space after which they both fall to the ground.

To perform this accurately and safely, the timing of trajectories is crucial. To make this mid-air collision work requires each ape to interiorise and 'interpolate' (Bower, 1974/1982) two coinciding vectors. When I write 'work' I do not just refer to the ballistics involved. It is not sufficient for the two chimps merely to hit each other in space. Their precision also involves a certain lightness of style which they both admire. My observations suggest that when an action-assembly fails to show perfect timing and precision (as does happen sometimes), a chimpanzee looks disappointed, somehow *diminished.* One is even tempted to add that he or she has the appearance of one who feels 'foolish'. (Darwin notes that Great Apes seem to feel 'ridicule' – Keynes, 2001, p. 262.) As in all their play-assemblies, from the most basic to the very complex, it is not enough simply to 'play the right notes in the right order', as comedian Eric Morecambe (1971) patiently explains to conductor André Previn. This is a subtle but crucial distinction and it is one to which we must return.

Representation precedes the real

Surely then, because of the difficulties and dangers involved with expressing an aeronautic aesthetic, it may be easier if the principle of the pounce is practised firstly in pretence rather than in real life. This may be one of the reasons why representation develops in the first place. In many cases, representation is easier than the real act because in the represented act one can denude it of unwanted extraneous features (faults in smooth execution, hurting one's partner or oneself, making a fool of oneself, getting killed etc.) whilst at the same time retain and polish its essence (timing, rhythm, fluidity). Thus, when we speak of representation this does not always mean a *re*-presentation of something real which has already happened (although this is sometimes what it is). Sometimes, representation precedes the real.

Explaining the imaginary

06 October 2005 (Part 2 continued)

Nkosi comes zooming in on a long pendulum-swing to make a virtual collision with his sister, Rah, who is sitting on a Horizontal Beam. Nkosi does not make physical contact with Rah but instead alights right next to her. This is a virtual collision. He certainly wants to surprise her with his sudden arrival right up close to her, touching her, but he does not want to unseat her. He knows that she does not yet know how to collaborate in these trajectory and impact plays.

If he really crashes into her he will certainly succeed in toppling her but will just as certainly fail to introduce her to this dynamic-structure – which is his main intention. He wants her to 'get the idea'. At the same time, he is giving his sister an *explanation of what imagination is*. This may be at the roots of language too. Rather than the dominant idea that language is hard-wired into humans (and humans alone), perhaps conscious beings find ways of demonstrating its structure to companions they care about. Alan Costall (2009, personal communication) refers to this as 'making language learnable'.

Nkosi's act is on the way to one of imagination because the intention to deceive is only partial and this he makes known to his little sister. He is teaching Rah about pretence. He does indeed want to surprise her but he also wants her to understand and to share the same idea with him; he wants her to be able to 'join-in'.

And for why? Because, to understand something properly and to get maximum enjoyment out of it, it has to be shared. Representation is rooted in and grows out of companionship. The secret of life is finding ways to pass the time pleasantly together (Trevarthen, 2003).

7 The end of the line

In the drawings of the human infant, the demarcation of the end of a line is pushed to a new developmental level when the child emphatically marks the end of a drawn line.

In some of their earliest drawings, very young children often mark the end of 'descending' lines with explosive squiggles. Sometimes they may synchronise 'ascending' and 'descending' vocalisations with their production of 'ascending' and 'descending' lines and, in a one-to-one correspondence with the explosive marking of the end of the line, synchronise an equally explosive vocalisation.

Children will also use the ascending and descending movement of the marker's point or the pencil point to represent the rising and falling flight, sometimes in the handheld flight of the marker itself; sometimes making lines on a drawing surface to represent the flight of imagined projectiles beyond the drawing surface. The doomed journey of an aeroplane is a classic example, with its catastrophic moment-of-impact demarcated by sound and vision in synchrony.

The dynamic-structure *trajectory and impact* appears frequently in adult human art too, a recent and excellent example being Anish Kapoor's *Shooting into a Corner* in which a huge cannon fires a projectile of red pigment into a corner of a gallery in the Royal Academy of Art, London (Kapoor, 2009).

In such cases, both child and adult add further levels of representation, symbolisation and metaphor to a simple dynamic-structure.

We simply do not know if the chimpanzee is capable of adding these further layers of meaning. However, at a structural or formal level – putting aside for a moment the question of representational values – there is ample evidence in the drawings and paintings of famous captive apes of the demarcation of the ends of painted and drawn lines. Later I will offer slight evidence of this demarcation of the end of a line made in the spontaneous drawing of one of my chimpanzees.

In human childhood, the passage of development from the *presentation* of rising and falling to its *representation* on the drawing surface is very striking. Here is an observation of a child drawing, or perhaps writing about, the idea of *descent*.

Going-down

> **06 December 1983 PAL VHS videotape-recorded child-observation and hand-written transcription (Hannah)**
>
> Hannah is 2 years and 3 months old. She is sitting at a table with a sheet of paper before her and some felt-tipped pens. Nearby her on the table is a pair of earphones for my video-camera. Hannah, holding a pen palmer-fashion, reaches toward the top edge of the paper. She places the pen-tip down on the paper and pulls the pen toward herself, making a long, vertical line down toward the bottom edge of the paper. As she makes this descending line she makes a descending, musical vocalisation. When the pen reaches the bottom of the paper she immediately picks up the nearby pair of earphones and drops them over the edge of the paper, continuing to make her descending vocalisation and looking down to watch the earphones as they fall to the floor.

Here, Hannah enacts a downward flight whilst synchronising an accompanying soundtrack in the form of a descending, musical vocalisation which underscores the event. When, as the pen reaches the bottom edge of the paper she runs out of space, she adds on to this *drawn* descending flight a *real* descent of a real object – the earphones – which she fluidly exchanges in mid-action with the pen, dropping the earphones to the floor. It is worth reminding ourselves that though we are referring to a vertical descent; the line drawn on the paper is physically horizontal, only the falling earphones describes a truly vertical axis. This may be considered one of those expressive and representational episodes in which the imaginary and symbolic is joined onto the physical and real. The episode starts off with a represented descent and finishes with a real descent which is imaginatively fixed onto the end.

In an observation recorded in Singapore, a Chinese 4-year-old represents the flight of an aeroplane, linking the real and the imagined in a similar way as does Hannah. First, the flight is represented by a handheld pencil flown in the air, then, second, as a whirling graphite line drawn on the paper and finally as the launching of the physical, graphite drawing itself, on the paper sheet, held by its corners and flown into the air. These different forms of represented flight are joined together in the imagination alone and by the boy's vocalisations which accompany the aeroplane's flight and its eventual impact. Like the Chinese boy, Hannah the Londoner may be linking the symbolic with the actual. Gunther Kress has noted similar linkages (Kress, 1997).

Chimpanzees do occasionally synchronise a little vocalisation to their actions in ways not completely unlike the way in which Hannah and the Chinese boy do.

The chimpanzees are able to produce percussive, plosive sounds and sharp screams and other sounds but they do not seem to use these in their interactions with each other. They do not use their voices very much to each other except in extreme situations, those which produce fear and conflict, for example, though, as Jane Goodall and Birute Galdikas note, chimpanzees sometimes cannot help but make chuckling noises of appreciation when they find food (Galdikas, 2005; Goodall, 2000). I cannot be certain but my chimpanzees may occasionally transport a very quiet, almost imperceptible version of such vocalisation over to accompany their play with handheld objects. Galdikas suggests this possibility but if I am to detect these quiet vocalisations (if they are made) I will need either to be nearer to my chimpanzees or else have better audio-recording capable of capturing such sounds.

A chimpanzee can however analogue the actions of another chimpanzee with one of his or her own. Earlier, in an observation of Ida breaking open a coconut, we saw two juvenile males echoing her series of downward strikes with their own use of a repeated downward thrusting (or 'nodding') of their jaws and heads.

Although some writers argue that apes are not capable of the representational thinking we see dawning in human childhood, the germinal beginnings of an incremental, polysemic layering is surely glimpsed in chimpanzees' use, organisation and permutation of natural, four-dimensional action-assemblies. The apes do not speak but they do seem to analogue the action of one body-part (or body-system) with an action of another. They make carefully controlled actions with body-parts not directly involved in the primary task or the chief play-action; actions which seem to echo or analogue the main performance. In human infancy, such synergistic reflections are the precursor actions for representation.

Falling

For example, the chimpanzee will make purposeful variations, elaborations and embellishments of the act of *falling*. The end of a linear trajectory may be punctuated in a variety of ways, including the simplest 'monosyllabic' 'full-stop', like a slap, clap or whack, to more complex ones including an ellipse through a horizontal axis, like a sudden, horizontally aligned rotation on the spot (a 'twirl'), or rotation through a vertical axis, as in a head-over-heels.

Another kind of demarcated linear ending I term the *tumble-fall*. Here the chimpanzee pretends to fall off a beam or branch and turns the fall into an elegant and sometimes comic downward tumble.

08 June 2007 (Part 2) Video-recorded observation (Edited DVD) Singapore

Rah performs a *tumble-fall* falling vertically downward whilst comically wriggling and writhing as she falls, her arms and legs winding and unwinding at high speed before she hits the grassy ground in a black bundle. This all happens within a fraction of a second.

Orang-utans are good at this sort of thing too. Human visitors to zoos find such antics very funny since they uncannily echo human actions. (This is actually rather paradoxical, since humans require considerable skill to perform some of these actions! It would seem that, though more complex than anything they could achieve themselves, humans recognise something in chimpanzee performances which reminds them of their own bathos.)

Chimpanzees and orang-utans do not learn this from humans. Although they recall the clowning and physical humour of great human comedians, such performances are not copied from people. Nor is it sufficient to remark that chimpanzees are just exhibitionists, for this too begs the question of how they understand that certain rhythmical sequences are entertaining and 'funny'. The basis of the humour may reside in a kind of script-violation in which the act of falling – in real life, life-threatening and to be avoided at all costs – is performed purposely and harmlessly as if to mock death. Because such falls are purposeful, primates find them pleasurable at a sensorimotoric and polymorphously sensual level too. It is worth comparing such intended falls with those which are accidental and unintended. In these latter cases, the chimpanzee's demeanour is quite different, indicating a sense of emotional discomfort, even embarrassment. (Later, I will give a good example in which Nkosi accidentally falls onto his mother, knocking her over!)

Perhaps, whilst watching the antics of the chimpanzees, the human brain, its ancestral memories twinned with that of its chimpanzee relative, similarly produces feelings of pleasure – which make us laugh.

Laughter

The beginnings of humour seem to go with the beginnings of symbolisation. Of relevance here is the research by Marina Davila Ross and her colleagues (Davila Ross et al., 2009) about tickling orang-utans, gorillas, chimpanzees, bonobos and human infants to make them laugh. It was until recently thought that because other apes could not control respiratory air-flow in the same way we do, through the larynx and pharynx to produce the necessary vocal-chord vibrations, then this makes both laughter and speech impossible for them. Because Davila Ross and her colleagues show that apes other than humans can produce laughter this may tell us something about the origins, perhaps over 15 million years ago, of human speech.

Charles Darwin was one of the first to wonder about the significance of tickling and the pleasure it produced in both babies and orang-utans. If humour, laughter and language somehow go together perhaps it is not by chance that Rah produces a great deal of emergent representation during these few weeks when she first makes me laugh.

Demarcated line ending and its variations

The tumble-fall is one of many ways in which the end of a line might be demarcated. The chimpanzees make numerous permutations of the basic assembly, *trajectory-and-moment-of-impact*. One version is *swing-and-touch*.

21 September 2005 Video-recorded observation (Edited DVD) Singapore (2nd movie Edit 2)

00:00:23 Hanging on an old, rubber fire-hose, Nkosi performs a pendulum-swing from a selected Point A to a Point B formed by his mother, Ida. At his moment of apogee he outstretches his arm and hand to touch her.

This variation of trajectory and impact I term *swing-and-touch*. On a subsequent inward swing to his mother, she reaches out to touch him at his point of arrival. Anticipated convergence of vectors or interpolated trajectories is involved here.

A variation of swing-and-touch is *swing-and-graze*.

01 September 2006 Video-recorded observation (Edited DVD) Singapore

00:06:34 Rah (1 year 8 months 23 days), swinging gently on a vertical strap above Level 3, describes a partial ellipse in the open air with her left hand, before finally grazing the surface of the platform beneath her with the index finger of this hand. She has earlier spilt mango juice on this surface and she now looks down, presumably at the trace-making effects she makes.

This activity is the beginning of mark-making and drawing and we will consider such incidents in detail later. Produced by the whole body in motion, this type of drawing has been demonstrated by several modern (human) artists. Carolee Schneemann's (1973–1977) performance *Up To and Including Her Limits* in which she draws on surfaces in a three-dimensional space whilst hanging from a rope, is a good example of drawing produced by a 'swinging' 1960s artist (Schneemann, 2002).

Zooming and looming

04 April 2008 Video-recorded observation (Edited DVD) Singapore

Rah (3 years 3 months 27 days) carries out a series of teasing pendulum-grazes targeted at Ainka (7 months 5 days) lying on her mother's stomach. On her final inward swing she looms into Ainka's face, provoking a snarl from her baby sister as well as an outward reach and swipe from her left hand.

A controlled *looming* into an infant's face is one of the many examples of trajectory and impact (or virtual impact) also practised by humans with their babies (Matthews, 1999, 2003, 2010; Trevarthen, 1995, 2003). The orang-utans in Singapore Zoo also appreciate this kind of play and will collaborate with their human Keepers to perform it. For example, Keeper Kumaren plays at looming his face in close to Merlin's, simultaneously startling and amusing this baby orang (08 September 2006, video-recorded observation, raw footage on DVD). Of additional interest is that, when young children gain some mastery of a video-camera and discover the function of the *zoom and wide-angle*, they use this device to amplify natural looming in and out (and even to videographically represent it) (Matthews, 2006a).

Another way of emphasising one's moment-of-impact is to select the medium for its particular expressive affordances. Consider the use of water in the following examples.

27 January 2006 (Part 1) Video-recorded observation (Edited DVD) Singapore

Water-spout

Makhulu (5 years 8 months 20 days) sucks up water into his mouth from the little Pool above the Waterfall and then, holding the water in his cheek, he launches himself on a vertical fire-hose into a long, arcing swoop toward Nkosi (5 years 7 months). As he looms up to his brother, so he squirts a jet of water at him from his mouth!

A spurt or splash of water demarcating the end of a line is used especially by Makhulu who, though like all chimpanzees cannot swim, nevertheless loves water. For the chimpanzees not so keen on the stuff, droplets may be splashed when the hand strikes a wet, wooden surface in the rain. Another method is to trickle water from the small pools trapped in leaves or down the runnels formed along the length of long stems. In the following observation, rain-water has collected in a long furrow in a rotting, fallen log.

Insolent sun and relentless rain have gradually chiselled out a long channel along much of the length of Bridge Log. Of course, the chimpanzees have done their bit to help along the depredations of decay. Sometimes, this narrow trough gets filled with water.

02 March 2007 (Part 2) Video-recorded observation (Edited DVD) Singapore

In the groove

Nkosi (6 years 8 months 5 days) carries Rara (2 years 2 months 25 days) on his back to Horizontal Log. They both drink from a puddle of rain-water which has collected in the groove in the big log. Poking an index finger into the water within the furrow, Nkosi traces a line in a single pulling action along this runnel.

02 March 2007 (Part 2 continued)

Rara also makes two pulling actions in the water collected in the groove on Horizontal Bridge Log. She pauses and makes a third pull stroke which causes water to spray up into the air in an arc.

02 March 2007 (Part 8) Video-recorded observation (Edited DVD) Singapore

Suzie is sitting near Twisted Tree and, in order to gain her attention, Rah swings in front of her on a Fireman's Hose. Perhaps feeling that she is still not getting the attention she deserves, Rah stalks off along Horizontal Log. She scrapes at the water in the little gulley, using an abbreviated pull stroke, knuckles downward. Then she looks at me and splashes water toward me using an underarm scooping arc. She returns her attention to Suzie, swings in front of her again and strikes the overhead Slack Strap with a vertical arc, boinging it just against Suzie's head and dripping water on her as it does so.

The big splash

Trajectory and impact developed

It is 1 year 4 months and 2 days later and Rah is becoming an expert in trajectory and impact in water.

04 July 2008 Video-recorded observation (raw footage on DVD) Singapore (Tape 1)

00:53:56 Rah (3 years 6 months 27 days) runs along Bridge Log from one end to the other and back again, making a big splash at each end of her trajectory by jumping into the pools of water there. From the cleft in the decaying timber clouds of silver droplets cascade, explosive exclamation marks at beginning and end of her linear route.

Rah also hones her swing-and-graze technique.

13 May 2009 (Part 2b) Video-recorded observation (Edited DVD) Singapore

Rah (4 years 5 months 6 days) hangs from a sack draped over a Slack Strap and swings out into empty space over the edge of the platform. She allows her one, free, left hand to hang vertically downward to sweep the wooden surface as she passes over it. As her hand grazes the platform she catches a green leaf up and discards it into the air. It flutters downward.

Her use of a hessian sack, slung over the Slack Strap and hung upon, is an example of an important non-human primate invention, the 'Hanger', which we mentioned earlier and which we will discuss fully later on.

In some of these examples, the way in which the point-of-arrival is enhanced is in the choice of target and the exploitation of its affordances. Some of our examples so far have been predominantly visual, dynamic and tactile but there are others which make more use of sound.

26 May 2006 Video-recorded observation (Edited DVD) Singapore

Nkosi (5 years 10 months 29 days) pendulum-swoops down upon the hanging, black, plastic bin, loudly announcing his arrival with a double drum-beat of his fists upon it.

18 April 2008 Video-recorded observation (raw footage on DVD) Singapore (Tape 1) (1 Tape only)

Rah (3 years 4 months 11 days) springs forward, launches herself into space and straight onto the trunk of Tremble Post making a jarring impact which sets it vibrating. She rapidly climbs up this post and sits on its sawn-off top, a black spot on the end of a black line like an upside-down exclamation mark.

Free flow

In the following observation, Nkosi extends the range of trajectory and impact and orchestrates it into a free-flowing chain of associations. Before I describe it you should know that the nails which usually secure Corner 4 of Word Platform have come loose.

19 January 2007 (Part 1) Video-recorded observation (Edited DVD) Singapore

Pluck

Nkosi (6 years 6 months 22 days) races up to Level 3 to pounce upon Word Platform yet in the very same instant as he lands on the board he contrives to lift up one of its heavy corners and let it fall with a crash, like a giant, wooden lid slamming shut. He stops, looks and listens. He races to Swinging Platform and thumps and then twangs one of its steel, cable-stays, which vibrates like a big, bass string on the vast sound-box formed by the wooden Construction.

Note that the Word Platform is a massive, heavy board and the steel cable is no slender guitar-string either, yet Nkosi finds them both 'pluckable'.

The chimpanzee's investigation of elasticity and pliability, its contrast and comparison with rigidity, is conducted in early infancy.

14 December 2007 Video-recorded observation (Edited DVD) Singapore

During the observation when Kimoni (4 months 1 day) high up on the Construction with Ganga, is experimenting dropping stems from his mother's spray of leaves, he pulls at a stem in the bunch in her fist but without withdrawing it, allowing it to spring back again. He repeats this. He is finding out that some things are 'twangy'.

Trajectory and impact in translation

As demonstrated in Nkosi's continuous series of trajectory and impact related assemblies, a trajectory does not have to end with the first impact but can be translated into a further, sometimes very different actions. Rah, for example, develops a variation of trajectory and impact by instantly translating her impact into a *slide* along a slippery surface (usually one of the shinier platforms).

Nor need trajectory and impact have to end with just one impact but may be repeated, the first impact acting as both pounce and *bounce*. We go back in time to when Nkosi is 5 years 3 months 9 days old and Rah is only 9 months and 29 days old.

Pounce bounce

06 October 2005 (Part 2) Video-recorded observation (Edited DVD) Singapore

Nkosi, hanging from a vertical fire-hose, and with baby Rah clinging to his back, makes long, sweeping arcs, launching himself toward one of the Platform Rocks embedded in the ground below. Instead of alighting and stopping here however, as soon as he touches down and without breaking the rhythm, he uses this rock as the next launching pad from which to propel himself and his passenger to a second rock.

This is a *single-bounce trajectory*. In fact, in this particular example, he does not stop there, but uses this second rock as the launching pad for a further trajectory, thus making this a *double-bounce trajectory*.

When children encode this dynamic structure onto the drawing surface I call it a *travelling 'm'*. In adult human design, a well-known example of this simple, reiterative trajectory and impact is the big, yellow 'M' in McDonald's. This is a *single-bounce trajectory*, or the first 'm' in a *travelling 'm'*. It is quite a thought that we might have started dreaming up such devices over 6 million years ago.

Human Olympic gymnasts like to play on variations of this structure, making a succession of such trajectory and bounce ensembles including highly elaborated flight-paths performed between bounces. Beth Tweddle's art is a good example; her complex, in-flight variations already dubbed 'twiddles' by some commentators (Kitson, 2009 in *The Guardian*).

Such *reiterative dynamic-structures* are common in human children's play and in their drawings in the form of *travelling waves, zigzags, 'm's*, and *'w's*. The *travelling 'e'* or *travelling loop* is another variation; the line this time repeatedly doubling back on itself before continuing on its forward journey. We find this form in chimpanzees' play too.

27 January 2006 Video-recorded observation (Edited DVD) Singapore

Nkosi is 5 years 7 months 27 days old when, in play with his father Togo, he careers over fallen logs and big stones as he forms a line which, at two points through its journey, double backs on itself as he performs two 360° spins before continuing on its way, forming a *travelling 'e'* in space.

The shape of the line Nkosi makes derives from its roots in his finding a way of running away (in play) from his father whilst also managing, without pausing, to look back at him, by spinning around, so checking his father's whereabouts and mood (which is today friendly and fun-loving) yet without losing outward-and-away momentum.

Toward the sign

In the following example, Nkosi adds a tumble-fall onto the end of a pendulum-swing.

14 December 2007 Video-recorded observation (Edited DVD) Singapore

Nkosi (7 years 5 months 17 days) performs a spectacular *swing and tumble-fall*, dropping, even pounding, to the ground. Of especial interest in this observation is that, some distance away, on top of Tower 2, Rah (3 years 7 days) echoes her brother's action when she allows herself to tumble to the ground. Remarkably, Rah's close copy of her brother's action is performed at some distance from him. Then Nkosi seems to echo her example and repeat it with another of his own but with the variation of additional twists and tangles.

Then he drops to the ground again, chases along for a few metres and gives a 'come here' gesture with his left hand to his sister, over on Tower 2; a hooking movement made in the air whilst he turns around to face her. Rah runs toward him.

When natural signing does occur, it seems to start when two chimpanzees are in closer proximity. That, in this observation, Nkosi makes his signal at distance of about 15 m from his sister – and that she responds to it – is remarkable. The background of this achievement is in fact seen in the moment earlier, when Rah was able to emulate Nkosi's tumble-fall. The depth of empathy between them is now great enough for him to know that she will be able to respond to his actions even over a considerable distance.

Spatial separation in itself seems to prompt the beginnings of signing, probably because it is economic to do so. (Nkosi does not have to physically traverse the space between him and his sister.) This is one of the reasons why its formation makes evolutionary sense. Such recognisable signs derive from abbreviations of natural actions made to physically coax or even tug another to come along in the same direction.

'Come with me' (go where I am going)

29 February 2008 Video-recorded observation (raw footage on DVD) Singapore (Tape 1)

00:18:31 With her right hand, Rah (3 years 4 months 12 days) grips her mother, Ida (34 years 4 months 16 days), by the scruff of her neck and leads her toward a new hessian sack she has espied and wants to investigate.

When approaching a new and unknown object, a youngster often wants the reassurance of an accompanying parent. Over time, Rah will reduce this physical tug to an abbreviated sign made in space alone.

Physical manipulation of another's head and face may even be used to direct the other's line of sight (rather than the other's direction of travel).

'Do you see what I mean?' (look where I am looking)

22 June 2007 Video-recorded observation (raw footage on DVD) Singapore

00:08:29 Nkosi (6 years 11 months 25 days) is sitting on Yellow Puzzle-Box spearing fruit through the slots with a stick. Rah (2 years 6 months 15 days) is sitting huddled up close to him, her face side-by-side with his, watching avidly. Something distracts Nkosi for a moment and he pauses in his work and looks away. Rah does not like this interruption however and wants him to resume fishing. She raises her right hand to his head, grasps hold of the lower part of his face below his eyes; that is, most of his jaw, and physically swivels his head back around so that he is once more looking where she wants him to look – back down at his work of spearing fruit! She takes her hand away from his face and they both look downward together, sharing the same field-of-view. (See Figures 7.1 and 7.2.)

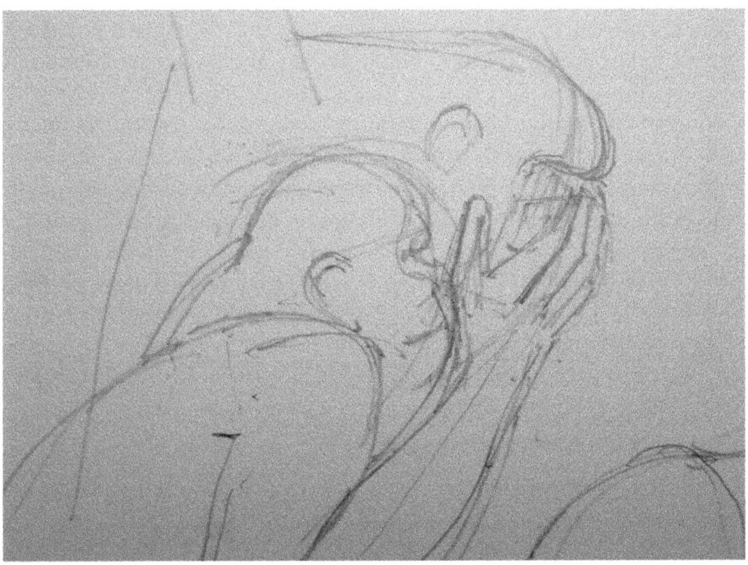

Figure 7.1 Rah pulls Nkosi's head around so that he is looking where she wants him to look.

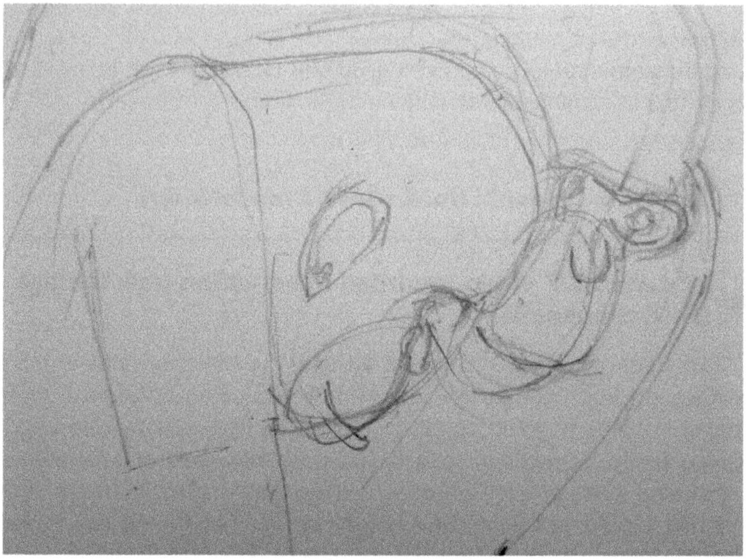

Figure 7.2 Now, they both share a similar field of view.

Rah sometimes manipulates her mother's head and face to make Ida look in the same direction as she (Rah) is looking (see Figure 7.3). As we noted earlier, a human could have pointed this out, with index finger alone, to another human – or even to a dog. However, there is a time in human development before which this is not understood. Early years educators sometimes report young children redirecting a teacher's gaze by forcibly turning the adult's face! (H. Matthews, personal communication, 27 March 2010). Chimpanzees can occasionally make use of the pointing gesture to indicate direction-of-travel to another but only in those instances when the gesture is supported by strong contextual cues. In Chapter 12 we will see Rah responding to a pointing gesture made by her father, but in the meantime her physical directing of her mother's line-of-sight is on an evolutionary path toward pointing as *sign*.

Deception, pretence and imagination

We now return to the observation of 19 January 2007 just after Nkosi has slammed down the heavy wooden platform and twanged a steel cable.

Figure 7.3 Shared moment. 'Look where I want you to look' (or 'look at what I am looking at'). Rah 'points' Ida's head in the direction she wants her mother to look so that she (Ida) sees what Rah sees.

19 January 2007 (Part 1) Video-recorded observation (Edited DVD) Singapore

Pluck (continued)

Without breaking the free flow of his passage of movement, Nkosi leaps to Hanging Branch which he levers swiftly into an upright position with front-hands plus heel of left foot-hand. He next seems to echo the arcing movements of the branch with a swing of an arm. He stops and looks down. Then he levers Hanging Branch back a few practice times before running suddenly to the left, holding the branch in his hands as if to launch himself into a pendulum-swing.

It is at this moment it occurs to me that he isn't going to make it. There just isn't enough room. The branch is not well situated to permit a launching manoeuvre. Sure enough, knowing he has insufficient run-up for his launch, and just at the moment when a spectacular swing would ordinarily commence, he lets go the branch and just makes a little hop, up and down, on the beam, as if *pretending* to pitch himself in the void.

Human children also do this.

22 October 2009 Child-observation, hand-written, London

Using a delicate finger-and-thumb pincer-grip, a 3-year-old girl carefully takes hold of the lowest end of a vertically hanging stem of a rose-bush and pretends to swing on this slender line. She moves with it through a tiny section of an arc making a 'Wheeee!' vocalisation. She cannot really swing upon it, as it is too thin, or even move it very much – because it is too thorny.

Pretence and placeholders

How best to interpret these observations? It is as if, unable to perform a real pendulum-swing, both chimp and child are obliged to fit in a notional representation of one, or at least provide a temporary 'placeholder' for a component of the assembly they know should occur at this moment in space and time but which they are unable, at present, to perform. Similar usage of the 'placeholder' occurs in both children's language (when the child is unsure of the exact word and so fills the gap with a 'babble') and in drawing (when the child is unable to draw a form he or she knows to be present so fills the space with a 'scribble') (Matthews, 1999, 2005; Willats, 1997, 2005; Winner, 1986).

The use of a placeholder (if this is an appropriate theoretical construct) is the use of a temporary, non-representational, or only notionally representational, stop-gap. The inventions of Nkosi and the girl are provoked by the physical impossibility of completing an intended action. Such in-fillings (the girl's minimal manoeuvring of the rose-bush stem; Nkosi's pretend swing) may be largely indexical but they still retain a tokenistic likeness to the real actions. They may be more like the patching together of pretence and reality, rather like ones mentioned before, when real flights made in real space are glued onto pretend flights made in drawn lines. The priority in all cases seems to be to preserve the flow of events.

24 March 2006 (Part 2) Video-recorded observation (Edited DVD) Singapore

Nkosi (5 years 8 months 27 days) hangs the sack over the Slack Strap, grasps both ends and hangs on it. Then he tries and partially succeeds in getting into the sack and almost swinging in it as if in a hammock. However, his exertions cause some of the hessian to slip away from the supporting strap so he has to fake it, pretending that he is hanging in his hammock by clenching sack and strap together, thus creating the impression that he is supported in the hessian hammock.

An interesting question is: For whom does he do this and why?

15 September 2006 Video-recorded observation (Edited DVD) Singapore (Version 2) (Interpretation 2)

00:17:09 It is 5 months 21 days later, Nkosi (6 years 2 months 18 days) has a hessian sack provided by the Keepers. It has been doused in the Waterfall so is sopping wet. He is using it to wrap up Look-Out Post 1 (a vertical trunk with sawn-off top) at Corner 2, Level 3. He wraps it tightly, and smoothes it down on the flat, circular top of the sawn-off trunk and over the sides of the trunk. Satisfied with the tautness and smoothness of the wrap, he then ascends the sides of the trunk.

His method of ascent is significant here. At first glance it appears that he is climbing up the sack as one would climb a rope-ladder. However, though he does indeed take hold of the sack in his front-hands he does not exert much force upon it but just pretends to climb up it, without actually pulling on it, allowing the precarious friction between wet sack and wet wood to hold it in place for the small jiffy of time it takes for him to get to the top. Here then is a careful balance between exerting sufficient leverage upon the sack so as to create the effect of ascent, but not so much that he pulls the sack off. Note that he has no one to fool by doing this. Perhaps he does not want to spoil the wet, stretched surfaces he has made and which so smoothly capture the form of the tree-trunk underneath.

Nine months and 29 days later:

22 June 2007 Video-recorded observation (Edited DVD) Singapore

Rah (2 years 6 months 15 days) wraps a strip of grass over a similar post and pretends to hang upon this fragile, green blade (see Figure 16.1).

19 September 2008 Video-recorded observation (raw footage on DVD) Singapore

Nkosi (8 years 2 months 22 days) holds a strip of sacking with both hands against a sloping vine. This scrap of hessian is a remnant of a complete hessian sack. It is about 80 cm long and about 12 cm wide. He touches one end of this strip to the vine and pulls the rest of it down into a taut, near vertical line. It looks like a strip of bandage hanging on a washing line except that it is not even a little bit folded over the line – he merely holds the sack at its top edge against the vine. He holds one end of the rag to the vine with his right hand, and pulls the other end down tightly with his left. He pretends to hang and swing upon this hessian strip though he is not in fact putting any weight upon it but is really holding it against the vine with his right hand.

This type of situation might induce precursor representation. I remember doing this sort of thing as a small boy whilst playing 'Tarzan'.

I think there is only a matter of degree between the two types of pretence here, the one that is merely a token; the other which is more like emergent representation. The word 'pretence' is used in two different ways in this book. One kind of pretence is really like a kind of deception; the chimpanzee pretends not to be watching; the chimpanzee feigns an attack and so on. Other kinds of pretence are not intended to deceive but rather to conjure. These require rather more critical alterations of actions, for example, techniques of curtailment, omission, deceleration and so on. I judge the present example more complex still because it requires sustaining an interiorised scenario; in a word, it requires *imagination.*

8 Repetition and variation upon a structural theme

In their action-assemblies the chimpanzees introduce repetition and variation upon a structural theme. They appreciate a well-constructed pattern of movement in time and space and will repeat it. I think they have an idea of the kind of perfection they want to achieve. Their instant correction of minor flaws is indicative of this chimpanzee aesthetic. For though they rarely make mistakes, that they occasionally do so (a barely discernible stumble; a missing half-beat) and that they take action to repair these, tells us a great deal about the 'patterns in their minds' (Jackendoff, 1994).

Making mistakes (and correcting them)

27 October 2006 Video-recorded observation (Edited DVD) Singapore

Nkosi soaks a bundle of leaves and then his entire body under the Waterfall. Soaking wet, and carrying the soggy sheaf of leaves, he makes a big pendulum-swing from Slack Strap. I think he intends to land upon sawn-off tree stump but, perhaps because he does not account for the extra weight of water carried in his leaves and in his drenched hair, he falls short of the stump, landing on the grass just in front of it. This is certainly not a 'personal best' and he seems disappointed, even glancing up to me as if to see whether I noticed this fault. I am tempted to write that he is embarrassed. Moments later, he makes another leap from a vertical fire-hose (Point A) and this time lands neatly on Level 3 (Point B). He looks at me as if requesting approval or praise and I say to him: 'Well done Nkosi.' He continues to look straight at me as he chews a wet leaf.

On other occasions chimpanzees instantly patch a perceived fault into a fluid sequence so that they are not really faults at all, or else they will incorporate what was initially an accident into the play-assembly.

This is reminiscent of the way a professional musician might recover from a mistake made in playing, or an actor on stage might improvise so that the audience fails to notice an error in the dialogue, or indeed, like an ordinary human child in deep play or in drawing might incorporate chance events. With practise, mistakes are no longer mistakes but part of skilful 'jamming'.

Not only this, chimpanzees will also make variations in the structure of a play-assembly whilst maintaining its overall form. In many instances, they will put in a detour or an 'aside', moving the action away from the main sequence, yet only to return to this core structure, without any hiccups in the flow of the action. Note that all this is spontaneous. None of it is pre-planned. Paradoxically these complex and dynamic passages of movement are at one and the same time improvised yet choreographed. It is this recursive form, in which a main theme may contain sub-themes, that has caused me to regard such action-sequences as linguistic structure. Roger Fouts (1997) reminds us that the linguistic need not be verbal.

The pivotal object and transitional phenomena

03 January 2008 Video-recorded observation (Edited DVD) Singapore

Rah (3 years 26 days) ascends Tall Tree cleverly carrying her stick-tool by wedging it between the top of her raised thigh and her body. On the way, she grabs a small bunch of leaves, one of which she lets fall. She looks down after it to see it fall to the ground. This purposely caused fall is a development of earlier, accidental ones. Then she swings down, carrying a remaining spear-shaped leaf in her left hand, her tool-stick still carried between her left side and her upraised left knee.

As in child development, what was once fortuitous is now performed purposely. This observation also shows that her early interest in falling leaves is not simply abandoned and forgotten. The leaf she carries is a favoured object which becomes a 'pivot' around which she can construct imaginative play. The idea of a *pivotal object*, deriving from Vygotsky's (1966) theory, is an essential concept for the present work.

The pivotal object acts like a lever with which the child separates words from objects and with which humans and (I believe) chimpanzees separate actions from meanings. Most people know of it simply as a 'toy' and would recognise that it need not be restricted to a manufactured toy but might include a huge range of objects. For some children, the drawing instrument can become a pivotal object. I believe this might have been so for my children Ben, Joel and Hannah with their felt-tip pens (Matthews 1984, 1999, 2003, 2010).

The concept of the pivotal object overlaps with another, related one, that of the *transitional* object from the work of Donald Winnicott (1971). Transitional phenomena occur within a space both physical and psychological formed between child and mother (or caregiver). This space acts like a buffer between interior subjective world of the child and the objective, outside world.

This space has a physical aspect too which is to be located in the interpersonal arena close to the mother; first of all in physical contact with her body and later, as the infant gains in ontological confidence, at gradually increasing distances. For a long period during the infancy of humans and chimpanzees the physical arena in which one may observe the beginnings of symbolic thought is in or near the lap of the mother. I have mentioned this special place earlier and call it the *lap-space*.

Iconic transitional objects in human childhood include the tattered teddy or the hugged blanket from which the child cannot be parted (Linus's treasured old blanket in the 'Peanuts' cartoon-strip by Schulz is a famous fictional example). For Rah, one pivotal and transitional object is the coloured leaf.

A coloured leaf

Like the human infant, the young chimpanzee also makes use of the 'pivotal object'; a further suggestion that the chimpanzee is on a potential evolutionary and developmental route to symbol-making.

I am going to concentrate upon Rah's use of one such pivotal and transitional object, the coloured leaf. From her early months of life Rah develops an interest in and an attachment to certain types of leaves. These leaves have to be spear-shaped, of about 7–8 cm long and 3–4 cm wide. They are brightly coloured, usually red or yellow. These, I believe, serve as *pivotal objects* around which precursor symbolic forms may be constructed. They may also act like *transitional* objects, in the sense intended by Winnicott, in that they help the infant construct a physical yet psychological space between the world of interior thinking and feeling and the world outside.

16 March 2007 (Part 1) Video-recorded observation (Edited DVD) Singapore

Rah (1 year 3 months 12 days) picks up a yellow and orange leaf and looks at it. Later on, I notice that she picks up another one, just like the first.

Her choice of a similar leaf means that she classifies the fallen leaves she sees around her and collects only those of a certain type. This is really like saying that one member of this type of leaf is equal to another of the same class; a little like a coin of a certain denomination in human currency. Later on in the book we see Rah possibly using it in exactly that sense. Note that the classifying of the leaf is made in terms of characteristics or properties selected by Rah. I am not here writing of an a priori or given class of leaf but one constructed in Rah's mind.

31 August 2007 Video-recorded observation (raw footage on DVD) Singapore (Tape 1)

Just over 5 months later, Rah (2 years 8 months 23 days) is lying on Level 2 whilst I talk to her. She is chewing a yellow-ochre-coloured leaf. She accidentally drops the leaf and takes a prolonged look down to where it fell.

Perhaps she is trying to distinguish it from the thousands of other leaves amongst which hers has fallen. When something interests them, chimpanzee discrimination, like that of human babies, is extremely sharp.

14 September 2007 (Part 2 – Editing notes) Video-recorded observation (Edited DVD) Singapore

Fourteen days later, Rah is still experimenting with *descent*. She pendulum-swings on Hanging Branch and, performing two swings, dislodges some leafy stems, her first causing the main mass of the leaves to fall, her second causing the remaining leaves to flutter down after. Then she scampers up to Look-Out Post 1 and hunches there as before but now with a single leaf in her teeth. She is deep in play.

26 October 2007 Video-recorded observation (raw footage on DVD) Singapore

Forty-two days later, Rah is sprawled over the top of a sawn-off post. She withdraws a leaf from her mouth and holds it away from her face, so that it is in focus at about 30 cm from her eyes. She regards it. She returns the leaf to her teeth and hops back down to Word Platform. She knuckle-runs leftward along Word Platform, the leaf still in her teeth.

She bounds down to Swinging Platform, lets go of the leaf and watches it flutter over the edge of the platform. She leaps after it but then stops for something to eat at the Grey Food-Box.

Rah's continuing experimentation with her coloured leaf and with the various ways it can be made to descend is rather like the formation of an *idea*.

26 October 2007 Video-recorded observation (Edited DVD) Singapore

(From the movie 'Smile')

Rah (2 years 10 months 19 days) ascends in a series of bounds up to the sawn-off top of Look-Out Post where she crouches, two leaves in her teeth. She allows one leaf to drop and it flutters downward. It is difficult to say whether this is intentional.

The remaining leaf in her teeth, Rah launches herself forward and down, following the first leaf's descent to Word Platform, one metre below.

When the palm-soles of her foot-hands impact upon the platform, she retains the rhythm set up by this sudden sound and makes a second beat of a foot-hand, after which she launches herself in an arcing trajectory, to land at her chosen Point B, which is this time Cable 3, Swinging Platform, adjacent to the Hanging Branch. From there, she swiftly climbs part way up Look-Out Post 2, before hopping back to Word Platform, the single leaf remaining in her teeth. She seems to be creeping up upon Look-Out Post 1, as though it is an enemy, and then she leaps onto it once again, as if pouncing on it.

In this leap, she reaches and grasps the far side of the sawn-off top and hangs there for a moment with her foot-hands clasped to the trunk of the post, before hauling herself up and lying on her stomach, holding the leaf to her mouth with her left hand and chewing it. She takes the leaf out of her mouth and holds it a few centimetres before her face, her eyesight aligned to it as she regards it again for a few seconds. She replaces the leaf in her teeth and again turns her head around to face me before leaping off and down to Word Platform, the leaf gripped by its stem in her teeth.

She scurries leftward and, grabbing the horizontal beam which supports Swinging Platform, swings down to this Platform, transferring the leaf from her mouth to her (right?) hand in flight. As soon as she hits the Platform she places the leaf before her on the platform. The slight breeze takes it forward along the surface of the platform and Rah pursues it. With both hands she makes the briefest of symmetrical horizontal arcs, knuckles downward, hands meeting momentarily at the midline of her body, but the leaf blows over the edge of the forward side of the platform (Side 1). It might be that the slight increase in air pressure caused by the coming together of her hands hastens the little leaf to its fall. Rah chases after it, launching herself into space. The leaf has become a companion; as in human childhood play, pivotal and transitional phenomena allow the chimpanzee to sustain the illusion.

It is worthwhile to note the role of chance and accident in Rah's play-assembly. Consider, for example, her dropping of the leaf and whether this is always intentional. In one passage of her play, the wind or change in air-pressure causes the leaf to fly over the edge of the platform. How easily such 'random perturbations' are incorporated into her free-flowing improvisation (Atkinson, 2002).

03 January 2008 Video-recorded observation (Edited DVD) Singapore

Rah (3 years 26 days) has a tool-stick but she carries it between her thighs because she wants her hands free to handle one of her favoured leaves.

Other leaves have been yellow or red but this one is a vivid green and it is interesting to wonder how she makes her selection. Although she usually chooses red or yellow leaves, this new, green leaf might also be considered a member of the same class of pivotal object perhaps because of its brightness. As the sun shines through it, it seems especially luminous and pure, like an immaculate piece of stained-glass. Perhaps this was why she was attracted to it. She and the other chimpanzees often lie on their backs looking at the light shining through translucent objects, including leaves, scraps of polythene and open-weave hessian. Remember the bright, red bag I once gave them?

06 November 2008 Video-recorded observation (Edited DVD) Singapore

The magenta-coloured bag has been reduced to a red rag and Rah (3 years 10 months 29 days) lies on her back, holding it up against the sunlight, before her eyes. Using sometimes four hands to hold it before her face, sometimes only her two front-hands, she stretches it open from its corners like a taut screen before her eyes, apparently enjoying the rosy red light which shines through it.

This is a popular viewing strategy for all the chimpanzees, young and old. Lying on their backs and using four outstretched hands to hold an object above them, chimpanzees construct a field-of-view in which an interesting optical array can be steadied for long enough to allow a prolonged gaze.

This is especially effective with the occasional piece of translucent material which can be opened up like a screen and held above the face to look at or through. In the present case, I believe that Rah is looking through the cloth and enjoying its red luminance as the sun shines through it. She is experiencing colour not just in its subdued form of *reflected* light, as when light is reflected from an opaque surface, but in light itself, for the sun is shining *through* the translucent cloth, as if through blood-red, stained glass or through an especially brilliant computer screen. Rah enjoys the action of looking and seeing itself. One is tempted to invoke the notion of *beauty*.

This supine position is part of the panoply of viewing strategies; some fixed, some ambulatory.

Does money grow on trees?

Sorting, classification and exchange

03 January 2008 (observation continued)

Returning to our observation about Rah (3 years 26 days) with the bright green leaf, she uses it to intensify one apogee of her swing-touch play; swinging the leaf outward before her in her fingers as she approaches the highest extreme of her pendulum flight. She still manages to keep hold of her tool-stick too, maintaining her grip upon it between her thigh and body until she eventually replaces it in her teeth. She thinks of something else she wants to do and adjusts cleverly and delicately the position of the green leaf to a place on top of her left thigh. It rests there only for a moment before the air casts it into space.

Though the leaf is lost, the fact that even for a moment she managed to rest it on the top her upraised thigh is still an extraordinarily skilful accomplishment, considering her swaying and yawing in mid-air.

Rah does not seem bothered about dropping the leaf. She knows there are plenty more where that came from. Because she has an idea about a particular kind of leaf, Rah can replace one leaf with another, very similar one. *Many individual leaves look alike, almost identical*. They almost replicate or duplicate each other. Therefore they can be *exchanged* for each other. Later, I will offer a little evidence of her making such an exchange, in the sense of a transaction. Other observations of the chimpanzees show them sorting different objects into groups, for example, long objects, including stems, hose-pipe and rope are gathered into a group for nest-building, whilst other objects, not saliently long, are discarded. Other primates sort and classify too. On one of my very first observations, I watch an orang-utan, sitting on the ground sorting leaves from twigs, putting leaves in a pile to his left and sticks in a pile to his right. (I don't know why.)

Play helps both chimpanzee and human baby capture information about the properties of objects and events which would otherwise remain indiscernible (Lorenz, 1944–1948). When Rah watches the sun shine through a leaf or watches it flutter downwards, this is at one and the same time an aesthetic and scientific experience.

Stone-age

The pivotal and transitional object need not be restricted to one type of object (a leaf or whatever) but can include a range of objects. It is not always readily apparent how the class is defined. One can perhaps understand why Rah might toy with a leaf in play but sometimes she enlists a decidedly unlikely candidate for a companion. Consider, for example, her use of a slab of concrete!

06 June 2008 Video-recorded observation (raw footage on DVD) Singapore

(Stone-age)

Rah (3 years 5 months 29 days) is on Level 3. When I call out her name, she gives me a cursory glance and turns away, seemingly engaged in something out of my sight; an object she is handling and looking at, upon the wooden platform.

00:08:22 Later, Rah is seen brachiating along a lateral vine, with her legs curled up beneath her in a sort of crossed-legged fashion, but in mid-air.

The only time I see the chimpanzees form this kind of locomotion is when they are carrying something with their legs. For a moment I cannot see what she has got; then I realise that on top of her upraised thighs she is carrying a large piece of concrete!

She places the concrete slab on a horizontal beam. It forms a rough and jagged triangle with two of its irregular sides about 27 cm long, whilst the 'base' of this 'triangle' is roughly 18 cm long. It is quite a thick slab of concrete, being approximately 7 cm in thickness. Its top surface – the one I can see clearly – is flat, though surely rough to the touch. Comparing it later with a slab of similar same size and shape I find outside the zoo, it must weigh at least 2 kg.

Rah, or one of her brothers, has managed to prise up this sizeable fragment from the old concrete path which runs through the void deck of the Construction.

Making plans for the future

Concealment and hiding

I did not see Rah carry the slab up to Level 3 today and since I have been watching from 09:00 this morning, when the chimpanzees come out of the Den, this means that she must have hidden it up there the evening before, just before they went to sleep in the Den. Perhaps planning to play with it the next day, Rah has hidden it out of sight of humans, anticipating that the Keepers will confiscate it if they find it. She knows a simple way of hiding it; placing it on the uppermost level of the Construction where it cannot be seen from the ground. She knows that humans are useless at climbing and she understands their *line-of-sight* from the ground cannot take in the top-side of the highest level of the Construction.

Primatologists have often wondered if chimpanzees can foresee and prepare for the future. Observations at Furuvik Zoo, Sweden, revealed that chimpanzee Santino secretly stored up stones in preparation for a planned attack on visitors (Osvath, 2009). Though, unlike Santino, Rah is not concealing a weapon of mass destruction, both chimpanzees know that they are forbidden these things by their Keepers and they know how to hide them ('out-of-sight') for future use. Santino's stones are pieces of concrete installation and bits of fake 'boulders' like the ones familiar to Rah and the rest of my chimpanzees. By tapping the concrete the chimpanzee discovers a weak area. Any hollow area in the concrete is betrayed by a difference in sound produced and a chunk might then be broken off. We have already noted that my chimpanzees also detect hollowness in objects by banging upon them, not only the obvious ones like plastic bins and so on, but also the less obvious. 'Hollow Rock' is discovered in this way, for example, and the chimpanzees soon learn to break holes in it. Perhaps Rah (or one of her brothers) has managed to prise up a section of the concrete path in this way.

Observation continues:

'You've got a big piece of stone, Rara?' I enquire. 'That looks very interesting', and a moment later: 'You're quite strong, aren't you?' I suggest with admiration, for the lump of concrete is clearly very heavy for her.

She gathers up a stick-tool and places it on the slab, seeming to align the stick with the slab's longer axis. Hugging the whole thing to her chest, she stands bipedal, ready to take a step upwards. Now obliged to use her left hand, she reaches up to the branch above and, holding the rock against her body but with her stick-tool held laterally in her teeth, she succeeds in ascending to the next level.

From below, Suzie seems to come after her, probably to confiscate the stone. To get out of her reach, Rah attempts to swing up to the top of Horizontal Beam, Level 3, where her mother Ida is sitting. However, with her heavy burden, she has not quite the strength to achieve this most difficult of manoeuvres. Failing to hoist herself up, she changes her mind, lowers herself down again and, carrying her slab, finds another route to the beam above where Ida sits.

Ganga has now taken over from Suzie in trying to catch the juvenile female to confiscate the heavy stone but Rah eludes her too, crouching hunched-up on the horizontal beam, huddling over the stone and clutching it to her chest.

Managing to deposit the great lump onto the Mezzanine Level and climbing up after it she then experiments in moving it around on the platform. She finds she can move it either bodily from one position to another or else she can slide or scrape it along.

The Mezzanine Level, unlike the other platforms, is both flat and smooth. It also slopes. The other platforms are made of separate timbers; thick planks put together like slats, upon which it is impossible to slide the slab. On the Mezzanine Level she drags the lump toward herself. From my position on the ground, I can hear the rough concrete being dragged grittily along the flat, smooth, painted surface. I cannot tell whether she is at all interested in the scratches it must make in the paintwork. Rah bends over the slab and licks it.

With her left hand holding onto the vine above, she pendulum-swings over the slab, touching it at each passing sweep, in a *swing-and-graze* assembly. She switches hands and now, holding on with her right hand, she swipes at it with her left hand as she swings over it. She develops these pendulum-swings and on each mid-point of her back-and-forth arcs, that is, just at the transient moment when her body is hanging vertically over the concrete slab, the palm-soles of her foot-hands passing just above it, she double-stamps its flat surface.

In this assembly, it is the *middle* of the trajectory – rather than its beginning or end – which is demarcated; the deepest 'trough' or 'well' of her sweeping arc, just above the slab, marking the middle of her passage. She hovers around above it, eventually swooping down to give it another lick.

The use of the *lick* (as opposed to a slap or a thump) to punctuate this middle point of her trajectory is also of interest because it betokens that the marking of time and space need not be a violent action but can be quite minimal.

Observation continues:

She dangles from her left hand and, for a few moments, her foot-hands tiptoe across the rock.

Now hanging upside-down from her two foot-hands, she drags the flat stone back and forth along the flat surface of the Mezzanine Level. She hovers, swaying upside-down; her face close to the surface of the piece of concrete. She adjusts her position so that she is hanging from her left front-hand and left foot-hand laterally over the slab, her body sideways on to it, whilst sliding and scuffing it with her right-hand and right foot-hand as she sways gently above it, like a black 'H' against the sky. She descends to the piece of concrete, tiptoeing lightly upon it but retaining her hold on the overhead vine with her left hand. The effect is now that of a swimmer, treading water and trying to reach an object on the seabed with her toes.

Leaving the lump of concrete on the surface of the Mezzanine Level, she pretend-stalks over to one corner and then to another, her shoulders hunched-up as if creeping up on an imaginary prey. She takes a short break from the stone and gently pendulum-swings from her right arm, dropping small pieces of chewed twig from her mouth upon it at the apogee of each swing.

Then she re-ascends to the Mezzanine Level and rushes up to her stone again, plonking herself down beside it and holding it with all four hands.

Keeping an eye on Nkosi, who is getting curious now and hovering slightly below on the other vine, she reaches up to the overhead vine with her left hand. She hugs the craggy lump to her chest with her right arm and pendulum-swings back and forth, running or stamping a few steps in the middle of each arc, where her foot-hands graze the surface of the platform. She varies this by sliding, rather than stamping, her foot-hands against the surface in the middle of each pendulum-swing, at the deepest point or well of the arc, whilst at the outward apogee of each swing she stamps the edge of the platform with her right foot-hand.

'That's amazing, Rara.' I comment, genuinely impressed by her synthesis of actions she introduced but an hour or so ago.

A moment later though, she drops (I think by accident) the heavy, concrete slab. It hits the wooden surface below her with a very loud 'thwack'.

Perhaps Suzie and Ganga had anticipated this eventuality and that is why they wanted to confiscate the dangerous object from her. (Perhaps I too, *homo-sapiens*, should have anticipated the danger and told the Keepers about it!)

The pivotal object occurs in many manifestations and any fixed defining characteristics are hard to pin down:

March 2000 Video-recorded child-observation (Edited DVD) Singapore

In a Singapore kindergarten, I ask a group of Chinese 4-year-olds to draw a green, plastic block measuring about 25cm × 10 cm × 10 cm. The Chinese teacher advises the children to handle it 'like a baby'. She uses this expression intending that they should be careful with the block but one of the children takes her words literally and actually picks it up in his arms, gently rocking it from side to side as if it were indeed a human infant.

The green block becomes a baby-in-arms for this 4-year-old child but how? Any physical likeness to a human baby seems at first non-existent. However, its rough size and proportion, coupled with the actions which may be performed upon it (hugging and rocking) mean that, for the young child, it can represent an infant in arms. At the beginning of the twentieth century, a girl in London's East End, too poor to own a toy doll, might carry around a common and worn house-brick wrapped in a cloth in such a way.

In our present observation, the fact that teacher and pupil share a verbal language brings in interesting complexities. Does the child really misunderstand his teacher or is he clever enough to seize an opportunity arising from the ambiguity of her words? In my hundreds of observations of the speechless chimpanzee world, I noticed that ambiguous situations arise continuously which might have easily been resolved with a few words (Morris, 1967). On the other hand, given millions of years, benefits may also accrue from the ambiguous situations continuously generated in this speechless world. This endless process of randomisation may constantly enrich the range of symbolic opportunities. The birth of meaning may be one of evolutionary selection.

The emergence of language may have channelled these symbolic possibilities in some ways but whilst human language can be used to make things clearer it can also complexify situations, a possibility which may further enrich symbolic affordances from moment to moment. The example of the Chinese boy's interpretation of his teacher's words illustrates this latter possibility.

The fluid indeterminacy of a life led mostly by movement and gesture can be disambiguated by language but, contrariwise and equally important, the same language may be used to randomise life too. The intertwining of these two features has had vast consequences and repercussions for human symbolisation.

Rah's block of concrete probably does not specifically represent a baby to her. However it does act like a pivot in a similar way as the way in which the green, plastic block acts for the human child; it allows her to perform actions of love and affection upon it.

Hence, the pivotal and transitional object makes possible quite complex play-assemblies which seem to contain an imaginative element. Consider the following observation.

Imagination

Trick the stick

27 October 2006 Video-recorded observation (Edited DVD) Singapore

Rah (1 year 10 months 20 days) holds a long, green stem in her hand and moves it through curves in space. She keeps her eye-focus upon it as she does so. She takes it to Nkosi (6 years 2 months) to show him, as if asking for his comment or advice about it. He glances at the stem and then leaves her to it, as if returning to his own thoughts. As if having been given the 'go-ahead' by her older and more experienced brother, she plays with it with one hand, moving it through coils in the air whilst with her other hand making synergistic push–pulls and arcings.

'Synergy' here refers to the empathic involvement of limbs not directly involved in the performance which is serving as the model.

She places the stick down on the wooden deck, then, moving away from it a short distance, finally turns around to look back at it. Then she creeps up on it, surely intending to pounce on it.

Surprisingly, instead of doing this, she walks straight over it, ignoring it, looking as if she has not noticed it or that she has forgotten it. The stick is now underneath and behind her and she seems to continue to walk away from it. She has one surprise left, however. Just as I assume that she is going to pass it by, at the last moment she suddenly retrieves the stick with her foot-hand.

As in other play-assemblies we have noticed earlier, this physical act looks like it has an imaginative dimension. It looks as if she is treating an inanimate object as if it is a living, thinking being and that she is trying to deceive it. She makes a surprise attack upon it. This is surely an act of *pretended* deception. The play-mode allows her to project an imaginary identity onto this dead object.

In a complex play of pretence, bluff and counter-bluff she treats this inert thing as if it is a conscious agency. Yet it is only a stick and surely never once was it out of her consciousness as just a stick.

Is this an instance of 'dual-function' of symbols and signs (Stetsenko, 1995; Vygotsky, 1966) in which an inert object (or, in human infancy, a mark on a piece of paper) is simultaneously just that thing and yet something else too? Or is it more like Judy DeLoach's (2000) idea of a 'dual-orientation' in which it is postulated that, in order to represent, one needs to hold together simultaneously in one's mind two quite different attitudes to the perceived world? (See Boysen & Kuhlmeier, 2002.)

Or is this not true representation at all but a case of substitution? Is the stick a *surrogate* for a playing brother? Some authors argue that what seems to pass for imaginative representation in chimpanzee play may really be the use of a *surrogate* (Mitchell, 2002b). The idea here is that whilst it might appear that the ape uses one object to represent another, the animal may merely substitute another object for it; to 'stand *in*' for it rather than 'stand *for*' it. This is not quite the same as representation.

According to this sort of interpretation, Rah, failing to get one of her brothers to join-in with her play, and yet unwilling to give up play, uses the stick as a surrogate for a playing brother.

To me, the distinction seems a fine one and if true it would still surely require Rah in some sense to interiorise an idea of the play-assembly in advance of performing it. Additionally and crucially, because Rah fails (according to this explanation) to get a partner to collaborate and compete for a real stick, this surely still sets the scene for emergent symbolisation in one way or another. Either she is obliged to perform upon the stick actions the like of which a partner was hoped to produce (the stick imagined to be moved by this imaginary partner), or else, she treats the stick as if it is itself alive and that it moves by itself. Either way, imagination seems involved.

The main difference between Rah's emergent representation and the tasks required of chimpanzees in many psychology experiments is that her act is spontaneous and self-chosen. She is not being tested. The representational dimension to her play emerges out of her own desires. When Rah represents (if for only a moment that is what she is doing), she can do so because this action is perched on the peak of a mountain of behaviours leading up to it. She only has to add 'one small step' (albeit a 'giant leap') to her original chase-and-grab play. We still don't know what this step is exactly but it surely constitutes what we think of as pretence and imagination.

15 June 2007 (Part 2) Video-recorded observation (Edited DVD) Singapore

Rah (2 years 6 months 8 days) is pulling at a long stem which has become wedged in a crack in the wooden platform. Suddenly the stem comes free and springs back. Rah responds with fright and jumps back.

Next, she regards it as if it is menacing and makes a horizontal arcing swipe in space at it with her front-hand, as if threatening it. Then she retreats from it, ascending Level 3 and peering down at the long stem from above, as if it were a foe.

Is this the beginning of imaginative abilities? Notice how the elementary-seeming action, the horizontal arc, is now incorporated into more complex action-assemblies. Like the stick itself, this action is an old friend. She can think of it in many ways. Her play-actions are initially the product of fear but she masters this fear by turning the episode emotionally around, so to speak, by altering the story in her imagination so that she becomes the winner, not the loser. She faces her fear by treating an inanimate object as if it were a thinking being – and a malicious one at that.

Rah constructs this play-narrative from that which is initially an accidental event. The sudden, natural whiplash of the branch really frightens her for a moment before she recognises what has happened. Now she turns this into an imaginative play scenario in which she 'takes her revenge', as it were, on the offending branch. In this way, like human children, she can assert herself safely, face-to-face with both the danger and her own fear of it.

Chimpanzees are easily startled by sudden sounds and movements but it could be that, after the initial genuine start, she then exaggerates somewhat, dramatising the effect, perhaps in order to understand and master her own fear. As we noted above, *exaggeration* and *dramatisation* are also techniques used by chimpanzees in their dawning expressive, representational and symbolic capabilities. It would be interesting to find out if any other mammal (or indeed, any creature) is able to do anything like exaggerate or dramatise.

Another feature of this incident is its accidental origins. What was fortuitous is incorporated into a conscious play-scenario. It is a characteristic both of the playing chimpanzee and the playing, human child that the fortuitous can be used in this way. Only the state of play makes this fluid kind of consciousness possible, into which anything and everything can be effortlessly incorporated. The use of chance and the fortuitous has been noted in human children's development too (Luquet, 1927/2001; Matthews, 1984, 1999, 2003, 2010).

In the above observation, it is an accidental event which triggers an imaginative proto-narrative. It is the chance insertion of a stem into a crack between the beams which prompts the scenario. Later on in development she will become fascinated with the ideas of *insertion* and *going-through* as interesting phenomena in themselves and will practise these purposefully.

We have seen how the early generations of structure are combined into more complex action-assemblies. We have also considered how these assemblies become layered with germinal expressive and representational possibilities. I have given examples which strongly suggest that the chimpanzees' play-assemblies have an imaginative dimension. Additionally, I have introduced the idea that chimpanzees may trace some of their actions onto flat surfaces in ways very similar to the manner in which early drawing starts in human infancy. This finding is in contrast to the assumption that non-human primate painting and drawing are unnatural phenomena contrived by human beings. We will study chimpanzee drawing in depth later on.

9 The giving of meaning

Showing something to another

06 October 2005 (Part 3 concluded) (Extra notes)
Video-recorded observation (Edited DVD) Singapore

The nearly flat surface (1)

00:02:30 Nkosi (5 years 3 months 10 days) is sitting on Platform Rock just to the left of baby Rah (10 months). He holds in his left hand a piece of melon husk which has a thin skim of fruit left on its inner side. Though it has a slight curve it is almost like a small, flat, roughly rectangular sheet about 18 cm × 15 cm and perhaps a centimetre in thickness. Nkosi is holding it steady against the rock in his left hand so that Rah has a direct line of sight to its surface at a good viewing distance from her eyes of about 35 cm. He even presses a knuckle into the back of the rind so as to straighten out its slight concavity. From Rah's viewpoint this leaves only a small remaining convex pucker in an otherwise generally flat surface. Nkosi has let Rah eat a little of the remaining flesh but I believe his main intention now is the positioning of the flattish rind before her so that she can look directly at its surface.

We don't know why he is doing this for Rara. Perhaps he is teasing her with it. Whatever the specific reason, he knows that he can make it interesting to her as a surface and he knows how to position it so that she can view it comfortably from her station-point. It is as if he is holding a book open before a small child or holding a picture up for her to look at.

Nkosi is interested in anything unusually flat (including very slightly concave and convex forms) and he is sharing this interest with his little sister. The following observation takes place 2 years 2 months 27 days later and is an instance of Nkosi displaying two key axes of an object to Rah.

03 January 2008 Video-recorded observation (Edited DVD) Singapore

Nkosi (7 years 6 months 6 days) holds a yellow leaf up before Rah's eyes. Rah is 3 years 26 days old. [Remember that she especially likes brightly coloured leaves.] Nkosi seems to adjust the yellow leaf for Rah's line-of-sight and in two ways: (1) along a lateral axis, that is, between his teeth longwise, so that, face on to Rara, he presents her with an area of yellow – a patch or *field* of this colour; and (2) gripped in his teeth and protruding outward so that, profile on to her, he presents a silhouetted, saliently long shape. This effect is further enhanced when Nkosi makes a minor adjustment to the leaf, turning it around in his teeth so that the leaf section is hidden in his mouth whilst the stalk sticks out like a black line from the black ball of his head against the sky, a little like the beak of a bird.

These are the primitives of the scene, natural signifiers which reveal key axes of the object and provide canonical views of these.

14 October 2005 (Part 4) Video-recorded observation (Edited DVD) Singapore

The nearly flat surface (2)

A week after Nkosi shows the flat piece of melon rind to Rah he is still showing an interest in the nearly flat surface. He is 5 years 3 months and 17 days old and he finds a piece of wood, roughly rectangular and approximately 13 cm × 10 cm. It is not quite flat, having a slight but regular concavity in its form. It could be a piece of bark or coconut shell. He puts it down on a flat stone, concave side upward, and beats upon it a brief tattoo of his knuckles. He then slides it, concave side down, over the back of his head, balances it on the back of his neck and knuckle-walks – or rather knuckle-strolls – off with it. [See Figure 9.1 for an example of Nkosi studying a nearly flat surface more than a year later.]

Humans make objects which have very good flat surfaces and Nkosi is interested in these when, once in a blue moon, he comes across one. On the 31st October 2008 he carries two small plywood boards, stacked congruently together, upon the back of his neck exactly as he did the piece of coconut shell or bark just over 3 years 1 month before. (I will tell you more about the plywood boards later.)

The act of sharing may start off with more basic commodities.

Figure 9.1 Nkosi (6 years 8 months 5 days) studies the nearly flat surface of a piece of bark he has found.

Giving something to another

14 October 2005 (Part 4) Video-recorded observation (Edited DVD) Singapore

During the same observation session, Nkosi sits patiently next to Suzie while she skilfully retrieves fruit from Yellow Puzzle-Box. She pokes her tool with her left hand down into the box, rolls the fruit along to the end of the box which has an opening and catches the fruit with her waiting right hand, bringing it to her mouth with this hand. She captures a segment of orange in this way and holds it to her mouth with her right hand. Nkosi pays close attention, his eyes steadily focused on the piece of orange, his face only a few centimetres from the older female's as she chews the orange.

After chewing away nearly all the fruit, with her right hand she passes the skin containing minute fragments of the flesh to Nkosi. He retrieves it in his right hand while, in a continuation of her gesture of passing the fruit, Suzie gently raises the same right hand to brush Nkosi away from her. He takes the hint and departs.

Typically, this high-ranking female makes sure she has taken nearly all the orange before passing it to Nkosi. Even so, I notice that the peel has not been utterly stripped bare. She has saved him some tiny morsels. Moreover, the way in which she passes it to Nkosi is of great interest; the gesture is slowed down, made gentle and she places it at his nose first of all, as if to see if he is inclined to accept the offering. He, in turn, smiles a little as he takes hold of it. There is very little food upon the rind but perhaps one might claim that 'it is the thought that counts'.

Anyway, he seems pleased with it for he takes it to show to his mother, Ida.

She has that famous metal pipe in her hands which she confiscated from the boys earlier on. She now allows him to take it back and he scampers off with it together with the orange peel.

Both these high-ranking females have offered him things which are but trifles to them but important to him.

A few minutes later, Ida is carrying Rah (10 months 7 days) on her back and Nkosi comes down to the pair and seems to be offering something from his hand to the baby, perhaps the residue of a piece of fruit. She declines whatever it is as she likewise does his offer to carry her (he gently puts an arm around her, motioning to her to climb onto him).

There seems to be an unostentatious level of sharing in the chimpanzees' interactions with each other. This sharing extends further than the exchange of material things to the sharing of ideas or understandings. Nkosi is in the process of learning how to be responsible when caring for a younger chimpanzee and, to a certain extent, Ida now trusts him. But Nkosi can still be rough and reckless which is perhaps why Rara chooses to stay where she is – on her mum's back!

Between ourselves

Mother and baby primate give meaning to an object or to an action

A parent or other senior chimpanzee may give meaning to an object and may share the giving of meaning to a young one. In the way in which they are handled within an interpersonal arena formed between by infant and parent, ordinary objects and the actions performed upon them become saturated in emotional values. Sometimes, when an object is passed between them, its inner meaning is passed on too. One might argue that act of sharing is itself passed on; sharing is itself shared. The act of giving is supposed not to occur much amongst chimpanzees but I am now wondering if this assumption might be wrong.

15 August 2008 Video-recorded observation (Edited DVD) Singapore

Old rope

00:09:32 Suzie (34 years 7 months 15 days) is nest-building on Level 2 of the Construction. She sits and bends long palm leaves around her bottom. Her 11 month 16-day-old daughter, Ainka, sits at her side. Interested in the nest-building, Rah (3 years 8 months 8 days) is coming down from the level above holding a single palm leaf in her teeth. She sits facing Suzie, watching her auntie at work as the older female plaits the leaves together and presses down the weave with the backs of her hands, the downward pressure issuing from arms held straight with elbows locked.

I notice that, amongst the lengths of palm, Suzie has hold of a piece of old rope; a thick piece of straggly, orange-brown sisal or hemp about 30 cm long and 4 cm in diameter. (It may have entered the enclosure unseen by the Keepers in the bundle of nest-building leaves they left for the chimpanzees this morning.) She is weaving it successfully into the lengths of green palm. She swivels a little to face me, and glances at Rah and at Ainka.

She swivels around to her right again and picks up the whole neatly woven sheaf and lowers it over and behind her head, so to curve it around her back. The hank of brown rope can be seen plaited into the structure. Rah swings across in front of her, clearly disturbing Suzie's thinking, for the older female pauses momentarily to decide what to do next. She even makes a little swipe of her right knuckles toward Rah who takes the hint and moves back a little.

Suzie seems unsatisfied with the piece of rope in the weave. She unhooks it from around the sheaf of leaves and is about to cast it aside on the deck but suddenly changes her mind and offers it up to Rah, who accepts it happily. Suzie's is a very clear act of *giving*.

Looking carefully at the sequence again, first of all Suzie weaves the saliently long length of rope into similarly long lengths of green stem. Here is a good example of the type of chimpanzee construction I call a *combine* – a term coined by artist Robert Rauschenberg (2005–2006) to describe works he made which combine different, found materials. Though Suzie's materials, stems and rope, are different from each other in many ways they share the dimension of length and the property of pliability. Suzie might well incorporate anything long and pliable into the weave. Moments later she decides that she does not, after all, want to incorporate the piece of rope into her work and so, with her right hand she unwinds it from the stems. She is about to cast it aside on the floor but changes her mind and hands it to Rah who quickly accepts it, taking hold of it with her right hand.

For her part, Rara, having been cautioned by Suzie not to get in the way, hovers nearby though leaving her hand hanging limply down into Suzie's field-of-view as an undemonstrative yet persistent reminder to the older female that she (Rah) is still there and interested in the hank of rope (should Suzie decide she can live without it).

This is a good example of thoughtful and sensitive support of the young Rah by the old female. Suzie is about to discard the hank of rope when it suddenly occurs to her that this is just the sort of thing which will interest Rah and so she hands it to her. For her part, Rah tries to remain unobtrusively in the back of Suzie's consciousness, with great subtlety reminding the older female of her presence by dangling her arm into the periphery of Suzie's field-of-view. Education has sometimes been thought of as a 'gift' (Luke, 2007) and this is an example of its beginning.

The creation of the pivotal object allows fantasy to take off. The idea of the pivotal object seems to me to overlap with Winnicott's (1971) idea of the transitional object and transitional phenomena. The infant, starting to become a little more independent, retains something of his or her mother in the objects which he or she treasures. These objects provide the link between an interior, imaginary world and an external, objectifiable one. The pivotal object is a lever to create meaning. All meaning is imaginary.

10 Sentences in the sky

Going-down and going-around

The stage-managed fall of an object, like other primitives of expression are choreographed into more complex action-assemblies.

04 April 2008 Video-recorded observation (Edited DVD) Singapore
(Tape 1)

Rah (3 years 4 months 17 days), on Level 3, moves along the edge of the platform from one corner to another. When she gets to this second corner, she trails an arc from her left to her right, with her left index finger, through a sprinkling of dry leaves, little twigs and dust which has gathered there. A few of these bits and pieces drop over the edge onto, or near to, her little sister Ainka (7 months 5 days). Rah looks over the edge to see the leaves fall down. With an outstretched right arm Rah this time swivels anticlockwise whilst with her index finger describes a large section of a circle in the detritus scattered on the platform.

This is an example of an early, natural drawing action. The near-circular movement is a Second Generation Structure in children's drawing. Rah's scattering of leaves and other detritus is seamlessly joined to the end of her drawing action and is really a continuation of the same action. In fact, the action does not even end there but includes her watching the fall of leaves and dust. As in human childhood, drawing actions are originally part of a bigger, all encompassing flow of multimodal play.

In most research on children's drawing, actions wrongly considered extraneous to drawing proper have been excluded from study. This is a shame because they often serve an important service of linking action to symbol. In both species, trace-making effects arise out of the natural rotations, swingings and swayings of the body. The general move from natural actions made in space and time, to their gradual encoding in purposeful action-statements seems true of human and chimpanzee. We should also note that Rah traces her actions on a flat, smooth surface. This smooth, flat surface alone is a product of human technology and thought. Would she be able do this in the wild?

Many people would say 'no' but in my opinion it is glib to dismiss chimpanzee drawing as the mere product of human contrivance (Janson, 1961). For one thing, biologists are sceptical of superfluity in the abilities of animals. Why would niggardly evolution give the animal unneeded capabilities? So the fact that, when provided by a carpentered, flat, smooth surface she makes drawing actions needs more careful explanation than this. Perhaps one may speak of *potentialities*.

08 August 2008 Video-recorded observation (raw footage on DVD) Singapore

Passing water

Rah (3 years 8 months 1 day), swings in and out on a long pendulum-swing to and from baby Ainka (11 months 9 days), who is with her mother Suzie on Level 3. At the end of each inward apogee, Rah strikes the edge of the platform with each foot-hand, one after the other, in a quick tattoo thus: swing, tap–tap; swing, tap–tap, and so on.

00:06:40 Later, Rah is upside-down, zooming in and out in a pendulum-swing from the same Slack Strap, holding on with three hands, including a right foot-hand (the leg curiously stretched around the back of her neck), to baby Ainka on Level 3. Exactly on the maxima of her final inward apogee, Rah lets fall from her mouth a piece of butter-coloured palm-stem and then, in the next instant, letting go the strap from all but her left front-hand, turns upright and with a sweep of right front-hand attempts to catch the falling morsel. She fails in this aeronautic feat but her confident style and speed suggests she has known success before.

00:22:48 Later, hanging upside-down from Level 2, she allows another piece of the butter-coloured palm-stem to drop from her mouth, adjusting the angle of her upside-down head so that the morsel falls plumb vertical. She watches it as it falls onto the platform beneath her. The sight and sound of its impact on the wooden floor seems to act as a trigger which sets her body into sudden motion; a mad, mid-air wriggle-dance. She flop-drops down to the platform below, rapping the wooden deck with her right hand as she lands in what seems an echo of the dropping lump of food. Then she slides in rapid circles in the vicinity of her discarded food remnant. She makes some graceful yet sprightly leaps along a vertical axis toward Ainka, directly above her. As she jumps upward at her little sister, she reaches up and grabs at her. In return, Ainka produces a little trickle of urine which descends in a silver stream toward Rah.

On this windless day, the trickle of liquid describes a more or less true, vertical axis. This is something the chimpanzees learn early on. The act of urinating on another ape from directly above is sometimes a way of creating fun. Rah, though, does not seem disconcerted as the trickle of wee just misses her.

In this complex play-assembly, Rah seems to be combining dynamic structures which were glimpsed in their emergent states only days before. One idea is about *going-down* (dropping), and even little Ainka is able to contribute a little to this – a little urine, that is. Another idea is about *going-around.* Going-around and turning in circles about a central axis I term *continuous-rotation.* This is part of Second Generation structure in human children's development. Going-around has its own rich developmental possibilities, leading in the acts of children (and possibly in those of apes) to the idea of the *circular*.

As in children's dynamic expression, Rah is joining together these two dynamic structures, *going-around* and *going-down,* to form a new play-assembly. Other ideas form subsets of this major dynamic structure; for example, *looking-down* is also incorporated into investigations of going-down.

Measurement (and other narratives)

Beginnings and ends (and the places in-between)

There is one other important structural principle we see in the last observation employed by Rah in her interactions with her baby sister. Rah is demarcating the *ends* of her inward pendulum-swings, first of all with two taps of each foot-hand against the edge of the platform and then with her dropping of the piece of palm from her mouth, which occurs on the apogee of her inward flight and is followed by a quick scoop of her right front-hand to intercept the falling food. This demarcation of the end of a linear route is palpably rhythmical; the first series of inward swings involving the double tap of her foot-hands at the end of each inward swing follow a strict pattern and the final demarcation of the final inward swing involving the dropping of the lump of palm and its attempted interception, although much more ambitious, nonetheless adheres to the same, strict rhythm. This time, the dropping of the palm has to be interjected a half-beat before the final beat of the bar, when Rah turns right-side-up to try to scoop up the lump. (That Rah this time fails to retrieve the falling particle does not matter – we can still see the rhythmical intention.)

Marking time

This is an example of how a chimpanzee demarcates the end of a linear route from A to B. Sometimes, the beginnings of such routes are demarcated in some way too, with a stamp or a slap, for example. Later, such linear passages may be interspersed with other staccato sounds or actions, as if the line of movement were being further subdivided. The placing of a sudden sound or sharp action within a fluid sequence seems to act as punctuation in action-assemblies whilst simultaneously lending them a rhythmic character. They are like beats in a piece of music. As well as demarcating space, the chimpanzee is marking time. First of all, as in human childhood, these punctuation points are placed at the *beginnings* and *ends* of linear routes. Here is an example from childhood.

20–21 December 2002 Video-recorded Child-observation (Edited DVD) London

(A–B route) (Poppy and Lily)

Lily (5 years) and Poppy (1 year 6 months) are playing on a mat; a longish oblong measuring 70 cm × 32 cm. It has the layout of the 'hopscotch' game printed upon it. Lily, the older child, understands the division of the area into squares and the numbering system which, by the rules of the game, allows movement along the length of the mat. Lily is trying to play the game properly but Poppy is trying to join in but is too young to understand the rules. Lily carefully throws a suitably sized object (a blue hair-band) onto one of the numbered squares on the mat and hops to that square. Poppy interferes, trying to take the blue hair-band whilst squealing with happiness and jumping wildly around the spot on which Lily is trying to stand. However, Lily does not mind too much and incorporates Poppy's lawless interpretation of the game into her own, legitimate one.

Chaotic as Poppy's play might seem, it does have one organising feature. She uses the two ends of the length of the mat as landmarks and she runs the length of the mat between these two ends. On one occasion she stamps on one end of the mat; on another occasion she slaps the back of a settee which happens to form a boundary at the other end of the mat.

The chimpanzees also demarcate the beginnings and ends of linear routes in much the same way.

21 March 2008 Video-recorded observation (Edited DVD) Singapore

Rah (3 years 3 months 14 days) knuckle-walks from one end of a fallen log to another, slapping each end of the log as she reaches it. This morning, Rah awoke to a transformed environment. Some of the timbers had been rearranged; older logs taken away and new ones added. Rah makes a linear route from one end of a fallen log to another, slapping each end of the log as she reaches it. Rah is noting that, although some of the environment has been changed, one of the older logs – this one – remains unchanged. She slaps each of its ends as if it is an old friend.

For both chimpanzee and human infant, the beginnings and ends of linear routes are the first salient landmarks. The positions in between are less immediately marked.

Counting

23 January 2009 Video-recorded observation (Edited DVD) Singapore

Kimoni (1 year 5 months 10 days) enjoys climbing up and down Oblique Log, a straight piece of timber which slopes at about 40°. To assist his climbing, he uses the short stumps of branches which stick up at intervals along the log's surface. I think of these as 'Handle Branches'.

He knows the whereabouts of the two ends of the log whilst the little branches provide information about positions *in-between* the two ends. It may be that he starts to notice the number of these familiar branches or perhaps the number of grasping actions he is required to make between beginning and end. Other observations support this hypothesis.

07 September 2007 Video-recorded observation (Edited DVD) Singapore

00:10:12 Rah (2 years 9 months) chases after Nkosi along Fallen Log 1 and with her right hand touches each of two hand-sized protuberances or nodules on the log, one in turn, as if counting them, one–two. After this she stops, sits and looks down at the two nodules as if considering.

We know that chimpanzees are even better than humans at remembering the visual configuration of a number of items (Cantlon & Brannon, 2007; Herrelko, 2010; Matsuzawa, 2007). Are they actually counting?

Young children likewise make note of any significant landmarks along a linear route. We go back to a previous observation to see an example.

22 October 2009 Child-observation, hand-written, London

A 3-year-old girl is with her mother walking along a London street. From over a high, brick wall, a single, wiry, rose-bush stem of a couple of metres in length hangs vertically, denuded of leaves but with still some thorns along it at almost regular intervals. The girl fingers the stem and her mum tells her to be careful. The girl notices the equal spacing of thorns up the stem and with an index finger indicates several of these thorns consecutively in a line. Then she takes the lowest end of the stem in a finger-and-thumb pincer-grip and moves it through a small section of an arc making a 'Wheeee!' vocalisation as if she is pretending to swing upon it.

We considered earlier the pretence (of flight) in this and other observations but now I want to draw attention to the way in which the girl notes the thorns which mark regular intervals along the stem. I think she is noting these in a similar way as Kimoni does the spiky stumps which stick out from Oblique Log and which he uses like handles or steps on a ladder.

These and other observations also suggest emergent measurement, in which long objects and possibly long intervals of time are divided and subdivided. We will return to these possible precursors of mensuration later on, when we consider further the chimpanzee's interest in shape and form.

In children's drawing too, the beginnings and ends of lines are demarcated first, with visual patches or marks, whilst positions in-between come a little later. Some of the child's first drawings of the human figure, the so-called 'tadpole figure', are also organised according to these landmarks. Very young children seem clear about the beginnings and ends of a sequence ('head' and 'legs', or 'hair' and 'feet', or even 'nose' and 'foot') but are less clear initially about the 'middles' ('bodies' or 'tummies'). This may be one of the reasons very young children tend to miss out the middle section (the 'body') in their early drawings of a human-being (or temporarily mark the spot where the body should be with one of those scribbled 'placeholders' we mentioned earlier) (Costall, 1985; Freeman, 1980; Freeman & Cox, 1985; Matthews, 1984, 1999, 2003, 2010; Willats, 1997, 2005). Shortly, we will consider another way in which the middle section of the human figure may come about.

In addition to demarcating different spatial positions, both children and chimpanzees mark direction changes.

18 February 2009 Video-recorded observation (Edited DVD) Singapore

Kimoni (1 year 6 months 5 days) leaves his mum, Ganga (16 years 10 days), whom we'll designate 'Point A', on Yellow Food-Box and runs on all fours along Fallen Log 1. He stops at a Point B, near the end of the log, and marks this spot with a downward tap of his heel, turns completely around and heads back.

Perhaps he feels this downward tap of the heel necessary to mark his change in direction. This is another device I have seen used in children's play and drawing.

Shortly afterward, an in-between point along the log is formed by the arrival of big brother Nkosi (8 years 7 months 21 days) who sits near the middle of the log between Kimoni and Ganga. Kimoni stays with his brother who, serving as a mother-substitute and temporary point of safety also marks a roughly half-way mark between Point A (Ganga) and Point B (end of log).

08 September 2005 Video-recorded observation (Edited DVD) Singapore

(Don't mess with Mr In-Between)

There is a big quarrel amongst the chimpanzees. There is a lot of angry shrieking with Nkosi (5 years 2 months 11 days) and Ida (32 years) at its centre. The young male has upset his mother in some way. Alpha male Togo (33 years 2 months and Nkosi's dad) sorts out the matter by getting into the middle of the group of quarrellers, separating them on his either side.

06 October 2005 (Part 2) Video-recorded observation (Edited DVD) Singapore

00:17:09 Makhulu (5 years 4 months 29 days) is approaching Nkosi (5 years 3 months 9 days) along Horizontal Beam Level 2, from Corner 2, with Ida (32 years 1 month 22 days) positioned in-between the two juvenile males. It is as if the two young males are each using her as a shield, so that they can creep up on each other, under cover of her big body, in its shelter, as it were.

02 March 2007 (Part 8) Video-recorded observation (Edited DVD) Singapore

This time it is Togo (34 years 9 months 6 days) whom Nkosi (6 years 8 months 5 days) has angered. The Alpha male chases him off. There is wild shrieking and screaming. Later, for protection and support, Nkosi positions himself between two high-ranking females (Ida and Suzie, I think). Sitting in a row on a Horizontal Beam, the three of them look like the Three Wise Monkeys.

This developmental sequence, beginning at the beginning (and sometimes at the end), and later on, with the addition of positions in-between, echoes one in human childhood. The human infant, having achieved self-locomotion makes daring excursions, outward from the known into the unknown and back again. Like the chimpanzee, the human infant starts to demarcate the beginnings and ends of these routes with actions of the body; a slap, a punch or a little tap, little hops or jumps perhaps; sometimes actions which make a sound; the stamp of feet at the end of a linear journey or sometimes just by visual and kinetic means, like the movement of a hand.

07 December 2007 Video-recorded observation (Edited DVD) Singapore

Suzie (32 years 25 days) is sitting with her baby Ainka (3 months 8 days) near the Yellow Puzzle-Box near the moat. Ainka knuckle-walks wobblingly within a passageway which winds its way between embedded rocks. She moves a couple of metres away from her mother and toward a small boulder. We can think of Suzie's body as Point A and the boulder as Point B. When the infant reaches this rock she stretches her arms up and along its smooth, sloping side before wobbling back to her mother. As Ainka approaches Suzie on the return journey, Suzie thoughtfully raises a hand as an easily visible landmark for the baby chimpanzee. Ainka makes one touch of Suzie's hand as she reaches this Point A. This single tap seems to give her a further burst of confidence, for she immediately sets off on another outward adventure.

This is very reminiscent of a human toddler venturing out on a wobbling course away from, and back toward, his or her mother (Bowlby, 1953).

07 December 2007 (observation continued)

Suzie flags the end of the 'home run' for the infant by raising her hand in the air. In another observation, made just a week later, Ganga too has occasion to raise her arm and hand as a 'sign-post' for her infant, Kimoni when he is around the same age as Ainka.

The mother's raised-hand 'stands-out', as we humans would say; presenting a distinct landmark to guide the infant back to safety. Additionally, the tottering baby chimpanzees can, if need be, hang onto the parent's raised arm as to a mast on a rocking ship. On this occasion Ainka's contact with Suzie's arm is just a tap, yet this gives the baby the necessary reassurance to renew her explorations. This is a good example of how rich symbolisation may emerge from what appears to be the most meagre source, like the merest touch of a finger.

The A to B routes described above are along bumpy terrain but we will think of these as roughly *horizontal axes*. In the next observation, an A to B linear route is made along a *vertical axis*.

29 August 2008 Video-recorded observation (Edited DVD) Singapore

Ganga (14 years 5 months 21 days) with her 1 year 16-day-old son Kimoni, sits at Yellow Puzzle-Box which is at the base of Right Tree (or Tall Tree) – the vertical support timber for Lateral Log and situated (if, for convenience, we continue to privilege my own station-point) at its right-hand side. Kimoni climbs confidently up away from his mother, up the tree to a height of about 1.5 m and then descends again along the same vertical route back to the safety of mother.

Scary, new objects too are investigated first of all within range of the safety of mother.

18 February 2009 Video-recorded observation (Edited DVD) Singapore

The Keepers have installed a new, blue bin hanging from a cable as a toy for the chimpanzees to play with. The chimpanzees are at first wary about it, as they are with any novel object. Nkosi is now 8 years 7 months and 21 days old yet even he is cautious, creeping over to it to briefly thump it and set it spinning before quickly running away from it. Now Kimoni (1 year 6 months 15 days) follows suit, copying big brother, leaving Ganga for less than a second, giving the blue bin a hasty tap with his little hand before, just as hastily, retreating to the shelter of his mother. Only experienced Ida approaches the bin calmly, certainly with caution but without fear, to check out the new installation.

The above observations show early adventuring away from the mother and the safe return to her. John Bowlby, in his (1953) 'attachment and loss' theory, writes about the human infant's excursions and incursions in relation to the safety and comfort of the mother and how profound is the effect of these on the child's future confidence.

My observations suggest that the chimpanzee mother favours starting her infant off on these adventures along the ground terrain, rather than high up in the arboreal environment, but I did see important exceptions to this. It seems to depend on particular circumstances and context. In the wild it is often safer to be off the ground and, if one is obliged to spend time in an arboreal environment from which one can fall, it makes sense to learn to use it safely.

Chimpanzee preparations for the future: Emergency procedures

14 December 2007 Video-recorded observation (Edited DVD) Singapore

Ganga (14 years 10 months 6 days) is with her baby, Kimoni (4 months 1 day), on Word Platform. Unusually for her, Ganga seems to be engaged in pretend nest-building. She has made a circle of leaves and mother and son sit on opposite sides of it. I think that she is encouraging Kimoni to manipulate the leafy stems. Kimoni picks up some and looks at them.

Soon I realise that the Ganga's pretend nest-building is just a pretext. She has set up an experiment which has nothing to do with nest-building. Knowing that Kimoni will become preoccupied in manipulating the leafy stems his mother creeps off away from the baby without his noticing. Keeping her eyes on her baby, Ganga moves to the far right of Word Platform. She leaves him sitting on the other side of the platform with his little back to its edge and a 10 m drop below. There is a distance of over a metre between the pair. I am worried lest he fall and she alarms me still more when she climbs off the platform altogether, at the opposite side from her son, and stands bipedal, still watching him, from an adjacent, horizontal beam.

Ganga maintains her eye-gaze upon Kimoni for a second or so and then, making a quick scurry across the platform, retrieves him and gathers him up in her arms, hugging him closely to her chest. She is practising what is essentially a 'safety drill' or 'emergency procedure' with her infant son.

00:34:48 Ganga again leaves Kimoni at the edge of the platform and moves to its right edge and watches him from there. This time, she lies down on her back as though ignoring him, though all the while looking toward him. She extends her left foot-hand toward him and he notices this and looks up at his mother.

Her raised foot-hand presents a highly visible and contrasting vertical axis to the baby chimpanzee. I believe that she is signalling the baby to climb onto her chest. Sure enough, Ganga reiterates her signal with the use of her two forearms, which she reaches forward to Kimoni and with them makes the barest suggestion of a pulling, pawing movement in the spread of leaves, as if encouraging the infant to come toward her, onto her chest. Then Ganga lowers her face right up close into the baby's face and, together, they study the leaves before them, their faces only centimetres from the platform's surface.

Mother and baby resume investigation and play with the leaves but after a short while Ganga again appears to abandon Kimoni and move to the opposite side of the Platform, watching over him for a few seconds whilst he investigates and plays with the leaves independently. Then, allowing a few seconds to pass, she quickly scoops the infant up into her arms and turns her back on me so that I cannot see him at all. Ganga practises with her son what I think of as 'emergency procedures' from time to time over the following eight and a half months.

The chimpanzee's tomorrow

Suzie likewise has rehearsed 'emergency' procedures with her baby, Ainka. We go back a few months to see the start of this.

14 February 2008 Video-recorded observation (Edited DVD) Singapore

Suzie leaves Ainka (now aged 6 months 1 week) intentionally at the edge of Hollow Rock where it plunges down to the edge of the moat. Hiding herself behind a log, 2 or 3 m away, Suzie waits a few seconds before rushing out and gathering up her baby in her arms.

This is similar to the contingency drills of Ganga and Kimoni but in the present case, the infant Ainka has additionally to cope with the disappearance, yet continued existence, of her mother. It is possible that Suzie is gently testing her baby's tolerance of her temporary disappearance. The work of John Bowlby (1953) and others shows that prolonged or endless disappearance of the mother leads to her virtual extinction in the imagination of the infant. In both human infants and chimpanzees this can lead to depression, despair and even death.

Earlier, we discussed Rah and Santino hiding stones for use on future occasions; the former for play, the latter to stockpile missiles (Osvath, 2009). This is evidence that chimpanzees have a concept of the future and make preparations for it. The observations above of Suzie and Ganga rehearsing contingency plans for the rescue of their infants in purely hypothetical emergencies is further evidence that chimpanzees possess what is termed 'autonoetic' consciousness; that is, the ability to both memorise past events and imagine the future. This ability is not only linked to real time but also allows the chimpanzee to distinguish between imagined events and current perception; an ability crucial in symbolic thought. Therefore, the existence of autonoetic thought in chimpanzees is further evidence of the likelihood of their representational and symbolic thought.

11 Point-of-arrival and moment-of-impact

We have seen that both human children and chimpanzees form an interest in *beginnings* and *ends* of linear routes and that they demarcate these important landmarks with a distinctive action or sound. Sometimes, to the young of both species the instant of impact becomes an event of interest in itself. This being so, I give this play-assembly a special term all of its own: *trajectory-and-moment-of-impact.*

Chimpanzees, like human infants, use special techniques to emphasise the point-of-arrival, as if exploiting this event for its expressive characteristics. Techniques used to *pronounce* the moment of impact may take many forms. An especially loud slap is a basic one but a range of surprisingly subtle and light caresses may also serve, as well as quite elaborate ornamentations – one example being Rah's dropping of a piece of food and her attempt to retrieve it before it falls, all within a single beat. The use of such different kinds and complexities of actions serving as punctuation suggests that the chimpanzee adopts an overarching principle which encompasses all these forms; the principle of *emphasis*.

'Wo de fei ji fei le!' (Mandarin-Chinese for 'My aeroplane flies!')

March 2001 Video-recorded child-observation (Edited DVD)

In a Singapore nursery, a group of Chinese 2–4-year-olds have been told they can draw anything they like. Tentatively at first, but soon enthusiastically, some of the children start producing First Generation mark-making Actions (including vertical arcs, horizontal arcs and push–pulls), whilst others produce more advanced, Second and Third Generation Structures yet to be discussed. The theme of aeroplane flight emerges, possibly from a suggestion by one boy who calls out excitedly: 'Wo de fei ji fei le!' (Mandarin-Chinese for 'My aeroplane flies!').

A 2-year-old boy stabs his pencil down furiously on the paper, synchronising a plosive vocalisation to each impact of his pencil. Clusters of heavy black dots and dashes (along with deep grazes and punctures in the paper) accumulate on the drawing surface.

Nearby, another child, of about the same age, very consciously and purposefully raises his pencil above and behind his head and aims it down in a controlled arc, the pencil-point touching the paper at right-angles to it and leaving a little dot. As the pencil touches the paper, he is seen to mouth a plosive morpheme. He repeats this successively many times, each time preparing the downward flight of the pencil by raising it out of his field-of-view-and-action before plunging it down again.

I have made many observations which show that when children come to appreciate that they are creating visual effects within their own field of view, they often withdraw their hands and handheld marker from this field of view so that they can more clearly survey its uncluttered appearance. Typically, the withdrawn hand hovers for a moment just outside the child's visual field, sometimes around and slightly behind the level of head and shoulder, as the child contemplates his or her work, before once more allowing the hand to enter the field of view to continue painting or drawing.

In contrast to the suggestion made by Thierry Lenain (1997), that chimpanzees do not, like humans, select a visual field (and therefore, according to him, their paintings and drawings are not 'art') my observations suggest that chimpanzees develop sensitivity to their own field of view and remove extraneous limbs or objects from it so as to see it as an immaculate visual sample. This way, they immediately notice any changes within it they then cause.

05 October 2007 Video-recorded observation (raw footage on DVD) Singapore

Rah (2 years 9 months 26 days) is pounding a lump of clay. The lump is about the size of a football. She pulls her right hand up and out of her field-of-view-and-action and just looks at the clay lump for a moment and the imprints she has made upon it.

It is also worth recalling that when she was only 10 months old, her big brother was sensitively showing her specially presented views of objects, carefully arranged for her unique viewpoint (see page 22, for example). This not only suggests that in certain situations chimpanzees show remarkable non-egocentricity, but that, even more than this, they educate their younger friends or siblings into this allocentric visual praxis too. But why would they do this for each other? I can only imagine that is out of *kindness*.

26 October 2007 Video-recorded observation (raw footage on DVD) Singapore

Ainka (only 1 month 27 days old) manipulates her left foot-hand with her front-hands. She looks at what she is doing whilst she does it. She presses her own foot-hand to her body so as to get it out of her field of view and then visually searches the surface of the log below her, reaching out her left hand again, so that it hovers over the area of interest with its fingers splayed out above the target area. She then raises this hand out of her field-of-view-and-action whilst she continues to survey the textured area of bark.

Is the insertion or withdrawal of the foot (or foot-hand) from one's field of view unique to the four-handed chimpanzee? Or will the human baby also insert or remove his or her foot from an optical sample too? I have some evidence that, before the hands (front-hands) come to dominate object-mastery, the human infant will also do this.

'Big boom!'

22 November 2008 Video-recorded child-observation (Edited DVD) Hong Kong

In a park in Hong Kong, Chloe (2 years 4 months 4 days) is gradually falling asleep in her stroller which is being pushed gently back-and-forth in a rhythmic and somnolent manner by her grandmother. Fountains softly splash nearby. Chloe murmurs a little invented song as she hugs her left foot close to her chest. She kicks out this foot a few times and after each kick retrieves the foot to her chest. This seems to give her an idea. Using both hands, she pulls her foot close to her chest and then suddenly lets it go, letting it kick out straight, in front of her. Then she pulls her foot back again with her hands very close to her face saying, 'Ready . . .' before letting the foot fly outward and away. As the foot reaches the fullest extent of its outward kick, Chloe synchronises the soft vocalisation, 'Boooom!' With both hands she recaptures her foot and draws it toward her again, pulling it back like a stone in a catapult, right up close to her face before releasing the foot like a missile. As her foot reaches the apogee of its flight, she sings again, 'Boooom!' She does this four times in its developed form though on the fourth go, as she pulls back her foot, instead of saying, 'Ready . . .', she says softly: 'Three, four, one . . .' and 'Big boom!' on the outward kick.

Chris Athey's (1990/2007) work has shown that a young human child is interested in two kinds of change – *changes of position* and *changes of state*. In the present observation, Chloe is interested in seeing her foot in Position A (close-up) and Position B (further away). She is interested in changes of position.

This observation shows a ludic trajectory and virtual impact. It may be that it is the restraining harness of her buggy which obliges her to insert and extract her own foot (as opposed to a hand) into and out of her visual field.

Armstrong et al. (1995) write that vision is the most fundamental sense for mammals and that it detects any changes in the pattern of light reaching the eye. They add that these changes may be extremely subtle and have to do with shape, location and movement.

06 October 2005 Video-recorded observation (Edited DVD) Singapore

Sitting in the hollow formed between her mother's open legs, the infant Rah (9 months) dangles her legs over the beam and lolls forward toward me, looking at me, her chin resting on Ida's right thigh and her right arm hanging over it. The baby swings her foot-hands up-and-down, and at the apogee of each upward swing, touches them with her hands before allowing the foot-hands to fall back down again in the downward-going arc. The contact between palm of hand and palm-sole of foot-hand is like a gentle brushing, very like the action performed by human baby, Chloe, with her hand and foot.

Changes of position and changes of state

17 October 2008 Video-recorded observation (Edited DVD) Singapore

Ainka (1 year 1 month 18 days) presses both hands together onto a big, green leaf of a growing plant, her left hand pressing on top of her right hand. [I have seen ceramists and cooks use this technique.] She is smiling with satisfaction. She holds her hands there, more or less motionless, for a second or so, the plant pressed down underneath them. She then releases her hands, withdrawing them to a waiting position, equidistant from each side of her body, just outside the boundary of her field-of-view-and-action. The crushed leaf slowly and naturally starts to uncrumple and Ainka stares down upon it as it seems to move under its own volition. The leaf slowly unfolds itself, nearly but not quite returning to its original form. The little chimpanzee looks at this effect with wonder.

In this example both *changes of position* and *changes of state* co-exist.

Trajectory and impact is further developed in other types of action and undergoes expressive variations. It is present within the form of a pendulum-swing in the following observation.

16 November 2007 Video-recorded observation (Edited DVD) Singapore

Hanging on a plastic vine, Nkosi (7 years 4 months 19 days) is making pendulum-swings, to and fro, to and fro. On his final 'to' swing, he lets go his hold on the vine and makes an arcing leap onto the wooden surface of Level 3. Both foot-hands slap the platform simultaneously at his moment of arrival.

In the following observation, Rah also uses the same type of combination in a swing-touch play with her brother, Nkosi.

Rhythmic sequences

03 January 2008 Video-recorded observation (Edited DVD) Singapore

00:05:09 Rah is pendulum swing-touching from a sloping vine to Lateral Log, reaching out with her right hand's fingertips to touch the Log at the point of apogee of her every inward swing. Then, on the final one, she makes a trajectory-and-impact-type landing onto the log with a follow-through downward hunch. This phrase is punctuated by a final downward knuckle-touch of her left hand against the surface of the log. This final moment of contact is not any sense functional; it is not used, as it is in other instances, to balance her point of arrival. Rather, the use of her straight-down punch against the wood is as a marker – a punctuation point only.

The chimpanzee develops subtle variations in this type of play-assembly capable of expressing and conveying different feelings or qualities. In fact, the demarcation of the terminus of a linear trajectory can take some rather surprising forms of physical contact.

The kiss

29 August 2008 Video-recorded observation (Edited DVD) Singapore

00:38:24 Kimoni (1 year 16 days) is learning how to dangle upside-down from a vine. He hangs on with right arm and makes little, oscillating swings, extending this movement by an additional swaying rhythm of his left arm at the apogee of each oscillation. He repeatedly makes vertically arcing swipes with his left hand, against a horizontal beam whilst hanging on with his right.

The top half of the horizontal beam is wet with rain whilst the bottom half is still mostly dry. The wet, upper half of the beam is dark in tone; the dry underside light in tone. It is toward the top, wet side of the beam that Kimoni strikes his hand, causing streaks of rain to run down below to the drier side of the beam, standing out as black lines against the pale wood below; straight, vertical and roughly parallel. Kimoni notices these lines and accentuates them by use of his vertically arcing swipes which help to propel the water downward, further strengthening and darkening the lines of water.

00:38:34 His original oscillations made whilst hanging from the overhead beam have now been amplified to a little pendulum-swing to and from the horizontal beam against which he swipes, each swipe emphasising his return swing. He increases the length of his swing until he brings his lips – instead of his hand – up to the wet beam on his every inward flight. At the instant his lips make contact, so he licks the water running down it. His point-of-arrival is sealed with a kiss.

The *kiss* seems to originate from the transference of food from the mother's mouth to the infant's mouth:

22 September 2006 Video-recorded observation (Edited DVD) Singapore

Ida (32 years 1 month 8 days) passes food from her mouth to 1 year 10 month 15-day-old Rah's mouth.

This exchange looks like a kiss. It acquires layers of emotion and meaning as the individual grows and perhaps through evolution too and as it is practised in slightly different contexts.

17 November 2006 (Part 5) Video-recorded observation (Edited DVD) Singapore

Rah is just 25 days older. She and big brother Nkosi (6 years 4 months 20 days) kiss each other and Rah then falls into his arms. No food is exchanged here, just affection alone.

16 March 2007 (Part 1) Video-recorded observation (Edited DVD) Singapore

It is nearly 4 months later and Nkosi and Rara are sitting on Yellow Puzzle-Box. Nkosi allows the little chimp to take his stick-tool and practise spearing fruit. [This will take her years to learn properly.] In return, Rah offers him some fruit, passing it to him from mouth-to-mouth.

03 October 2008 Video-recorded observation (raw footage on DVD) Singapore

00:02:41 It is 1 year 6 months and 17 days later. The Keepers have left some tasty food-pellets in the enclosure this morning. Ganga (16 years 7 months 25 days) has some in her hand and one in her mouth. As she passes Suzie (33 years 9 months 3 days) she uses the mouth-to-mouth method of exchange to pass the food-pellet to Suzie.

This transpires in a single, seamless movement, with Suzie anticipating Ganga's impending arrival and seeing what she is intending. As she becomes in range, their faces looming up to each other, Suzie puts her left arm carefully around Ganga's shoulder, so as to maintain stability and to allow for accuracy of docking with Ganga's lips.

In this example, together with the kiss, is the *embrace*, which may have its roots in the actions of a couple using hands and arms to steady each other, a little like lovers, so that the mouth-to-mouth transaction can take place (see Figure 11.1).

The 'kiss' then is emotionally laden and acquires polysemic meanings. Elliot Eisner (2006) makes the important point that one cannot speak of an action being 'merely' motoric; for sensorimotoric actions are loaded with meaning. He cites the 'kiss' as a prime example.

Many of our examples of the impact at the end of the trajectory have been sharp, aggressive, sometimes involving a sudden action and a loud sound but they need not be so. Because the kiss is so laden with meaning, Kimoni's silent and softly planted kiss made on arrival at the wet wood surface is as effective as a crash-landing.

I see other instances of the minimised and softened moment of contact. For example, in an observation of Rah when she is 2 years, 1 month, 19 days old (page 206), after watching leaves falling down from her mouth and rain falling from her fingers she performs a spiralling descent to Suzie and as she lands she points an index finger at Suzie's head (26 January 2007, video-recorded observation, edited DVD, Singapore).

What may have started its developmental life as a more solid strike is here restrained to soft touch, perhaps the evolutionary predecessor of the indexical or

Figure 11.1 The kiss and the embrace perhaps start from the transfer of food from mouth-to-mouth. Ganga (carrying Kimoni ventrally) passes food to Suzie.

'pointing' gesture. As in other cases too, such soft-landings, really more like tokens of physical contacts or even signals, may arise from the wish to avoid hurting a companion – especially a high-ranking one (Suzie, remember, is not Rah's mother so Rah must be careful not to offend her).

In some kinds of demarcation of line-ending, action, sound and vision are naturally united. In the above example and the one following, no sound is produced; only action and image are used in manners which are both iconic and indexical.

21 March 2008 Video-recorded observation (Edited DVD) Singapore
(Tape 1)

00:06:13 (3 years 3 months 14 days) Rah makes her way again along the Slack Vine to her mother Ida (33 years 7 months 7 days), commencing her brachiation with an upward, underarm arc or 'wave' with her left hand and marking the finish of her journey, at the other end of Sloping Log, with an almost identical, upward motion or 'wave' of her hand toward Ida. Perhaps the signal, *wave*, starts life as an ornament on one or both ends of a walked line, only attaining semiotic status later.

In this and other signals of the body, including of course those that do make sound, the ape uses the principle of *contrast*. Later, we will see the development of the wave and of other signals which rely on an understanding of contrast.

Trajectory and point-of-arrival may operate at various scales too, one form at one scale seeming to reiterate, echo or possibly represent another form at another scale.

06 October 2005 (Part 2) (1st cut) Video-recorded observation (Edited DVD) Singapore

00:30:00 Rah is nearly 10 months old and is sitting in Ida's lap, dangling her legs over the beam on which her mother sits, watching her two brothers, Makhulu (5 years 5 months 29 days) and Nkosi (5 years 3 months 9 days) play. Their play-assembly includes graceful and dramatic pendulum-swings, swoops and collisions, high up in the Construction. They end their play taking turns to zoom into their baby sister, touching her feet on each incoming moment of apogee. She does not yet have the skills to join in so she participates in the only way she can. First of all she swings her own little legs rhythmically back and forth from the knees whilst in matching fashion swinging her forearms toward them, her front-hands stroking her foot-hands each time they meet.

She turns to look me straight in the eyes. She swings her right forearm (which is hanging over Ida's thigh) in an upward arc, the pivotal point of which is formed at her elbow joint, so that her palm strikes her own forehead at the end of its upward trajectory. She repeats this natural upward arc of her forearm, again striking her own forehead with her palm. I notice that she closes her eyes against both of the impending collision.

This observation is very similar to the following one I make of a Chinese 4-year-old in a Singapore nursery.

02 October 2001 Video-recorded child-observation (Edited DVD) Singapore

'Wo de fei ji fei le' ('my aeroplane flies') (2)

The teacher is telling the class about the 9/11 disaster (which happened 3 weeks earlier) and she is showing them a photograph of the dreadful impact of a terrorist hi-jacked plane against the side of the tower. Whilst staring at the photograph, the boy, gripped by this image, makes an upward arcing movement of his right forearm, the movement issuing from the elbow and finishing with the palm striking his forehead. In synchrony with the impact of his palm, the boy vocalises a softly spoken, 'boom!' Apart from this vocalisation, his action is very like Rah's and, like the chimpanzee, he also repeats it.

His action is an *action-representation* (Matthews, 1984, 1999, 2003, 2010). Of course, in his case, the action is replete with levels of meaning and makes a direct reference to an event which has happened recently in the real world – it is a *re*-presentation.

This kind of action-representation is used time and time again in human culture. Examples of trajectory and impact operate at many levels and span a vast range, from, say, Tom the Cat being banged in the face with a frying pan (his face instantly flattened into frying-pan form) (Hanna & Barbera, 1940), to comic genius Bill Hicks' enactment of the moment in which the sniper's bullet hits John F. Kennedy in the side of the head; Hicks striking his temple with the palm of his hand, very like the way in which the Chinese boy does, except that the skilled comedian makes conscious reference to the cinematic representation of this event, by repeating it, over and over again in 'slomo' (Hicks, 1991).

The Chinese boy has not yet formed the layered representational understandings and skills of the American comedian but he is well on the way. One way in which he brings to the surface of his mind the structure and meaning of his action-representation is by the act of drawing. Shortly after he slaps his own forehead he goes on to represent this graphically, with felt-tip pen on a sheet of paper. The trajectory and impact initially represented with arm and hand in empty space is now encoded onto the drawing surface in lines which end with explosive markings.

Rah does not take this further step of drawing but we can see that in *formal* terms, her action of hand and arm is similar to that of the boy; that is, leaving aside for the moment the question of representational content, the basic structure is the same. Physically and dynamically it is almost identical to the 4-year-old boy's self-slap of forehead (which, for him, emulated the impact of an aeroplane).

Dare we suggest that perhaps Rah's action is akin to his in the further sense of having an extra layer of meaning beneath it? It may be that she too is re-presenting a recent event, though in this case, one that occurred only moments ago – her brothers' trajectory and impact play. The contiguity of the two events, her brothers' actions and then hers, may help her achieve this precursor representation. She has been watching them avidly and now seems to echo, on a miniaturised scale, their cycle of trajectory and impact.

To remind the reader, *miniaturisation* is when an event seems to be reiterated on a smaller scale with a part of one's body. Rah's upward rising arm and hand ends in a slap of her own forehead which emulates, on a smaller scale, the pendulum-swings and moments-of-impact of her brothers' play.

Is her action also a communication with her brothers? Is she trying to find a way of 'joining in' with them? As Rah performs her second forehead slap, Nkosi comes zooming in on a long pendulum-swing to make a virtual collision with her. He does not make physical contact but alights on the Horizontal Beam next to her. This is very like an observation we discussed earlier in which Nkosi makes an almost identical trajectory-and-impact leap which ends right up close to his little sister but without crashing into her. As with that previous example, at one level of interpretation it is an act of deception made to tease Rah by causing her to think he really is going to crash into her. There is more to it than this, however. Although it is true that he is bluffing his sister, partway through the event (which lasts only milliseconds), he also wants her to *know* that he is bluffing. I would argue that this is a pretend or virtual collision; a performance on the way to becoming an *imaginary* collision. It is an act of pretence because he inhibits complete execution of the action by use of the technique of *curtailment.*

As we have noticed before, an act of imagination may commence life as a desire to 'explain' something to another. Nkosi wants Rah to 'get the idea'; to understand by *joining-in.* Getting others to 'join-in' is important because in order to obtain maximum enjoyment from play, it has to be shared. Companionship is at the heart of representation (Reddy, 1991; Trevarthen, 1995). The play-assembly is an embodied and shared *idea.*

At this present moment in her development, Rah's inability to participate fully, safely, with her brothers in their play may be behind one of the ways in which evolution encourages and sustains the representational act. Because she cannot make full-body movement in mid-air like them (and she knows it would be dangerous to try), a bottleneck of desire forms within her. She has to find a way out of this cul-de-sac. She must not jump because she knows she will injure herself but has to remain sitting on the beam. The desire to participate and be part of the action excites parts of the body which may be moved safely. Her desire is expressed in her miniaturised translation of her brothers' swinging motions and moments-of-impact. This translation takes the form of her swinging forearm, pivoting from her elbow and the collision of her palm against her forehead.

Nor can Rah participate in her brothers' complex variations which follow but they nevertheless show her the elements of which these are composed. They break down the action-assemblies into watchable component parts. They do this to entertain their little sister and also to introduce her to this play-mode. It seems very like the learning of language.

The actions and interactions from here become very complex and the description which follows only captures part of its structure:

06 October 2005 (Part 2 continued)

Makhulu, hanging on a vine, performs a series of pendulum-swings in relation to a group comprising Ida, Rara and Nkosi, with Nkosi being the main target. Makhulu swings inward outstretching arm or legs at the moment of apogee. His assembly is made more complex by a contribution by Nkosi, who, sitting on the end of a corner-post, reaches out with his left hand to Makhulu's vine each time it approaches and tugs it exactly one beat after each of his brother's moments of impact.

The rhythm is one–two–three, one–two–three, comprising: (1) inward swing and impact by Makhulu; (2) tug on vine by Nkosi; (3) outward-swing by Makhulu. Variations occur when Makhulu turns upside-down and zooms in toward the group head-first.

Nkosi's hand-contact with the vine is also varied; sometimes he grabs forcibly, sometimes he just touches it with his left index finger alone. So, the intention behind each touch is as variable as is its emotional content. Of great interest to us is that on one inward-swing, the much older Ida and mother of Nkosi also contributes by a quick outward touch with her left hand at Makhulu as he arrives at the innermost point of his apogee. She understands the meaning and structure of the boys' play and is supporting it. Sometimes too, Ida will also offer a gentle touch of restraint, also at the inward apogee of the chimps' flight, especially when they get too rough; too close to the baby Rah.

This entire action-assembly then is complex; filled with subtle variations and what seem to be representations within representations. It is also a group effort, resulting from companionship.

A desire for an item of food may have been all that set it off (when one of the brothers pursued the other in order to grab it), but this is quickly followed by reciprocal teasing between the chimpanzees. From there, a deeper motivating force and organising principle emerges. This is the wish to organise one's actions according to an aesthetic. As Vygotsky says, play creates a new form of desire (Vygotsky, 1966).

12 Helping hands

Learning

Our last observation contains many instances of the chimpanzees' learning and their provision of learning opportunities for their young. We will look now at other ways in which the learning of the young is supported by the older or more experienced chimpanzees. This educational process in chimpanzee culture may have been misinterpreted and underestimated. Again, this is hardly surprising as it seems to have been all but forgotten in human, technological societies (if not actively suppressed by their controllers). It is often assumed that chimpanzees do not give very much to each other, nor do they support, share or 'scaffold'. This notion is very misleading. One way in which they share and give is in the education of their young and inexperienced.

27 January 2006 (Part 3) Video-recorded observation (Edited DVD) Singapore

00:20:13 Nkosi (5 years 7 months 27 days), entrusted to look after Rah age (1 year 1 month 20 days) carries her ventrally as he plays a high-speed, aerobatic play-chase with his brothers, Manzu and Makhulu, through the Construction. Makhulu is soaking wet from the Waterfall and cascades of water enhance the explosive direction-changes of their three-dimensional, harum-scarum chase. Rah clings on to Nkosi's chest as she is whirled through this high-velocity choreography, inducted safely into a complex play-assembly.

00:21:01 Later, when the play-chase is over, we see Nkosi escorting Rah along Fallen Log 1. He is walking backwards in front of her, encouraging her to come towards him. She knuckle-walking along the log toward him whilst he keeps a constant eye on her, smiling at her as he walks backward before her, monitoring her forward movement. He twice allows her to pass through his legs to overtake him but each time circling in front of her again, the second time by swinging from an overhead beam, to face her again, resuming his backward shuffle in order that she can continue her forward movement unbroken.

When they get to the end of the log she takes hold of his right arm in her right hand as a support down which to climb to the ground. The interchange here is smoothly reciprocal; as she takes hold of his arm so he anticipates her descent and is already in the act of lowering her. She continues this motion by trailing her little fingers through his big ones, before letting go of his hand, a residual trace of her fingers in space forming a request for him to follow her.

Before we leave this observation, of note is that the configuration Nkosi describes in space, moving along the straightish line of the log whilst circling twice en route in front of forward-moving Rah is a *travelling-loop* or *travelling 'e'* formation similar to the one he performed just minutes earlier today when running from his father in play (see page 99). (Other travelling 'e's made by Nkosi can be seen illustrated in Figures 18.1–18.4 and 18.10–18.16.)

In the last observation a baby chimpanzee is helped to walk along a straight route near the ground. Because they lead a partly arboreal existence it is necessary to learn how to pursue a straight course high above the ground too.

18 April 2008 Video-recorded observation (raw footage on DVD) Singapore

00:19:26 Leaving Kimoni (9 months 5 days) several metres away at the top-right end of Lateral Log, high above the ground, Ganga (14 years 2 months 10 days) backs away along the log to the opposite end, the top-left end, though maintaining a constant line-of-sight to the little baby as she withdraws. However, unlike Rah, who was happy to follow her backward-walking brother along a log on the ground, Kimoni will not be induced to follow his mum along a log 10 m in the air! Ganga races back along the length of Lateral Log to rescue him.

Helping little sister to play 'peepbo'

11 April 2008 Video-recorded observation (Edited DVD) Singapore

Rah is now 3 years 4 months 14 days and is trying to entice her baby sister, 7 month 12-day-old Ainka, to play 'peepbo' with her but Ainka does not understand the signals Rah is making. These include looking at the baby chimpanzee through a gap in the sawn-off stumps of branches, reaching her arm through the gap towards her and then quickly dodging back out of sight again. Unable to comprehend Rah's promptings, Ainka fails to respond.

Helping little brother to spin (1)

11 April 2008 Video-recorded observation (Edited DVD) Singapore
(Observation continued)

Eventually, Suzie rescues Ainka from Rah's attentions. Realising it is 'no go' with Ainka, she then tries her luck with her brother Kimoni. She tries to teach him how to spin around whilst hanging on a vine but again, the little male cannot take advantage of her cues because he does not understand what she is trying to 'tell' him.

We will discuss this particular observation further when we consider the emergence of the Third Generation Dynamic-Structure *continuous-rotation*.

Here is another example observation of a learning situation in the chimpanzee world:

Helping Kimoni to climb

It is 9 months and 12 days later. Ganga is allowing her infant Kimoni (1 year 5 months 10 days) to climb unassisted along a branch just below her.

23 January 2009 Video-recorded observation (Edited DVD) Singapore

00:03:15 She is ready to help Kimoni should he need it. She tentatively reaches out her right hand toward him as if to take hold of him but immediately realises he does not require her help so withdraws her hand.

Instead, she provides very watchful support. She lowers her hand once more but again hesitates and withdraws it. After allowing Kimoni a few minutes solo climbing, she finally makes a decision and uses her right hand to lift him onto her back as away she climbs.

There is an interesting moment in which Ganga signals, in a coiling gesture in space made with her only free hand, for Kimoni to climb onto her back. He does not yet understand the signal so he cannot cooperate.

That Ganga is obliged to use her only free hand in this manner is important and may tell us something about the necessity of inventing the sign. Then, because Ganga knows that Kimoni does not understand what she wants him to do – to climb aboard – she accordingly adjusts her level of teaching to his level of understanding.

00:07:00 Ganga lowers her right hand from Level 3 down to Level 2 to physically hoist Kimoni onto her back.

It is worth noting that, even with a functional, physical strategy like this, instigated and carried out by the parent, reciprocal exchange is nonetheless required between the pair of them. As Ganga lowers her left hand, so Kimoni senses its arrival and accommodates to this event, bringing his left hand around into a shared field-of-view-and-action so that the hands of each chimpanzee meet and hook together. She then hauls him upward and he swings over onto her back.

Over the following months, when this manoeuvre has been practised over and over again, it will become sufficiently interiorised in Kimoni's brain to allow Ganga to abbreviate it to a *sign*. The coiling gesture tried by Ganga in the above observation is one example of the abbreviation of a functional, physical movement (physically taking hold of her son) to a sign, a visual and kinaesthetic image which refers to an idea. Whether such signs are equivalent to words is a controversial question (Savage-Rumbaugh & Lewin, 1994).

Although it might seem that Ganga is not doing anything much, she is in fact doing a great deal. Helping an infant with a task involves very subtle levels of communication rarely mentioned in current educational debate (though there are important exceptions including Chris Athey 1990/2007; Tina Bruce, 1987, 1991; Worthington & Carruthers, 2006, Blenkin & Kelly, 1988/1996; Atkinson, 2002; John Jessel, Gregory, Islam, Kenner, & Ruby, 2004). In the present case, there exists a very subtle yet palpable empathy between Ganga and Kimoni in which cues are offered; some taken up, others not. The sensitivity of mother to baby causes her to follow up certain avenues of communication but abandon others. When she knows it is to no avail she does not persist in overt teaching. Of significance to those interested in children's education is that, whilst opportunities in chimpanzee culture for learning are plentiful, many of which are supplied to the young by the older or more experienced chimpanzee, there exists nothing in their culture which one might term 'rote learning'. Perhaps this is one of the many advantages of *not* having language.

Here is another example of a mother helping her infant learn to climb.

05 October 2007 Video-recorded observation (Edited DVD) Singapore

Suzie (34 years 1 month 21 days) is trialling Ainka (1 month 6 days) to see if she is able yet to hang independently on an overhead horizontal beam. Suzie hoists Ainka over her head by gripping the baby's wrists and trying to place the baby's hands onto the beam (it is circular in cross-section). Though the baby seems to have the necessary physical strength to do this she is unable to. The reason is that it is beyond her comprehension at this time. Ainka seems not to understand what is required of her. It is illuminating to analyse Suzie's response to her baby's inability.

It tells us important things about the mother's care and support for her infant's development. She is finding out how strong her baby is whilst also introducing her to a task at the very threshold of her skills and understandings. Vygotsky (1986), writing about the cognitive development of the human infant referred to what he termed the infant's Zone of Proximal Development. This 'ZPD' means the anticipated forms of understanding toward which the child is moving; the knowledge 'nearby' in the developmental distance, as it were, of the infant's present understanding. Vygotsky writes that an optimal learning experience is one in which the adult companion anticipates and predicts to a certain extent, the understandings towards which her infant may be heading and supports or 'scaffolds' the task accordingly. In the case of the chimpanzee mother and baby dyad in the present observation, it turns out that the skills of climbing solo are just out of range of Ainka's 'ZPD'. It is fascinating and salutary for educators to note that, like Ganga, as soon as Suzie realises her baby's level of understanding, she does not persist in any further teaching at this time.

Much current educational discourse gravely misunderstands *scaffolding* – it is not a one-way transfer of information from teacher to learner. It is multilinear communication with the traffic of information travelling in many directions, including from child to adult. Whilst many a curriculum hack is forced to boringly churn out formulae about 'scaffolding', as if it was a series of instructions in an IKEA assembly kit, Mitchell Resnick notes that *it is extremely unlikely that an adult can predict what the child's ZPD will turn out to be* (Resnick, 2002). Alan Costall (2009, personal communication) too points out that the word itself, 'scaffolding', implies a framework erected for an edifice already built or, at least, planned.

John Jessel and colleagues' (2004) term 'synergy' better encapsulates the complex support provided by the 'more experienced learner' (to use Geva Blenkin's better, 1990 term for 'teacher').

It is important to understand that the issue for Ainka is not just physical competence. Though Ainka may have the physical strength necessary to accomplish the task of climbing up, it is her lack of cognitive knowledge and communication skills which prevent her doing so. She does not take Suzie's hint to climb because she does not understand what her mother is trying to do (or to 'tell' her). For an interesting comparison consider how a human parent might try to put a baby on a swing before the baby has the idea of swinging. One thing that can happen is that every time the parent lowers the infant down onto the seat, instead of the child bending his or her legs and forming a sitting position, the child persists in holding his or her legs stiff and straight down.

In the next observation we see a young chimpanzee who has both the physical strength and the mental understandings required for climbing solo. All she needs is just a little help from dad.

Father supports baby's play

01 December 2006 (Part 2) Video-recorded observation (Edited DVD) Singapore

Togo (nearly 34 years) and baby Rah (nearly 2 years) are at the base of Twisted Tree, Togo sprawling on his back on Bridge Log and wrestling in play with his infant daughter. He cuddles Rah closely to his chest and playfully mouths at biting her. A fire-hose hangs almost vertically from the top of Twisted Tree and Togo allows her to climb solo up it while he holds the strap taut and steady for her. To motion her to climb up he uses an upward pointing gesture of his left index finger which he lines up with the vertical rubber strap and in conjunction with emphatic directional up and down nodding of his head. Rah ascends Twisted Tree and when she gets to the top, Togo looks briefly around at me.

This is the example promised earlier which shows that in some contexts, chimpanzees can make use of the pointing gesture to indicate direction-of-travel; that is, in situations supported by contextual cues (a main one here being the nearby vertical fire-hose) they can use a physical gesture as a *sign*. In the current observation, Togo has one hand occupied holding the strap straight and taut for his daughter, leaving his other hand free to align a straight, upwardly pointing finger immediately adjacent to the straight, upward axis of the rubber strap. He has one remaining free body-part or body-system which he can enlist to emphasise his wish that Rah proceed upward along the vertical fire-hose. This spare body-system is his head which he nods, up and down, very distinctly, twice; its motion also in line with the direction of the rubber strap. (See Figure 12.1.)

Shortly afterward, Nkosi, who has been hanging upside-down from the nearby Slack Strap, watching them play, takes over the care of his baby sister, gathering her up in a cuddle on the contorted limbs of Twisted Tree. The use of the gesture as a sign is seen again shortly afterwards, on the same day and at the same site, again demonstrated for Rah's benefit but this time by Nkosi.

This observation does not actually show any physical mark-making but it seems to me to be on a developmental route somewhere between play and something very akin to drawing.

Figure 12.1 Togo (bottom right) holds hanging strap taut for Rah and, using an index finger pointing in the same direction as strap whilst at the same time nodding his head up-and-down, indicates to her that she is to climb up it.

Four-dimensional language: A trace in space

01 December 2006 (Part 2 continued)

On one of the boughs of Twisted Tree Nkosi cuddles Rah in his arms, nuzzles her and playfully bites her fingers. Then he climbs down below her on the trunk of the tree, a trunk stripped so bare of bark by the chimpanzees that it appears almost white in the bright Singapore sun.

Rah looks down at her big brother as he fingers the pronounced grain of the tree which stands out like black lines winding in a set of concentric spirals obliquely down the tree. Such grooves of grain are often gripped by the young chimpanzees and used for climbing. Nkosi fingers one particularly pronounced fissure, deepened and widened by sun and rain; sharp and clear like a hard, black line on the white wood.

Perhaps it is to entertain his sister that he makes the very intriguing action which follows.

He pretends to lose his balance, letting himself tumble-fall to the left whilst at the same time tracing the fingers of both hands in a descending arc on the trunk, following the line of grain with his fingers.

Still simulating an accidental fall, his fingers continue to trace a line through empty space in the same direction as that set up by their tracing of the grain in the wood of the trunk.

Finally, at the end of this whole-body trajectory, he allows himself to tumble into the grass, nearly, but not quite, performing a punctuating head-over-heels, so adding the full-stop to his performance. The tracing of his finger through space, perfectly in line with the line of wood-grain in the trunk, underscores his graceful descent and looks like a mime of a drawing action; a theatrical event.

In this observation Nkosi adds a new and imaginary line to the end of an old and physical one. With fingertips, he carefully follows an oblique line derived from the pattern of the grain and as this line comes to its end so he continues it with fingertips in space as he falls sideways, ending it with his final collapse to the ground.

This observation recalls that of Hannah at 2 years 3 months (Hannah, 06 December 1983, PAL VHS videotape-recorded and written observation, page 91) in which she adds, to the end of a drawn, 'descending' line, a real descending 'line' in three-dimensional space, with the fall of a real object (a set of earphones); or like the Chinese boy, who after enacting the flight of an aeroplane with a moving line on a piece of paper, 'adds on' to this graphite trace the movement of the piece of paper itself (bearing the pencil line upon it) through the air. Like them, Nkosi seems to be giving the same 'denotational value' (Willats, 1997, 2005) to a line made up of two, quite different parts, one physical and visible, one invisible, imaginary.

Additionally, as well as configuring a trajectory in space, as if in a mimed action of invisible drawing, Nkosi's trailing hand bears a signalling function too; he is beckoning Rah to follow him. To help you picture this I will risk a bad pun and write that he is using this gesture to 'draw' her after him; he is enticing her to follow him – and she does just that.

13 Shape, form and symmetry

Like us, chimpanzees are clearly interested in the shape and form of all sorts of objects they might find. In their interactions with them they uncover their main characteristics, features and axes. Earlier on we noted that some of the actions they perform upon objects might form the precursors for strategies of *measurement*.

The calculated action

The primate body in itself embodies a symmetrical geometry which gives shape and form to its actions and interactions within the environment and with its fellows.

Chimpanzees learn about the *midline* of their faces and bodies. Their use of the paired halves of their bodies naturally invokes a midline. The midline seems to be an important reference point for some primates. Human mothers hand their babies toys at the midline of the baby's chest (Gray, 1978). It would be of interest to see if chimpanzee parents do this.

At the end of the last chapter, it is the midline of his face which Togo exploits when he uses his up-and-down head-bob to encourage Rara to climb upward. An imaginary vertical line down through the centre of his forehead, between his eyes and down his nose to the centre of his chin is lined up with the nearby, salient vertical axis of the hanging fire-hose (01 December 2006, video-recorded observation, edited DVD, Singapore, page 156. Also see Figure 12.1.)

The facial midline is used as a reference point in other ways too.

14 December 2007 Video-recorded observation (Edited DVD) Singapore

'Kimoni . . .!' I call out to the 4-month-old chimpanzee. He dodges his head back behind Ganga's leg, allowing only the left side of his face to remain visible. Then, as I continue to call out, he peeps out his left eye and looks down at me, straight into my face. After a while Kimoni averts his gaze away to his left. The baby chimpanzee is already curious about this constantly watching human. Nevertheless, taking cues from his mother, he remains sensibly wary.

The balance between fear and curiosity is an embodied form of measurement in itself, involving a psychological symmetry. We will encounter again this symmetry between opposing feelings expressed in the *agonistic play* of chimpanzees.

Measurement

At present, Kimoni's calculation of the relationship between fear and curiosity is expressed in the strategies he uses. His actions are very much like the way in which a mixture of curiosity and fear is expressed by a human infant when approached by a stranger. He or she uses the mother's body as a shield yet at the same time still wants to peek out from this safe vantage point. Kimoni uses the symmetry of the face to hide as much as he is able whilst retaining the all-important view of me with one eye. The other baby chimpanzees also use this strategy.

There is an important optical consequence of this practice which we will discuss fully later on, but briefly, by viewing a scene with a single eye, the usual disparity between the two images which result from binocular vision (essential for capturing information about depth and three dimensions), is cancelled out and one obtains a single, monocular image of the scene.

Kimoni's strategy recalls the 'Peepbo' play performed by human babies with the help of their mothers or fathers. As both human and chimpanzee grow older they perform this with increasing skill.

Nkosi, for example, has become good at it by the time he is 6 years 8 months 27 days old (24 March 2006, video-recorded observation, part 1, edited DVD, Singapore) and sometimes uses it as a viewing strategy when regarding me (see Figures 13.1 and 13.2).

Figure 13.1 Nkosi (6 years 8 months 27 days) partially hides but still wants to see what I am doing so retains a single eye-gaze upon me (obtaining a monocular image).

Figure 13.2 A moment later Nkosi looks directly at me with both eyes (obtaining a binocular image).

In their interactions with the environment the chimpanzees seem to display an embodied knowledge of geometry and measurement. One act of mensuration continuously performed by the chimpanzees is the estimation of distance. They gauge distances every time they make a perilous leap from branch to branch. Being able to estimate a distance is necessary to protect someone else, too, especially one's offspring. As we have seen earlier, it could be vital to know how far one's baby can move in *x* number of seconds. Here is another example of estimation used in the care of a young one:

29 February 2008 Video-recorded observation (raw footage on DVD) Singapore (Tape 1)

Ida (33 years 6 months 15 days), with Rah (3 years 2 months 22 days) on her back, carefully negotiates a passage underneath sharp, pointed Spike Branch, which juts down dangerously from above, so as to avoid banging her daughter's head on it. The old female estimates the extra height needed to allow Rah, mounted on her back, to pass safely underneath.

Earlier on, I suggested that when chimpanzees are young they locate and mark in some way the beginnings and ends of linear routes. Later, they take note of any significant landmark 'in-between'. Calculating more exactly the half-way mark takes more experience. One way in which chimpanzees gain the background experience necessary to make this measurement is when they swing on vertically hanging vines and fire-hoses. When they describe pendulum-swings, they sometimes come close to marking the centre of an arc, which, in this type of swing is at its lowest point, or the *well* of the arc. This knowledge shows up clearly when the chimpanzee forms a *swing-and-touch*, or *swing-and-graze* assembly just above the ground or another surface. Remember when Rah (3 years 5 months 29 days) hung from a vine above Level 3 carrying a 2 kg slab of concrete in her foot-hands? Hanging like a plumb-line she deposits the slab on the platform just below her toes. When she pendulum-swings over it she stamps it rapidly, twice in succession at each pass, which occurs exactly in the central 'well' of her pendulum-swing, at its deepest point, halfway through her arcing swoop.

Another way in which the chimpanzees find the middle of a length is when they sit or hang from a Slack Strap or Slack Vine. Unlike the vertically hanging lines, these, you will remember, are plastic vines or old fire-hoses tied (by the Keepers) at both ends laterally between two posts or trees, though not tightened into straight lines but allowed to sag.

When the chimpanzees swing on these Slack Vines or Slack Straps they like to find the middle of these drooping lines and swing from the lowest point of the deep 'V' formed by their weight. The range of the pendulum-swing formed in this way is greatest; the flight-path satisfyingly balanced. They 'gravitate' toward this midway-point in order to achieve a special sensation.

This is a form of *embodied knowledge* perhaps based upon an interiorised representation of the primate body in the primate mind. Such embodied knowledge is amplified in actions the chimpanzees perform upon objects and in their locomotion through space.

Non-accidental structure

My chimpanzees seem especially intrigued by atypical, natural structure. Both unusual straightness (including flatness) and curvature is of interest to them. They investigate what I think of as *non-accidental structure* (Biederman, 1987; Costall, 1993, 1995). Non-accidental structure includes human-made form but it also refers to certain spatial relationships noticed by the chimpanzee which are not necessarily human in origin. This may seem confusing but there is a good reason for my using the term in this way. Non-accidental structure includes properties like co-linearity (things being-in-line with each other, or which run side-by-side), as well as unusual roundness, flatness, smoothness and hollowness and also regular *slopes* (which we noted earlier, exploited by young Rah). These phenomena are non-accidental in the sense that, though they are cued by natural phenomena, they are only fully resolved in the consciousness of a thinking being. Other non-accidental structure includes uncommon juxtaposition of forms or associations between them, as in chance aggregates, collections and combines as well as self-supporting, gravity-assisted structures or *polypods*. The apes themselves cause certain of these non-accidental structures to come into being whilst others they notice readymade because they are on the lookout for them.

Some non-accidental structure comes about from the transient visual relationships between forms which result from one's line-of-sight to objects. Non-accidental structure is thus supplied by the mind of the thinking primate and so when we try to imagine the chimpanzee world we should try to imagine a psychological as well as physical environment in which there is a gradual change (rather than a sharp division) from that which is generally considered 'natural' to that which is 'made'.

Being in close proximity to humans, my chimpanzees of course have close encounters with 'made' structure in the form of poles, rods, plastic bottles, hessian sacks, lengths of thread and rope, as well as the great, carpentered, wooden Construction on which they spend so much of their lives. They encounter 'made', flat, planar objects – the wooden platforms being a prime example. Additionally, they have more distant encounters with other objects and machines which they espy beyond the moat and electric fence, the outskirts of their territory. My video-camera on its tripod; my ladder and my umbrella are familiar ones to them but there are other objects, slightly more distant still, including cars, lorries and mechanical excavating machines. The chimpanzees must be making inferences from these sightings.

Saliently long objects (long stems, branches, plastic bottles, lengths of thread, etc.) are of interest to them because of their straightness or linearity. These include objects which are *fixedly* straight, like rods, sticks and poles and so on, and those which can be *made* straight (or into other forms too), including long, malleable stems and lengths of thread.

Another important pivotal or transitional object, the hessian sack, can be *straightened-out* carefully on a flat surface so creating the *straight-edge*, or hung out flat on a vine like a washed sheet on a line. Ganga likes to perform both these strategies. The actions the chimps perform upon these special objects not only uncover the main axes and characteristics of the objects themselves but simultaneously demonstrate the natural possibilities of the movements and configurations of the primate body. The chimpanzees' interactions with objects reveal their thorough understanding of the relationship between action and shape; a prerequisite for the use of many forms of semiotic system. In the following observation Rah has encountered an amorphous material which, though it is not straight, can be made straight.

Horizontal, vertical, curved (or just droopy)

10 October 2008 Video-recorded observation (Edited DVD) Singapore

Rah (3 years 10 months 3 days) is on Mezzanine Level with a piece of canvas (measuring approximately 1 m^2). She pulls out a length of cotton-thread from the edge of the canvas and, holding one end of the thread from each hand she outstretches her arms laterally to fullest extent to make the line of thread straight. A similar action occurs moments later but this time she stretches her arms upward toward the sky, pulling a length of thread with them. Thus she makes both horizontal and vertical straight lines. Then she pulls the thread around her neck, making a curved line rather than a straight one. There is an expressive dimension to Rah's techniques which will gain full power as she grows.

Shape the result of action

We can see this expressive dimension developed very clearly in Nkosi (5 years 10 months 29 days) in Figures 13.3 to 13.8 in which shapes made by his body seem to embody deep, emotional feeling.

Nkosi uses the hessian sack to make *bights*, *loops*, *closures*. In tyers' and rope-workers' terminology, the folding-over of the sacking, this hessian 'rope', in half, the two ends of equal length and touching but not crossing, is called a *bight* whereas the crossing-over of the ends of the rope is a *loop* (Budsworth, 2004) (in my terms, a *closure* or *closed-shape*).

Finally, he screws together in his front-hands the two ends of his rope or strap and puts his weight upon it, hanging in the air. He has made a *Hanger* which he uses to slide down Slack Strap from Level 2 to Level 1.

Minutes later, up on Word Platform, he makes other forms by exerting measured, directed forces upon the sacking's warp and woof. These shapes are produced in a sequence which runs as follows and can be seen in Figures 13.4 to 13.7. A *curved line* formed by pulling the sack over his upper back (Figure 13.4); a *right-angular*

Figure 13.3 Nkosi (5 years 10 months 29 days) makes a Hanger from a strip of hessian sack. Hanging upon this he slides down Slack Strap, thus combining two dynamic-structures – the *hanger* and *slide*.

junction by pulling and tearing one length of sack directly upward whilst the other directly to his right (Figure 13.5). Notice the important supporting role played by a foot-hand in Figures 13.4 and 13.5. In Figure 13.6 he makes a strong horizontal line in which the direction of his left upper arm and elbow and even his closed eyes all follow the line of stretched sacking. Throwing the strip over his head whilst holding

Figure 13.4 Curved lines can be made by stretching lengths of material (in this case hessian sacking) around the shoulders and other parts of the back as *formers*. Up on Word Platform, Nkosi experiments with the sacking.

Figure 13.5 Shape is the result of action. Nkosi pulls on the sack along two opposing axes. The left leg and foot-hand push directly away from the self; the right front-hand and arm pull with about equal strength to his right, so forming a right-angle.

Figure 13.6 A strong horizontal line made by Nkosi. The direction his left upper arm and elbow and even his closed eyes all follow the line of stretched sacking.

Figure 13.7 Another way of making a curved line in space.

on to each end with each front-hand is another way to form a transient curve (Figure 13.7).

Finally, he offers one end of the hessian strip to Ganga and, in an interesting act of cooperation, they make between them a straight line. (See Figure 13.8.) Cooperation between chimpanzees is noteworthy and will be discussed more fully, later on.

Figure 13.8 Between them, Nkosi and Ganga form a straight line with the stretched 'rope' of hessian.

Figure 13.9 Later, Rah (1 year 5 months 19 days) becomes interested in the sack played with by her big brother and, by investigating it, makes shapes out of the hessian thread.

After they have finished with the sack, little Rah (1 year 5 months 19 days) comes along and investigates a hessian thread, pulling it into triangular and other shapes (see Figure 13.9). Later on, near the end of this book, we will see this *agonistic* play (play in which the struggle is with oneself) developed to an extraordinary level in Rah.

Chimpanzees notice the properties of straightness and of the saliently elongated. They build up knowledge of the contrast between straightness and curvature. There seems to be an aesthetic dimension too. For example, a droopy stick is picked out by Nkosi (at age 6 years 8 months 5 days) and used to comical effect. It is seemed to be found interesting (or 'funny'?) simply because it droops.

Earlier on we saw instances of embodied or embedded measurement, when, for example, the chimpanzees learnt to place themselves in-between two persons or objects. Now we consider some of the ways in which this embedded knowledge might be disembedded to form what might be forerunner acts of measurement.

Division

The division of something into two roughly equal parts (be it a stick, a squashed plastic bottle or a conversation) arises from the actions of the body performed upon the object, or within the event; actions which themselves originate in the natural symmetry of the body. The beginnings of the practice of measurement are sometimes seen in actions of *halving* an object or simply *balancing* it.

15 December 2006 (Part 1) Video-recorded observation (Edited DVD) Singapore

Spear

00:03:42 I am alerted by the sound of a stick being snapped. I pan the camera around to find Nkosi (6 years 5 months 18 days) holding a long, straight rod in his right front-hand. His grip on the rod is made more-or-less equidistantly along its length. I know this because when he momentarily loosens his grip on the rod it does not drop from his hand but rather balances on his palm.

He wants to balance the rod and as a consequence of achieving this wish he finds its middle. At the risk of stating the obvious, note that he does not in any way purposely set out to measure or find where its middle is. As in human infancy, such knowledge is a by-product of fulfilling a desire. This explanation avoids the mistake of crediting evolution with the bestowal upon the chimpanzee of superfluous cognitive gifts. The chimpanzee's knowledge is both embedded in, and released from, certain activities he or she finds interesting.

15 December 2006 (observation continued)

Nkosi's rod is saliently long, rigid and straight. He carries it differently from other sticks, holding it roughly horizontally from its mid-section as if weighing it in his hand, like a spear. Carrying it in this fashion means that neither ends drag along the ground.

Perhaps the rod's weight and straightness feel 'good' when balanced evenly in his hand and perhaps, as a consequence of his balancing-act; is even 'measured' in the hand. Whether we think of this as a functional or aesthetic requirement is an interesting issue. The aesthetic is never very far away from the functional.

Embodied measuring occurs along both vertical and horizontal axes.

15 December 2006 (observation continued)

Nkosi takes the rod to the stream and pokes it vertically downward into the water, touching the bottom of the shallow stream.

It would be misleading to claim that Nkosi is 'measuring' either the depth of the water or the length of the rod but his action might form the background experience for such calculations.

Rah also performs the same kind of action when she is 2 years 6 months 12 days old.

29 June 2007 Video-recorded observation (Edited DVD) Singapore

Rah dips a long twig into the pool near the Waterfall as deeply as it will go vertically. She touches the bottom of the pool with the stick.

As with the observation of Nkosi, a 'by-product' of her action is a proto-measurement. In looking at it this way we have again avoided the invocation of superfluous knowledge. This is embodied knowledge awaiting its evolutionary opportunity.

Note that, when a stick is inserted into the water, the depth of the water is automatically marked somewhere along its length by a waterline. Here is another, rather more developed example, again from Rah, but just over 18 months later.

18 February 2009 Video-recorded observation (Edited DVD) Singapore

Rah (4 years 2 months 11 days) sits atop Tower 2 with a leafy stem she has found. Rah has an impatient personality and is in a hurry to use her tool. She does not take the time to strip the waving headdress of long leaves it wears at one end but inserts its other, sharpened end without delay deeply downward into a hole in the Tower. She thrusts it completely home so that it inevitably gets stopped at the end of the tunnel.

It is what she does next which is fascinating. She clasps a hand around the section of the stem just at the position at which it disappears into the hole. She retains her hand at this place on the stick as she withdraws it from the hole. She contrives to maintain her grip on the same spot on the long stem even while manoeuvring it around in the air and it is from this location on the stick that she strips off the leaves.

Using her clenched hand, she has noted the position at which she wants to strip the leaves from the stick. It is also worth mentioning in passing that, if Rah's hand had been wet or dirty, a mark would have been left at a measurable position along the length of stick.

Beginnings, middles and ends

Learning how to demarcate the beginnings and ends of routes and how to estimate the half-way point or middle could confer an evolutionary advantage. This kind of knowledge, linked with navigational understandings might provide some of the basis for rational thinking and planning and, possibly, even language.

In human development, the child also starts off locating the beginnings and ends of a route, line or length, whilst for a while, like the chimpanzee, being less sure about the whereabouts of the middles.

The mystery of the middle explains the popularity with young children of certain songs, stories and nursery rhymes (in human culture, in language) which celebrate these measured stages – Beginnings, Middles and Ends along vertical, horizontal and oblique pathways – or which refer to the Top, the Bottom and Only Halfway-Up (as in the 'The Grand Old Duke of York'), or even changes in volume in three stages from Big, Middle to Small (consider 'The Three Bears', for example). Nursery stories are made for adults too, though with slight but important changes in terminology, for example; 'Grande', 'Tall' and 'Regular' ('Regular' because no one likes 'Small') and even 'Fun-Size' (which means, of course, 'Extremely Small'). A long anxiety in mathematical cosmology about the nature of infinity also testifies to our affection for beginnings, middles and ends (Cooter, 2010).

A difference between the species then is that children (and grownups) like hearing about beginnings, middles and ends in language (and number). Humans also like encoding these divisions in writing and drawing:

06 January 1984, PAL VHS videotape-recorded child-observation, London

At 2 years 4 months, with pencil and paper, Hannah makes series of vertical lines on a sheet of paper and, as each line is completed and just before she moves on to draw the next, she marks, with a little squiggle, the end of each line.

Here, she has demarcated the terminus of a line. In other drawings of this time, the beginning of a line is also similarly marked. Of interest to us is that captive chimpanzees who like to paint also produce paintings in which the beginnings and ends of lines are demarcated by a blob of colour. Desmond Morris (2009, personal communication) showed me an interesting example by Congo. In my own study of the Singapore chimpanzees, I saw just one possible instance of a chimpanzee marking the end of a drawn line but I will describe this later.

In human childhood, the middle section between two termini appears a little later in drawing development. We noted earlier on that child's early drawing of a person, the 'tadpole', is a two-unit figure possessing beginning and end ('head' and 'legs' or 'head and feet') but only later on in development does it acquire a third unit, a middle section (or 'body'). We have seen that when the chimpanzee finds the middle of an object this is as a consequence of enacting an event. This is often true for the child too, even when the object is a two-dimensional one, like a drawing. For example, putting the clothes on a 'naked' tadpole figure may entail shading or colouring over the join between 'head' and 'legs' and as a consequence arriving at the location of the middle part of a 'beginning, middle and end' sequence (Matthews, 1984, 1999, 2003, 2010).

Such examples entail an almost narrative mode not restricted to the representation of the human figure.

06 April 1985, 35 mm colour-slide film-recorded child-observation plus hand-written transcription, (Hannah) London

Hannah (3 years 7 months) draws a line along a lateral axis on the paper and as she draws it she gives a running commentary about a man who walks along, bends over and finally falls into a dustbin. In this drawing, the undulations of the line show the various stages within the event between the starting and ending point. This is a visual and linguistic description of the misadventure of the poor man in which she demarcates beginning, middle and end of his story in time as well as in space.

In several of the paintings and drawings executed by Congo, in the company of Desmond Morris, it does look like this chimpanzee is measuring space and demarcating at least the beginnings and ends of lines, and sometimes the halfway position within shapes (Morris, 1962). This also seems true of Alpha, the chimpanzee whose drawings were studied by Paul Schiller (1951). I have seen nothing as complex as this sequencing in the drawings of my chimpanzees but you might agree with me that their play-assemblies, locomotions and some of their object manipulations constitute passages with beginnings, middles and ends.

Now follows a good example of measuring by a chimpanzee. This is performed by Nkosi (at 6 years 8 months 5 days) and combined within a complex, repeated, circuitous play-assembly with his baby sister Rah (2 years 2 months 25 days).

Running alongside

02 March 2007 (Part 6; being an analysis of Part 5)
Video-recorded observation (Edited DVD) Singapore

At one point on the circuit around which Rah and Nkosi travel in their play lays a log of about 2 m in length which is sloping down the grassy bank to the moat. Each time he passes it, Nkosi either runs along it from its bottom to its top, or leaps up its length, in one jump from bottom to top, or else runs up the grassy slope alongside it. On his final circuit he makes yet another and telling variation, which is to leap up alongside it, on the grass, making no physical contact with the log, starting from a position level with the log's bottom end, and landing at about the level of the top end of the log, again at its side, on the grass. It is as if he is measuring the length of his jump against the aligned gauge or 'ruler' formed by the log, a bit like the way in which the distance jumped by a long-jump athlete is measured.

This is an instance of *measurement* made possible by the chimpanzee's use of the non-accidental structure *co-linearity*. (See Figures 13.10 to 13.13.)

One way the chimpanzees find out about how to divide and subdivide distances is by folding or bending sticks, stems, twigs and even the occasional plastic bottle.

Figure 13.10 At one section of a repeated play-circuit, a fallen log is used as a gauge for leaping. Nkosi jumps from the lower end of the log (Point A) to Rara who waits at the higher end of the log (Point B).

Figure 13.11

Figure 13.12

Figure 13.13

21 September 2005 Video-recorded observation (Edited DVD) Singapore (2nd movie, Edit 2)

Bending and halving

Nkosi (5 years 2 months 24 days) bends an empty, plastic mineral-water bottle in half before jumping up to Level 3 and subjecting the now crumpled object to a single, overarm, flailing, vertical arc.

Hanging up the washing (and doing the ironing)

Ganga (11 years 7 months 13 days) folds and unfolds a hessian sack with some care. She straightens it out and hangs it over the top of a sawn-off log. This action results in two, flattened sides roughly equal in length. It recalls the actions of someone hanging washing up to dry. A third flat side can be seen if we count the section of cloth stretched in a flat horizontal plane over the sawn-off trunk, as if over an ironing board.

The Hanger (1)

13 May 2009 (Part 2b continued) Video-recorded observation (Edited DVD) Singapore

Rah (4 years 5 months 6 days) takes a hessian sack in her front-hands, stands up bipedal and leaps up to the overhead slack vine. She expertly passes one end of the sack over one side of the plastic vine and pulls it down on the other side so that the two hanging sides are of roughly equal length. She grasps the two near matching ends together in her hands, puts her weight upon the sack, lifts her palm-soles off the floor and swings.

'That's a very nice Hanger', I comment into the camera's microphone.

We will return to this interesting invention, the *Hanger* later on.

Half-way round

Young chimpanzees also find out about the midway position in their social and interpersonal relationships and in their play. For example, in a play-chase, two or three chimpanzees will find out how to remain equidistant to the others. Sometimes they will incorporate another object or another chimpanzee as a central pivot around which to rotate.

Rotation around an axis

Round and round the Rotten Stump

08 December 2006 Video-recorded observation (Edited DVD) Singapore

(Rotten Stump)

Makhulu (6 years 7 months) and Nkosi (6 years 3 months 11 days) chase each other around Rotten Stump. They sometimes hang onto any suitable protuberance high up on the old, crumbling trunk in order to propel themselves more quickly through a greater segment of arc. They grin, with top teeth covered and lower teeth revealed.

Sometimes, to trick the other, they each feign a false movement, an unexpected clockwise turn instead of the expected anticlockwise one, for example. Perhaps they can hear my gentle laughter which may support and encourage such actions.

Whatever variations of this rotation-around-an axis they might perform they maintain their relative positions to each other on opposite sides of Rotten Stump. This symmetrical relationship contains within it information about beginnings, middles and ends.

Another way in which the middle might be discovered is in the natural geometry formed in acts of cooperation between two chimpanzees. Earlier on I mentioned that Ganga and Nkosi cooperated on the task of making the strip of sack straight. Now it is time to consider the significance of cooperation in chimpanzees.

Cooperating chimps

Peter Reynolds (2002) claims that no non-human primate has been observed cooperating in a task but the following observations challenge this assertion. Nkosi makes a discovery on his own, tries to fulfil a task by himself, but then seeks assistance from his mother, Ida. Though only momentary, their cooperation seems to suggest their grasp of the 'role complementarity' necessary for two beings to work together on a task (Reynolds, 2002, p. 205). We start with Nkosi's independent discovery which is especially interesting since it shows a cross-species empathy, this time between a chimpanzee and a tiny amphibian.

23 February 2007 (Part 1) Video-recorded observation (Edited DVD) Singapore

The frog

Nkosi (now aged 6 years 7 months 26 days) comes down to Rotten Stump. He looks into a little hole in the old tree-stump and sees something there which interests him. He now looks around in the grass until he finds a stick of suitable thinness and length.

He inserts a stick into the hole until its non-working end is flush with the bark of the old stump. Now he withdraws it with his teeth; carefully removing a little, green frog, which then attaches itself to the side of the Rotten Stump. Nkosi, fascinated by the frog, touches it gently until it eventually leaps off into the grass.

Standing up, Nkosi now shakes the tree stump with a levering action, sometimes with right foot-hand braced against the trunk, and then from the opposite side of the trunk with his foot-hands planted on the ground, whilst he pulls at the top of the trunk with his front-hands.

Suddenly, a piece of rotten wood breaks off in his hands. This startles him and he scampers off, up onto Level 2 to safety, to reconsider the situation.

He regards the Rotten Stump warily from his perch before finally returning to it. Whilst the breaking wood startles him it also tells him that the old timber can indeed be broken. Once back at the old tree-stump, he breaks off another chunk of decaying wood but this time purposely.

He goes to his mother, Ida (32 years 6 months 9 days), who is sitting on Level 1. He coaxes her to come down, by touching her left wrist with his left hand and then pushing gently from behind her head. With her usual patience, Ida slowly wanders down to take a look.

Cooperation and symmetry (push and pull)

23 February 2007 (Part 3, continued)

Rotten Stump

00:13:28 Once they are at the Rotten Stump, she pushes it from one side whilst he returns a push from the other side. In this fashion, taking turns to push alternately from each of their opposite sides, they try to uproot the old stump. They do this only for a couple of seconds or so but it is an act of cooperation nonetheless; one which makes use of symmetrical, oppositional forces.

Similar acts of cooperation between chimpanzees have been observed before (see for example de Waal, 2000, 2005).

Nkosi then stands in front of Ida and pulls fiercely at the top of Rotten Stump. Suddenly, another big chunk of wood comes off in his hand and he falls backward, comically, straight onto his mother, accidentally knocking her over!

'Whoops! You okay, Nkosi?' I cannot help laughing from the other side of the moat. Do they both feel foolish, even *embarrassed*? Or am I projecting this idea into their actions? Any comedy here is certainly unintended. This is quite unlike Nkosi's other purposeful comedy routines (if that is what they are) discussed earlier. Does the fact that this tumble is unintended account for the marked change in demeanour in the pair?

Nkosi tries different strategies to loosen the old stump from the ground; shaking it furiously from one side and then the other and standing on its top and rocking it. He also tries swinging from an overhead Horizontal Beam and crashing into the stump.

Now sitting by the stump, Nkosi listens to two squirrels talking to each other in the trees. He makes as if to pick up the piece of rotten wood which broke away from the old stump as if to throw at them but then, changing his mind, he slaps his own wrist and squeezes his arms tightly around his body, trapping his hands between armpits and upper arms, as if in self-restraint.

Later, he tries once more to shift the stump but with no success, so he swings himself on top of it. Poised with foot-hands and front-hands on the rim section, he meditatively rocks the stump from side to side, wobbling the rotting trunk like a giant, decaying tooth which stubbornly refuses to come loose.

14 Empathy and imagination

The observations in the last chapter illustrate how cooperation, reciprocity and sharing may have a geometrical structure formed in the spatiotemporal relations between chimpanzees and things and between two or more chimpanzees when they find they are working together on the same task. Perhaps, over time, ideas 'congregate' (as it were) at a particular site frequently visited (the decrepit timber wobbled by anyone who happens to pass by; the piece of loose concrete routinely tugged by all); or else collaborative solutions suggest themselves in the geometrical structures formed between individuals when a pair or a group happen to work on the same task.

Hanging out together

Now follows another observation in which empathy and understanding emerges from the formal structure of the relationship between a pair of chimpanzees doing something together.

07 September 2007 Video-recorded observation (Edited DVD) Singapore

'Getting the sack' (Part 2)

00:02:10 Facing Nkosi (7 years 3 months 10 days), Rah (2 years 9 months) puts a hessian sack over her head which prompts her brother to chase her. A fast and furious play-circuit ensues.

Gaining possession of the sack is only the pretext for a turn-taking play between brother and sister.

00:02:28 There occurs a momentary tug-of-war between them which ends with Nkosi pulling the sack free and dragging it over the shiny surface of the log. The chase and tussle continues high up in the Construction, where both use the various posts for momentary concealment, peeking and teasing.

00:02:52 Rah hurls herself onto sloping Slack Strap and slides down it with the sack cuddled up to her chest. Nkosi is in hot pursuit.

00:03:07 He leaps at her in mid-air and snatches the sack from her. Scampering to the Void Deck, he lurks half hidden behind a corner post. There is a lull in proceedings as Rah sits atop of the sawn-off top of a Look-Out Post, pretending she has lost interest. This is a pause or 'rest' in an almost musical sense.

Such interludes of stillness form part of the structure of the play-assembly and I will discuss them later.

00:03:22 Nkosi, hidden behind the post, dangles the sack provocatively by its corner into Rah's field-of-view until she suddenly leaps toward him and it.

00:08:23 The noise of human builders around the enclosure causes Nkosi to pause and Rah follows suit. Gunfire from the nearby military base adds to Nkosi's unease. To further complicate the cacophony, siamungs begin to sing their strange song, unseen, somewhere in the mysterious distance, in the remains of the rain-forest.

00:08:41 Nkosi is forced finally to concentrate on the play when Rah cannonades head-first into his chest! She wants to ensure that she regains his full attention. Now in deep play once more, there follows a beautiful sequence in which Nkosi, on Horizontal Beam Level 1, the sack held tightly in his hands, dangles Rah beneath him, hanging onto the other end of the hessian. This hanging technique is a relative of the tug-of-war which preceded it a few minutes ago. (See Figures 14.1–14.5.)

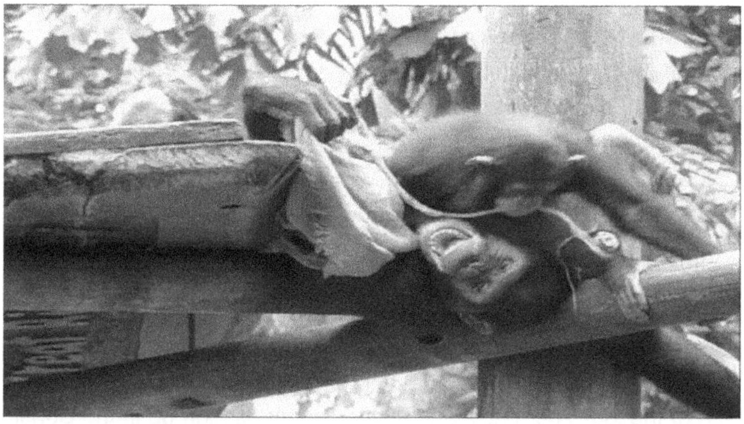

Figure 14.1 Sibling support for learning. A complex play-assembly with Nkosi, Rah and a hessian sack.

Figure 14.2 Later on in the play, Rah and Nkosi collaborate in making a hanger. After a while, Rah (left) entrusts her safety to her big brother.

Figure 14.3 Later on in the play, Rah swings from the hessian from Level 3, supported by Nkosi's strength alone. For extra purchase, Nkosi holds onto little nodules ('foot-rests') on the log with his foot-hands.

Figure 14.4 Nkosi and Rah take turns to hang from their hanger.

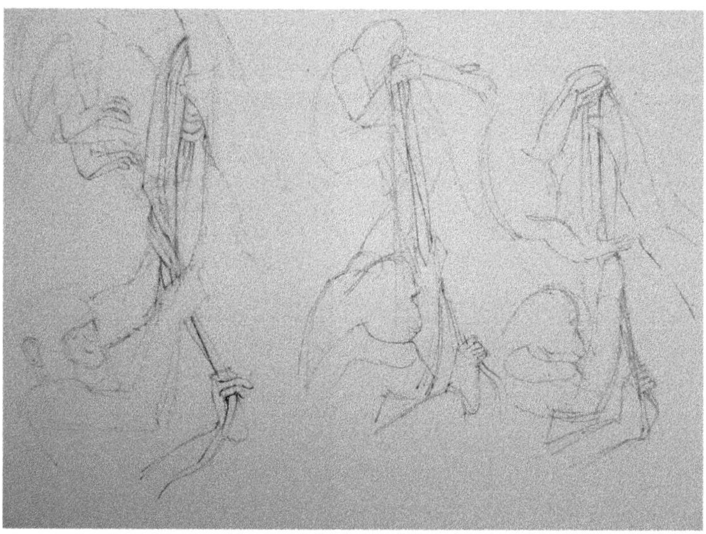

Figure 14.5 They experiment with different ways of holding on and swinging.

Nkosi hugs Rah to his chest and play-bites her. He winds her up in the sack and then she hangs upon it, taking part in its winding and unwinding until it becomes a vertical screw of hessian.

Winding, unwinding and other kinds of *continuous-rotation* form a family of important dynamic-structures and will be discussed in the next chapter.

Foot-rests

After tangle-dangling in this way and thrashing about in space, Nkosi pauses for a moment and so Rah is obliged to, too. Nkosi sits down on Fallen Log 1 and rests his foot-hands on two little lumps or nodules in the bark which are aligned horizontally on the log at a distance from each other of about 30 cm and which act as readymade 'foot-rests'. As he sits down he looks pointedly toward me because (I believe) he knows full well the quality of his performance and that I am likely to be interested in it.

00:09:04 Additionally (I believe), he notes that a sudden shrieking of machinery has not alarmed this familiar human-watcher and he is reassured by this.

00:09:31 Play resumes when he is satisfied that it is safe; taking his foot-hands off the two projecting nodules and spinning around to set off the play-chase again.

Counting (2)

00:10:12 Rah chases after Nkosi along Fallen Log 1 and there occurs the moment when, with her right front-hand, she taps the two foot-rest nodules one in turn as if counting them, one – two.

00:11:08 Nkosi passes a corner of the sack over a high horizontal beam to Rah on the other side of it. Is he hoping that she will take hold of it and pull from her side? Possibly. In any case I am interested to see that she does indeed get the idea midway through the event. As the sack's corner passes over to her side of the beam she realises that she is to take hold of it. (See Figures 14.2–14.5.)

They are cooperating together on the construction of a 'hanger' and Nkosi is making apparent to his little sister the principles involved. They now both dangle from it, hanging each from opposite ends of the sack, crashing and colliding into each other in empty space; hanging from the hanger they have built together; a device which exists only as long as each chimp, brother and sister, keeps tight hold.

Counting on you

Later, making off, up and into the Construction, the sack trailing from a corner in his hand, Nkosi meets up with Rah on Word Platform again.

00:16:08 Rah and Nkosi are wrestling and rolling on Word Platform, Level 3, taking turns to attack each other.

Rah dives head-first into the hessian screen and he wraps it over her. This develops into a repeated sequence with Nkosi now throwing the sack over his sister's head; she pummelling him again in a repeated rhythm of alternating vertical arcs. She bashes him some more but this time with two fists coming down together in double vertical arcs. When, moments later, she grasps two corners of the sack and throws it over Nkosi she is really amplifying what is essentially a double vertical arc.

00:19:10 At one point, Rah takes a corner of the hessian in her teeth and, knowing that Nkosi is holding the other end, carries it with her as she plunges over the edge of Word Platform. As she falls she takes hold of the hessian in her front-hands. Arrested in her flight by the strong, young male at the other end of the hessian rope, she climbs back to tussle with him once more on Word Platform.

00:21:33 During a breathing space, Nkosi dangles the cloth over the edge of the Word Platform. Rah once more seizes its lowest corner and with the other end gripped by Nkosi, she hangs upon the cork-screwing hessian rope (though keeping another hand on a nearby fire-hose for security!).

00:22:34 Nkosi, leaning over the edge wrestles with her and though she still uses her left hand to hold the fire-hose, her weight is now largely supported by her brother who has tight hold of a corner of the sack and is tugging her up and down with it while she hangs onto it with her right hand and both foot-hands.

00:24:31 After some more rolling and wrestling in the sack on the platform, Rah seizes a corner of the hessian and leaps off the side of the platform and this time finally lets go of her extra life-line, the fire-hose, and hangs from the sack alone, swinging upon it; her entire weight supported by Nkosi's left hand and arm. He looks down at her to encourage her and Rah, after looking down to check the depth beneath her, swings upon this hessian rope, supported entirely by her brother. (See Figures 14.6–14.8.)

After making one trial run, not putting her entire weight on the hessian but tip-toeing a little across the horizontal beam, Rah now entrusts Nkosi completely and swings in a sweeping pendulum in empty space. Nkosi takes the strain with his left hand and arm as she swings a wide pendulum arc from one side of the Construction to the other.

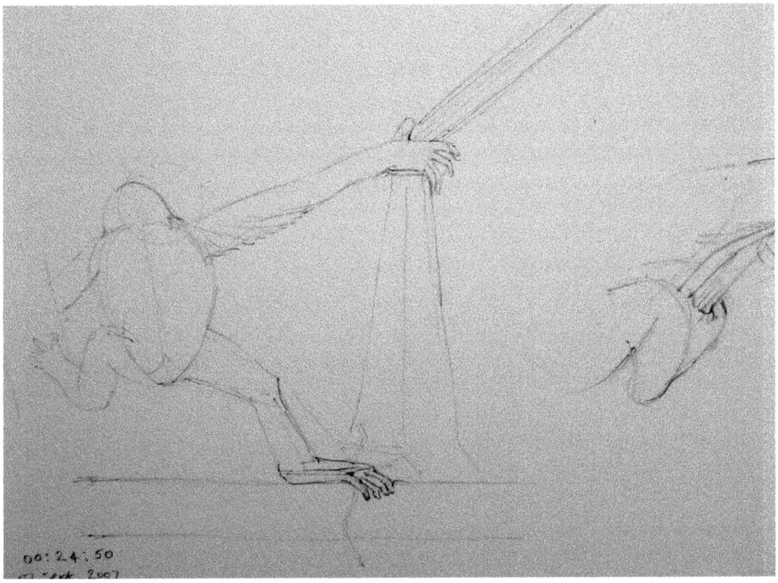

Figure 14.6 The higher end of the hessian is gripped by Nkosi, high above on Level 3, whilst down below Rah puts her trusts in him entirely when she performs a pendulum-swing from the lower end.

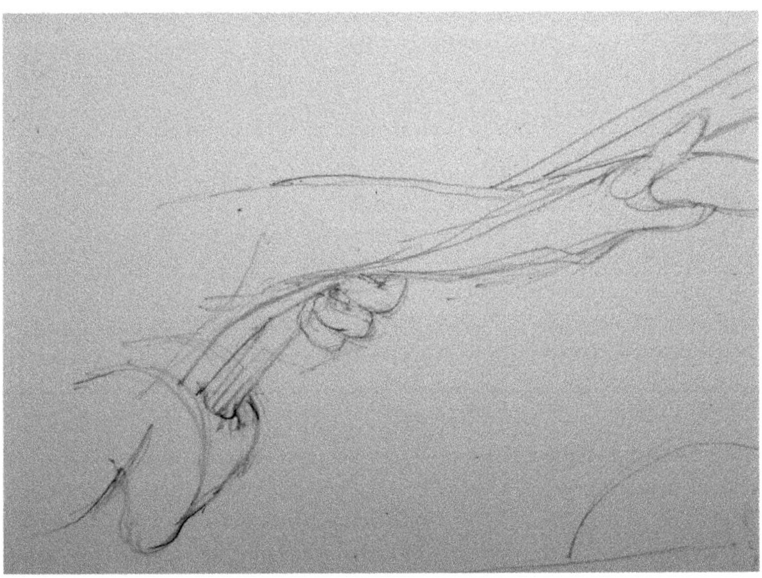

Figure 14.7 The moment before lift-off. This drawing shows a detail of a typical launching position for a pendulum-swing in which two front-hands plus a foot-hand grip the hessian 'rope'.

Figure 14.8 We have lift-off. Rah pendulum-swings in mid-air on a strip of sacking held by Nkosi in his left hand, high above on Level 3.

They have both participated in a task requiring total and complete collaboration, otherwise serious injury could befall. Did they both have the same project – the same idea – in mind? Or (a slightly different interpretation, this) did each one realise what was required of the other as the process played itself out? Of additional interest is that some human passers-by, including a mother and her young child, stop and look, fascinated by the chimpanzees' play. There happens an interesting moment when both non-human primates and the watching human kind are silenced, gripped; all interested in the same thing, though neither are able to put it into words.

15 Dialogue

Suzie's bad eye

01 December 2006 Video-recorded observation (Edited DVD) Singapore

Suzie's bad eye (wrongly labelled 'Bianca's bad eye')

Suzie is lying on her back on Level 3, Word Platform.

Togo comes over to the edge of the moat to see me.

'What's occurring, Togo?' I enquire of him. [Human colleagues have pointed out that whereas other researchers' chimpanzees have merely learnt English – at least in sign – mine have learnt Cockney.]

He nods at me.

Today, all the chimpanzees seem preoccupied with watching for Keepers and I wonder why. The chimpanzees watch expectantly as a few of them appear in a truck but these men go by without stopping.

Togo climbs up to Level 3, Word Platform, to sit next to Suzie who is still supine.

Then I notice that baby Rah is also sitting by Suzie's side too. Later, Nkosi climbs up to join the small group, all clustered around Suzie, who is still lying down.

Later on, she rises and makes her way toward me and at the edge of Level 2 Platform she sits and faces me. Suddenly, she opens her mouth extremely wide, stretching her great jaws cavernously, revealing huge teeth. It is like a gargantuan yawn except that her face bears the unmistakable signs of distress.

Gradually, I realise she is making this yawning type action in an attempt to open her right eye. Through the zoom-lens of my camera I see that this eye is very sticky-looking as if she has some kind of conjunctivitis. She tries to clean it with an index finger – *or could it be that she is pointing to her eye?*

'Are you in pain, Suzie?' I ask her, and then, to the camera: 'Is there something wrong with her eyes, I wonder?'

This time she dips her index finger in her mouth and attempts to wash her eye with this wetted fingertip.

Nkosi comes to sit next to her.

I continue my narration into the microphone: 'One of the boys seems to be a bit worried about her. He's gone up to see Suzie, to see what's wrong.'

Nkosi continues to sit next to her. He looks serious.

'I think he knows that she's not well.'

Suzie makes more tortured yawns, and then Nkosi makes some too, joining-in, his jaws gaping open-wide in an excellent imitation!

I switch off my video-camera. Packing up my gear I trudge off in the blazing sun to find the zoo vet.

From that day on, Suzie treats me as a friend.

Such acts of empathy, sympathy and cooperation are multilayered in meaning and necessitate communication (and in this case, cross-species communication). Perhaps they even form the background for language.

Such episodes are complex but they start off simple enough.

11 April 2008 Video-recorded observation (Edited DVD) Singapore

As Rah (3 years 4 months 4 days) climbs onto Lateral Log she slips a little and Suzie (33 years 7 months 27 days) gives her a little helping hand by cupping her right hand underneath Rah's bum, though without actually touching it.

Remember that Suzie is not Rara's mother. I read this action as friendly, companionable gesture rather than a strictly necessary, functional, survival one.

Even a simple, empathic 'hands-on', quicker-than-reason action like this soon gets layered in meaning.

30 November 2007 Video-recorded observation (Edited DVD) Singapore

Togo (34 years 5 months) climbs up the Construction toward Ganga (13 years 9 months 22 days). Big and bulky as he is, the old fellow wobbles slightly and she, noticing this, offers him her hand, which he accepts. She hauls him up to her side.

This empathic hand-holding is of the type that even human strangers might offer each other before they have time to think about it. Today, Ganga is very agitated about the building work going on just outside the enclosure (preparations for 'The Coastal Adventure and Children's Playground' by Seletar Lake, a few metres from the chimpanzee enclosure). Where once there were luxuriant trees at the edge of the shore, now there is a hole in the sky and a huge pile-driver pounds concrete into the ground in repetitive, mechanised, vertical arcs; harsh, sonorous sounds which often herald the phenomenon of humans making money.

Observation continues:

> Togo has gone up to sit by and comfort the agitated Ganga. Welcoming his attention, Ganga gives him a hand up.

Now follows another example of an adult supporting the development of an infant chimpanzee's skills and one which again involves induction into a world of signs.

> **13 May 2009 (Part 2b)**
>
> Kimoni (1 year 9 months) climbs up a steeply sloping log. Ganga (15 years 3 months 5 days), on the grass nearby, raises her right forearm and hand, matching his rate of ascent with this hand so that it remains aimed at around the middle of his chest despite his movement upward. She places her hand and arm there as a guide to him and it will remain there should he require it. When he is as high up as he will go (or as high as she wants him to go) she upturns her hand as an option for him. He notices this action and climbs down toward her. As he climbs off the tree to her, she gently makes to support his right hand which in turn reaches toward her hand in reciprocation. She lets him walk for a moment on the grass and then she gathers him up, places him on her back and knuckle-walks off with him.

This is a learning and teaching experience in which empathic handling, both physically and metaphorically 'supportive', is being encoded into natural signs.

Conversation

23 February 2007 (Part 3, continued)

Later on in the day when Nkosi (6 years 7 months 26 days) has tried, with his mother, to uproot Rotten Stump, he and Rah (2 years 2 months 16 days) play swing-touch at Twisted Tree; a play of cooperation requiring each player in turn to reiterate or paraphrase the action of the other.

This is a conversation-like structure.

Cooperation is also involved when I speak to Nkosi later on. This happens during one of the rare occasions when he seems to appreciate my company. I don't fool myself that he has any deep feelings for me. He comes to see me when no-one else will play with him or when he has nothing better to do. He just sits near me listening to my words, using from time to time a gentle, waving motion of his right hand, palm uppermost, as if to keep up his end of the conversation. He seems to fit these little waves in between my short bursts of speech. One might object that his is merely a 'begging'-type gesture but I don't think so. Although his action may be based on this action he knows I do not give food. I am pretty sure his movement is used to communicate with me. It is like a 'wave'. He is turn-taking and he has to find the simplest, most economical movement to insert into the temporal passages I leave for him. The upward-moving forearm, pivoting from the elbow has the necessary brevity and ease of production. It is familiar to us from earlier observations. Desmond Morris notes its use by human babies too (Morris, 2008).

16 Dynamic-structure

Going-over and going-around

We have been discussing patterns of movement which I think of as *dynamic-structures*, dynamic in the sense that they are formed not only in space but in time too. Dynamic-structures seem to emerge in the primate's actions and interactions within the environment and become grouped together to form classes. Members of different classes can also combine with each other in many ways and form hybrids. We have considered *going-along, going-up, going-down* and *point-of-arrival*. These also emerge in human infancy and I refer to them as Second Generation Structure. They will reappear shortly in this chapter but entwined with the developmental trajectories of two other important Second Generation Structures explored by child and chimpanzee – *going-over* and *going-around*. Like the other structures we have considered, both these seem to form overarching, emergent concepts containing a variety of related structures.

Going-over

22 June 2007 (3) Video-recorded observation (Edited DVD) Singapore

Rah (Breakthrough) (Edit 1)

There is a leaf resting on Horizontal Beam, curling over it. Rah (2 years 6 months 15 days) pushes the leaf along the beam and then allows it to drop. Later, Rah bends a branch over the beam. Later on still, she bends a long blade of grass over the sawn-off top of a Look-Out Post (see Figure 16.1).

Rah notices the similarity between these different instances of *going-over*. She is also forming an interest in objects which are saliently long. Only a long object can go-over or wrap around another.

Figure 16.1 Rah practises the dynamic relationship *going-over* across a range of contexts.

08 September 2005 Video-recorded observation (Edited DVD) Singapore

00:22:40 Nkosi (5 years 3 months) has a big bunch of leafy stems on a platform. He toys with one of these stems, putting it over his head and then, still holding it by its two ends, allowing it to drop behind his back. Retaining his hold on each end of the stem he pulls at these so that the stem is stretched around the girth of his back.

This recalls human infant behaviour when a baby will pick up a long object, raise it up above his or her head and lower it or drop it over the back of the head and down behind him or her to the floor. I saw examples of this many times with my own and others' children. (This is why polythene bags are dangerous for babies during this phase of their development.)

Going-around, going-over, wrapping and curving

Such behaviours incorporate a cluster of emergent concepts including *going-around, going-over, wrapping* and *curving*. With chimpanzees, these types of action derive from nest-building behaviours. Nkosi is near the beginning of a long apprenticeship in these skills, one of which is to coil a stem over the back of one's head and neck and, pulling from its two ends, wrap it around one's back in order to curve it (rather than bend or break it). Like a human child, his next action is to drop it behind him. In Nkosi's case, the curved shape so formed could become part of a nest of woven sticks, stems and leaves making a roughly circular nest on which the chimpanzee can sit or sleep. In the wild, chimpanzees will make a fresh nest every day. In the zoo, nest-building frequency is a little more erratic, depending as it does on the Keepers' provision of suitable nest-building material and also because these captive chimpanzees do not spend the night out in the open but in the Den. Nevertheless, the older chimpanzees often make proper, serious nests and show considerable skills in weaving natural – and sometimes unnatural – material together.

It takes the young ones many years to learn these skills and, as in all tasks of object and material-mastery, they are assisted in having access to the play-mode. In the observation under discussion, Nkosi's use of nest-building strategies is only minimally practical. He is here using his nest-building techniques in play.

In play, the young ones uncouple the skills involved in a task from their usual end-point (in this case, making a nest) and investigate these actions as of interest in themselves. This is what is happening here. Nkosi is interested in the process of bending the stem against his strong back which serves as a mould (or *former*) so that it makes a *curve*. Earlier, I offered evidence that the chimpanzees differentiate between curved and straight lines and are interested in this difference. The present observation shows one way the chimpanzees form a curve or even a near circle, as opposed to a straight line. In the case of the straight line, the chimpanzees find 'readymade' sticks and stems which are already straight, or else they find sticks which are nearly straight and then improve upon them. Large-size, near-circular shapes are less common in nature (you may find more under the sea; barrel-sponges – *xestospongia testudinaria* – for example) but chimpanzees can make them, often using their own bodies as moulds or formers.

The observation continues:

> 00:25:57 Nkosi sits and then lies down on the leafy branches, squashing them flat. Then, he pulls them as a complete bunch over his body as he lies back.

Here, the action and emergent concept of *going-over* leads him to cover his entire body with the leaves. The action and concept of *covering* grows out of investigation and play with *going-over.* The actions of sitting-on or lying-down-upon are procedures used in nest-building proper. They are made with the purpose of squashing and flattening the nest-building material. Yet though Nkosi performs these actions upon his leafy branch this is not serious nest-building. Rather, it is play and this is testified by his next action which is to take the product of all his hard work – a nest half-built – and throw all of the leafy branches right over himself in a green shower.

Play allows him to divorce means from ends; to both separate and recombine actions, sometimes putting together actions of totally different classes to form unique action-assemblies. In this observation, most of his actions are derived from functional, adaptational procedures yet the entire assembly is non-utilitarian. He has no fixed end-point in mind. He is quite happy to mess up a potentially good start in building a nest and then see what these effects might suggest to him. Human artists often work in this way.

Covering

Covering is also a key concept in human infancy and again, I have many observations from my longitudinal studies of children's play and investigation.

> **20 December 2002 Child-observation hand-written, (Poppy) London**
>
> Poppy (just over 18 months old), lying down on her back on a cushion pulls a blanket over herself. She retains her grip on the top edge of the blanket and lies still; partially hidden by the blanket (though her feet are sticking out from its lower edge). Suddenly she pulls the blanket away from her face and body with a sharp downward tug of her hands. I respond to this action with dutiful surprise and she shrieks with laughter.

In this example, *covering* is linked to the concept of *hiding*. The act of covering is performed by the infant in relation to an adult companion who is watching interestedly.

The action of *covering* in order to *hide* is performed by the chimpanzees in serious situations when they want to avoid conflict but also in play with each other and, occasionally, in relation to me, a human observer (see Figures 16.2 and 16.3). Additionally, as we will shortly see, a chimpanzee will hide from what I believe to be an *imaginary other*.

Earlier on, I described the *Hanger* made, collaboratively or individually, by both Rah and Nkosi, and promised more about this invention. In the following example observation, we can see that its use grows out of natural action amplified by an artefact of chimpanzee technology.

The Hanger (2)

27 January 2006 Video-recorded observation (Edited DVD) Singapore

Nkosi is now 5 years 7 months. He is standing on a horizontal beam (circular in section) whilst holding onto a similar, overhead beam. He uses both his front-hands clasped together over the beam with fingers interlaced. He hangs from his hands like this, lifting his foot-hands from the beam on which he stands. At other times, he stands on the beam and slides his clasped hands back and forth loosely along the smooth, curved surface of another beam overhead.

Later, he finds what looks like a strip of torn bark, or perhaps it is a long, strong leaf, and he hangs this over the beam above. Gripping both its ends together, he hangs and swings from this *Hanger.*

His Hanger is a technological development of his earlier, double-handed grip on the overhead beam. He has invented a device which will extend an action performed originally with his bare hands. Somewhere in that short period when hanging by his hands upon the beam, the idea occurred to him that he could improve this experience with an improvised device.

I also made observations of the orang-utans at the zoo making Hangers from hessian sacks. Please note that the orang-utans and the chimpanzees never see each other and do not know of each other's existence. Therefore, the invention of the Hanger is made independently by each species.

Now, holding onto each of its ends, Nkosi slides this strip of bark (or long, tough, leaf) back and forth along the smooth, curved surface of beam in much the same way as he did with his crossed, clasped hands just a moment ago.

Figures 16.2–16.3 Rah (3 years 10 months 29 days) enjoys covering and uncovering her face with the wet, magenta cloth. She likes looking through the cloth too. Perhaps she likes the colour against the light.

Figure 16.3

I comment into the camera's microphone: 'That's the invention of the strap, or the hook, or the hanger ... based on the concept of going-around or going-over'.

Later on I see him trying to hook a big branch over an overhead beam. After some effort, he manages to achieve this. Holding onto each end of the branch he manages to hang there. The rigidity of the branch, as opposed to the pliability of the long leaf, creates interesting differences in what is, after all, the same strategy.

Right-angular cross-over and junction

This big branch crosses the overhead beam at right-angles to it and rests upon it. The *near-right-angular junction* is an important structural innovation in early human childhood development. In childhood, the near right-angular join is a Third Generation Structure; that is, a new structure arising out of a differentiation and combination of earlier First and Second Generation Structures.

The chimpanzees take note of any example of *right-angular* or *near right-angular join* from early infancy (see Figures 16.4 and 16.5).

27 January 2006 (observation continued)

Nkosi's makes a near right-angular crossover when he rests his big branch on top of the horizontal beam. The beam on which the branch rests is circular in section which means that the branch rocks a little as Nkosi hangs upon it.

'Well done Nkosi! Keep trying!' I say to him.

He has made a significant variation to the *Hanger*, which he has been developing for the last few weeks. Whereas, with pliable material, the two ends of the Hanger hang down and touch each other, here, this rigid hanger maintains a horizontal axis over the supporting beam.

In the present observation, Nkosi responds to my encouragement and support. I have no illusions about my importance to him however. He may welcome my interest only because no other chimpanzee seems to pay any attention to what he is doing. However, it turns out that I am not, after all, the only one interested in what Nkosi is doing.

27 January 2006 (Part 6) Video-recorded observation (Edited DVD) Singapore

Togo (Nkosi's father), like me, has become interested in his son's efforts and sits down beside him to watch. He lends support and encouragement to the young chimpanzee by touching his son's arm and the branches and other materials which Nkosi is testing for their possible use as 'Hangers'.

Figure 16.4 Rara (1 year 5 months 19 days) notices a small branch sticking out at a striking angle from the larger branch.

Figure 16.5 She can't resist touching it with an index finger.

In this observation, Togo clearly takes an interest in his son's activity and supports him with it. Parent or older chimpanzee may offer more support of their offspring's or friend's efforts than is sometimes assumed.

> Nkosi goes on to try out other long objects as Hangers. One of these is a strip of grass which of course breaks and he tumbles to the ground. He performs a head-over-heels when he hits the ground.

This head-over-heels may have been a way of cushioning the blow of his accidental fall or perhaps, as I have suggested earlier – it may be a way of covering up a mistake. Perhaps it was even a *joke*. For whatever reason it was instigated in this particular instance, a *head-over-heels* is another way of demarcating or perhaps adorning the end of a trajectory. In the next chapter we will see this form of demarcation used to signify the end of rotational trajectories too.

17 Continuous-rotation

Both human babies and chimpanzees learn to perform the dynamic-structure, *continuous-rotation*. This is a Third Generation Structure arising out of both differentiation and fusion of earlier generations of, side-to-side and push–pulling oscillations. In whole-body play it entails spinning on the spot, essentially using one's own spinal column as a central axis. In close-up, field-of-view play in human childhood the rotation is minimised with the use of hand or fingers (and perhaps with a handheld tool or toy). When empty-handed, it consists of invisible traces made in empty space, whilst in drawing it manifests itself as visible, elliptical tracings of pigment on a surface. This begins with the child 'opening up', as it were, what were originally highly condensed push–pull and arcing movements into elliptical movements so that empty spaces are gradually left in between elongated closures. Congo the chimpanzee does this too in his painting (see Figure 1.1). Gradually, both child and painting chimpanzee make rotations which are more circular in shape. There are a couple of such very roughly circular closures in Congo's painting.

We will look at whole-body rotational play first of all.

When continuous-rotation starts to emerge in human infancy, it is noteworthy that the child becomes interested in all that spins and moves in ellipses and circles. The rotating child might become interested in the rotation of clothes in a washing machine, of electric fans, helicopters, lion-dances or even hurricanes seen on TV. The child also sees older siblings or friends performing elliptical movements and tries to emulate these.

First attempts at continuous-rotation may lead to giddiness and falling over.

20 December 2002 Video-recorded child-observation (Edited DVD) (Poppy) London

Poppy (1 year 6 months 11 days) spins on the spot before toppling onto a pile of cushions. (She has just watched her older sister rotating expertly.)

06 June 1983 PAL VHS video-recorded child-observation plus written transcription, (Hannah) London

Hannah (1 year 9 months 8 days) tries to spin on the spot and falls down after a couple of revolutions.

17 March 2006 Video-recorded observation (Edited DVD) Singapore

Chimpanzee Rah (1 year 3 months 10 days) tries to spin around on the spot. She has been watching her big brothers Nkosi, Manzu and Makhulu performing rotation. Her attempts are not as expertly performed as those of the young males and she sometimes wobbles to a standstill or falls over.

In the next observation, made just over a month later, Rah has now mastered the continuous-rotation. Even that which was initially an accidental fall at the end of the rotation is now added purposely to punctuate its conclusion. We noted earlier how, in both human and non-human primate development, what started off as a fortuitous occurrence is later incorporated purposely into a play-assembly.

Rotation and sudden descent

05 May 2006 Video-recorded observation (Edited DVD) Singapore

(Going-around and going-down)

Standing bipedal on Platform Rock, Rah (1 year 5 months 18 days) performs anticlockwise spinning-on-the-spot and sits down suddenly at the end of the rotation.

This is an example of continuous-rotation combined with a descending axis. Rah's whirling ellipse is terminated with a sudden, vertical drop to the ground which acts as a punctuation to the action-assembly.

I made many observations too of the orang-utans at Singapore Zoo performing this play-assembly; a rotation followed by sudden drop or spiral to the ground. (Remember that the orangs and the chimps do not know of each others' existence and thus have independently invented the same ensemble.)

It's unusual to cite evidence from a feature film but the following observation is too good to miss. The humour of rotation and sudden descent is well represented by humans in films from the early days of cinematography. This example is from Jon Amiel's (2009) film about Charles Darwin, *Creation*. In one scene, actor Paul Bettany (Darwin) is in a cage with a young orang-utan. The orang rotates on the spot and drops to the floor and Bettany, brilliantly improvising, imitates this rotation and descent very well indeed. (This orang-utan also does a little drawing, but I will save this for later.)

Continuous-rotation and descent is spontaneously generated in human infancy as an expressive combination of what started life as two different dynamic-structures.

Spinning and falling

06 June 1983 PAL VHS video-recorded child-observation plus written transcription, (Hannah) London

Hannah (1 year 9 months 8 days) rotates clockwise on the spot, flailing a tape-measure out as she does so. She keeps hold of one end of the tape-measure causing the rest of its length to fly out wildly away from the central axle of her spinning spine, according to the law of centrifugal force. Whilst she spins she makes ululating song: 'OOO – ooo, OOO – ooo, OOO – ooo . . .!' After several rotations on the spot she slows to a standstill, loses her balance and sits down suddenly on the carpet. Video-recording the event, her older brothers, Ben and Joel (11 years 10 months and 9 years 1 month, respectively) laugh.

Her use of the tape-measure to enhance her rotation is just like the apes' use of palm leaves, sacks or other amorphous materials which can be made to fly outward in one's whirling hand, to enhance and make visual the dynamic forces of *angular-momentum* and *moment-of-turn*.

11 August 2006 (Parts 4 and 4a) Video-recorded observation (Edited DVD) Singapore

00:25:25 Nkosi (6 years 1 month 14 days) swings wildly in a rotation around a vertical plastic vine with small scrap of muddy polythene gripped in his teeth. As he whirls around, his head thrown back in ecstasy, so the polythene describes an additional concentric orbit, flung outward from his teeth by centrifugal force.

00:29:00 He continues to develop this, trying it in both clockwise and anti-clockwise directions, with varying degrees of outward lean and at ever-increasing speed until his mother, Ida, sitting nearby, restrains him with a touch of her hand.

In the observation of Hannah, the combination of rotation and sudden descent is accidental; the child losing her balance after several rotations. What happens next in the same observation is interesting because it shows one way in which different types of action are combined to make new action-assemblies.

Winding and unwinding (1)

06 June 1983 (continued) PAL VHS video-recorded child-observation plus written transcription, (Hannah) London

Responding to her brothers' laughter, Hannah gives an encore, repeating the spin but this time rotating in the opposite direction (anticlockwise).

Hannah the Human, like Nkosi the Non-Human, knows that the act of reversal alone will make a difference to the effect of this dynamic-structure yet (paradoxically perhaps) without essentially changing its shape and form.

As she comes to a stop, Hannah lets the tape-measure uncoil vertically from her hand to the floor whilst holding onto one end. She synchronises a descending, musical vocalisation to the tape-measure's unwinding descent.

Chimpanzees don't sing but they do become interested, as Hannah is, in *winding* and *unwinding* and, as we will later see, they make many variations of these actions.

Hannah rewinds herself the opposite way, this time very excitedly stamping her feet whilst her now purposeful fall at the end of her rotation is this time preceded by a funny, somewhat stagey stumble and a laugh, before she lets herself drop to the carpet on her behind.

The chimpanzees also seemingly 'play-act', making melodramatic or comic-looking endings. In such cases, that which was initially a product of accident is appreciated for its expressive, structural and dramatic values. Sometimes, as in Hannah's second performance, the original accidental action (her fall at the end of the rotation) is repeated but this time purposely; purposely added on to the end of the ensemble. In such cases she is appreciating the expressive character created when rotation is followed by abrupt descent and moment-of-impact.

Going-around and going-down

I have seen the chimpanzees produce essentially the same action-assembly composed of the elements *going-around* and *going-down.* The sense one has of their satisfaction with this ensemble seems the same as that felt by children. As we noted earlier in a different context, although excellent gymnasts, chimpanzees do occasionally miss a beat or they stumble slightly. However, just like Hannah, such errors sometimes give them a new idea which they then incorporate into a similar performance, perhaps a pretend stumble added to their high-speed improvisations 'next time around'. This on-the-spot revision seems important in that it suggests an awareness of structure and what is required in purely formal terms.

The object for the player, either chimp or child, is to not break the rhythm. In this sense it is a bit like a professional musical performance, where the process of recovery from a mistake (wrong note; botched beat) is made as seamless as possible.

Aside from the absence of vocalisation, the chimpanzees' use of rotation and sudden descent seems more or less identical in both structure and expressive intent to that of the human child.

05 May 2006 Video-recorded observation (Edited DVD) Singapore

Baby Rah (1 year 4 months 28 days) performs an anticlockwise rotation on a flat rock and suddenly sits down at the end of the rotation.

Sometimes, with both species, the ludic and the non-ludic are combined in seamless sequence. The same observation continues:

Rah spins anticlockwise again but this time she incorporates into the action of dropping downward the skilful retrieval of a piece of coconut shell which is lying on the rock's surface.

In this case, the same action-assembly is used firstly in a ludic mode, expressively, and then repeated but with an extra action added on; ostensibly a non-ludic one, that of gathering up the piece of coconut and inserting this new action into the whole action-assembly.

Insertion

Perhaps one may think of the above example as a more metaphorical act of *insertion* or at least the physical and dynamic correlates for metaphor. She thinks quickly, on-the-spot (physically and metaphorically), without interrupting the flow of the play. She wants to play the play fluidly but she also wants to retrieve something 'outside' of the play. She does what human children do when they wish to preserve the play-mode: she inserts the outside, inside.

Retrieval of the piece of coconut may be considered part ludic, part adaptational; she certainly wants to eat the coconut but at the same time she does not want to break the flow of play.

This rotation and sudden descent is similar in construction to that of Hannah, with rotation punctuated by an abrupt drop to the floor. In Rah's case, however, it is the following ascent back to the vertical which forms the salient 'full-stop' to this performance. Instead of finishing, as did Hannah, on the downstroke (as it were), she incorporates an additional upstroke into her performance. She adds an extra beat.

It is at the precise moment in time, just prior to this added beat – the sudden return to the upright – that Rah builds into her play the retrieval of the piece of coconut. She does this without missing the final beat into which fits her concluding ascent to the upright.

In addition to the seamless nature of this dynamic-combine we should also note the more subtle combination of different states of mind; of the adaptational and the ludic. The gathering up of the coconut is a practical, adaptational action requiring object mastery yet it is invisibly stitched into the ludic ensemble.

So far, in the expressive development of both humans and chimpanzees, we have seen a developmental procession in which actions are initially coalesced into an undifferentiated whole, then gradually differentiated and separated, to be practised and perfected singly. Once this is achieved, they are once again combined together but this time purposely.

14 October 2005 Video-recorded observation (Edited DVD) Singapore

(Continuous-rotation continued)

00:24:30 Makhulu (5 years 5 months 7 days) makes a semi-rotation in the stream culminating in underarm scooping of water and splashing it at Nkosi (5 years 3 months 17 days).

Nkosi makes off but appears a moment later, baring his teeth (though in play) and splashing Makhulu.

They play-fight rotationally clockwise, each assuming opposite sides of an imagined moving circle, splashing in the stream, linked together for a moment by hooking of arms, so rotating in a centrifuge.

00:01:27 Later on, Nkosi is digging at something near the stream. Something seems to startle him and interrupt him in his task, as if someone has crept up behind him. He jumps into the stream and makes a very fast and perfect continuous-rotation in water, silver droplets flying off him in the centrifugal force.

Does anyone or anything real really approach him? Perhaps his sharp ear detects a sound which my dulled one does not. Or is it an imaginary interloper; one whom he mentally provides because he wants to pretend to be frightened and to whirl very fast in water?

We have considered how the combination of two contrasting actions might be arranged for expressive effect and noted that rotation ending with sudden descent is one of these combines. Here is another kind, also involving the contrast of two oppositional axes:

The observation continues:

> 00:01:53 Suddenly Nkosi stops rotation and in a single movement stretches upright very taut and straight, as if standing to attention, spine absolutely vertical and with head and neck craned back.

In this case, the structure is similar to that of rotation and sudden descent, involving the juxtaposition of two contrasting axes of movement but here, instead of a concluding descent, the final full-stop is a sudden vertical ascent. The entire action-assembly is an elliptical statement concluded with a black exclamation mark.

Another way going-around can be combined with going-down is in a *spiral-descent.*

Spiral-descent

> **26 January 2007 Video-recorded observation (Edited DVD) Singapore**
>
> Rah (2 years 1 month 19 days) performs a spiralling descent to Suzie below her on the Construction and as she lands she points an index finger to the top of Suzie's head. It is difficult to determine whether or not Rah's finger actually physically touches her aunt's head or falls short of it by a few millimetres.
>
> I advise Rah: 'Don't annoy Suzie'.

A spiral-fall descent is when the chimpanzee drops whilst at the same time spinning around on his or her own axis. It becomes possible when the chimpanzee has learnt to combine *descending* with *rotation* to form a new action-assembly. In childhood development, this new ensemble is part of what I think of as Third Generation Structure. The interest Rah displays in things which turn around as they fall occurs in human childhood development too, when *going-around* is combined with *going-down.*

One difference between Rah's spiral-descent and those performed by children is that children eventually encode this combination of actions onto a drawing surface. A continuous-rotation is translated through a vertical axis (descending or ascending) in a spiralling line made on a two-dimensional surface (Matthews, 2003).

Additionally, children may synchronise a musical vocalisation to their descent. When Hannah, gripping the tape-measure from one end, allows the rest of it to uncoil from her hand and fall in a vertical line downward, there comes from her lips a musical, descending vocalisation.

Unlike Hannah, chimpanzees cannot speak so the capacity to make cross-modal associations between physical descent and musical descent is denied them. This capacity in humans has had far-reaching effects on the development of symbolisation and may have set the stage for the advent of verbal language (Mithen, 2005).

But there is at least one way in which Rah shows indications that she is part-way translating a real descent into a represented one and that is in her use of her index finger which describes her downward spiral to stop gently on, or very near, Suzie's head. Rah knows that she must not really crash into Suzie (which is after all the idea behind her fall) so she minimises action into the barest whisper of a trace, ending in an indexical touch of her finger at the Point B of her trajectory (Auntie Suzie's head). This terminal touch, or virtual touch, might be considered a precursor *pointing* gesture.

As for chimpanzee vocalisation; perhaps we should not dismiss it altogether. Birute Galdikas notes that chimpanzees do sometimes vocalise in conjunction with making other, quite different actions (Galdikas, 2005). Perhaps useful research might be conducted about this. In any case, though one might argue that chimpanzee vocalisation is limited compared with that of the human, there are other ways in which chimpanzees make analogues between events occurring within different sensory channels and different body systems.

18 In a spin: Continuous-rotation continued

08 September 2006 Video-recorded observation (Edited DVD) Singapore (continued)

Nkosi (6 years 2 months 11 days) has become an expert in spinning round and round on the vertically hanging fire-hose. He leaps onto it with great force and winds the rubber-hose up as tightly as he can. Then, still holding onto it, lets it unwind, slowly at first and then with ever-increasing speed. Like some kind of hairy ballet dancer or ice skater, he alters his speed by either hunching up along his vertical axis with his knees bent and his legs tucked in (to accelerate), or else lengthening himself out and splaying out one or other leg (to decelerate). He does this several times. One time, he carries a small piece of crumpled polythene in his teeth and lets it fly off in the centrifugal force he produces. With his mouth wide open, his head arched back and his eyes closed, he looks like he is in Heaven.

Like human children, chimpanzees detect the same dynamic-structures despite differences in their manifestation and they systematically investigate these, exploiting their expressive affordances.

02 March 2007 (Part 7) Video-recorded observation (Edited DVD) Singapore

Today, Nkosi (6 years 8 months 5 days) is interested in all things spinning. First of all, with Rah (2 years 2 months 25 days) on his back, he performs a double, anticlockwise rotation whilst moving rightwards across the picture.

This is a *travelling-loop* or *travelling 'e'* formation. Note that he holds a stick in his right hand with which he describes an ellipse in the grass. (See Figures 18.1 to 18.4.)

Figures 18.1–18.4 02 March 2007 (Part 7) Spin. With Rara riding dorsally, Nkosi starts the first ellipse of a double rotation whilst moving leftward and, with a stick in his right front-hand, describes an anticlockwise *travelling-loop* or *travelling 'e'* in the grass.

Figure 18.2

Figure 18.3

Figure 18.4

After he has rotated with his sister on his back he then spins around, solo, on the Hanging Dustbin [see Figures 18.5 to 18.9], and after this he finishes off with a high-velocity, quadruple travelling-loop across the grass again, but this time without the burden of Rara. (See Figures 18.10 to 18.16.)

Figures 18.5–18.9 Immediately after his continuous-rotation in the grass, he spins anti-clockwise on the big plastic drum hanging from its chain.

Figure 18.6

Figure 18.7

Figure 18.8

Figure 18.9

Figures 18.10–18.16 As if reiterating his rotations on the plastic drum, after dismounting from it Nkosi makes the first of a series of quadruple *travelling-loops* or *travelling 'e's* moving rightward.

Figure 18.11

Figure 18.12

Figure 18.13

Figure 18.14

Figure 18.15

Figure 18.16

In some of the observations of dynamic trajectory, be they rotational or other configurations, a single finger sometimes echoes the moving body. At other times, the effect is amplified with a handheld object, a stick or a piece of crumpled polythene are two examples.

The flight-path of the flower

26 January 2007 (observation continued)

It has been raining. Rah drops one leaf over the edge of a platform and watches it fall. She rests her head on its side on the wet, wooden surface and then spies something else that interests her. It is a small flower or blossom. She lifts it up and gracefully zooms it up though the air before her face. She maintains eye-gaze upon the flower as she makes it ascend.

The flight-path of the flower has a clearly marked beginning and end and perhaps one could additionally argue that the trajectory in between, with its graceful deceleration at the moment of apogee, made directly before her upward-turned eyes, forms a middle section.

In the above example, control and mastery of the handheld object has been achieved and coordinated with visual tracking. Of especial interest is the fact that, though the chimpanzees can use four hands, for special acts involving careful manipulation and viewing of a handheld object, the preferred hands are the front-hands. In examples like this and the previous one the chimpanzee focuses down to an even smaller region, a single finger on a hand, or even a flower held between the fingers. In an action as fine as this the foot-hands may also play an important role.

This is an example of a chimpanzee's handheld flight of an object made in the close-up mode before the chimp's own watchful eyes. It looks very like the pretend flights human children create in front of their eyes with toys. With human children of about age 3 years and over, this act often represents the flight-path of another object, an aeroplane or space-craft being common subjects.

Trace in space: The moving finger

26 January 2007 Video-recorded observation (Edited DVD) Singapore (continued)

Still on the same day, Rah (2 years 1 month 19 days) with her left hand plays with a handheld bunch of wet stems, raising them high above her head and flying them through graceful curves in the air.

She holds the bunch by her second, third and fourth finger plus her thumb, whilst leaving free her index finger which, like the nose-cone of an aeroplane, leads the flight of the fuselage formed by the handheld stems upward through a graceful trajectory. The leading index finger might be likened to a *pointing* finger. Rah visually tracks the ascending and descending flight.

In human childhood, this use of the handheld object is often made to simulate flight. With the child's eyesight constantly monitoring the hand-made trajectory, this is a precursor *action-representation.*

Until comparatively recently, in studies of the development of children's semiotic thought, this form of representation was concealed by definitional problems about what constituted a 'pictorial' representation (Wolf & Fucigna, 1983). Such a representation does not record the *shape* of an object but rather its *movement* from A to B. In childhood, this becomes an idea about movement which soon finds its way onto the drawing surface where it is encoded in physical lines.

Rah's actions with the flying stems are very similar to those performed by very young children and, like theirs, perhaps is the start of represented movement.

Winding and unwinding (2)

As we have seen, *winding* and *unwinding* is another way in which rotation is investigated and played with. We have already encountered it in some of the observations above about rotation, for example, when Hannah coils and uncoils a tape-measure, and when Nkosi whirls around on a vine and lets go a piece of polythene from his teeth. Although rotation is based on natural movements of the primate's muscular and skeletal frame its potential is only fully realised within the culture of the primate. The following observations show the emergence of winding and unwinding and how its development is encouraged and fostered by friends and relations.

08 September 2005 Video-recorded observation (Edited DVD) Singapore

Rotation

Nkosi (5 years 2 months 11 days), a small twig in his teeth, does a rotational swing on a downwardly drooping branch causing the branch to twist around on its own axis, as if he is 'winding up' the branch like an elastic band. Then, when it can be twisted no more, he allows the branch to unwind itself, making use of its stored or kinetic energy to set his own body spinning.

Other Great Apes independently discover, explore and enjoy this kind of rotational, winding and unwinding play too.

18 November 2004 Video-recorded observation (Edited DVD) Singapore

Orang-utan winding and unwinding

An infant orang-utan swings from a tree holding onto some long leaves which he holds together in a bunch so that they form a sort of rope. On the platform below sits his adoptive father. [This is an orphaned infant, his mother shot by poachers in Sumatra. Rescued by Singapore Zoo, this old male has chosen to look after him.] The infant sends himself into a spin on his rope of leaves and in doing so winds up the leaves into a cork-screw. Then he allows his leafy rope to unwind, gradually at first, and then ever faster, using the kinetic energy he has supplied to the leaves with his 'winding up' of them. At one point he swings close enough to his adoptive father to reach out his right hand and pull the old male's long, orange hair. Momentarily, the old male wraps a tolerant right hand around the youngster, giving him a quick cuddle, as if to share the moment with him.

The baby orang-utan now spins in an anticlockwise, rotational movement and he gazes up happily into the canopy of the tree. He climbs up the bunch of leaves which he now holds together in all four hands. He hangs there for a few moments and then exerts a sudden downward force upon the leaves, breaking them from the tree and falling, with leaves in hand, harmlessly to the floor next to his adoptive father.

Like the orang-utan's, the chimpanzee versions of winding and unwinding, and of rotation with sudden descent, are not 'natural' or instinctive actions. They have to be learnt and, as in the case of the baby orang-utan, learnt in a supportive community of more experienced siblings and elders.

11 August 2006 (Parts 4 and 4a) Video-recorded observation (Edited DVD) Singapore

Learning to swing

Rah (1 year 8 months 4 days) is hanging high up on one of the old, vertical, rubber fire-hoses. Nkosi (6 years 1 month 15 days) is at the bottom of the strap and he takes hold of it and sways it. He wants to give his little sister the maximum swinging experience. Suddenly, Makhulu (6 years 3 months 4 days) zooms into the scene and dive-bombs Nkosi, taking over command of the strap and swaying Rah even more vigorously. At this point, Ida steps in and lays a hand on Makhulu to stop him from being too rough with Rah.

In the above observation, Nkosi was content to get Rara swaying in any old fashion at all on the fire-hose. Twenty-seven days later he raises the standard – he wants to help her to rotate properly:

08 September 2006 Video-recorded observation (Edited DVD) Singapore

Learning to spin

Today, Nkosi (6 years 2 months 12 days) is enjoying spinning wildly upon the old fire-hose from its lower end whilst Rah (1 year 9 months 1 day) is hanging upon the same length of perished rubber but much higher up, near its top, where it is joined to the Construction. Her hanging upon the same strap as her brother is her way of 'joining-in' and benefiting from his exertions down below. Then Nkosi stops spinning and stands bipedal on the horizontal beam, as if a thought has struck him. He grips the bottom of the strap and, using a circular motion of his hand and arm, tries to whisk his small sister into a spin like his own of a moment ago.

Rah is not able to do this yet. Though her brother's circular swirling of the strap's bottom end is intended to prompt her into cooperating in the rotational movement he is trying to set up, she cannot yet respond to this cue. The older chimpanzee is trying to support the younger one's play but he is still in the process of gauging her level of competence whilst she herself requires more independent experiment and practice.

Describing apparently premature attempts at giving assistance to another illuminates just what is involved in successful learning experiences in terms of the empathic awareness and the gradual fine-tuning of actions necessary to support such learning.

'Getting the idea' (grasping something)

29 September 2006 Video-recorded observation (Edited DVD) Singapore

Twenty days later Rah (1 year 9 months and 21 days) is now winding and unwinding solo on the rubber strap. Perhaps Nkosi's teaching was not so premature after all.

Eight months twenty-three days later she is proficient at the continuous rotation and can combine it with other action-assemblies:

22 June 2007 (3) Video-recorded observation (Edited DVD) Singapore

Today, Rah (2 years 6 months 15 days) combines the *slide* with the *rotation*, making a *rotational-swivel*. On a shiny, wooden platform, she rotates around and around on her ischial callosities.

We glimpsed the forerunner of this in the early swivels she made on her with her brothers on their backsides when she was only 1 year 11 months 17 days old.

14 March 2008 Video-recorded observation (raw footage on DVD) Singapore

Another 8 months and 22 days later, Rah (3 years 3 months 7 days) is very good at rotation though she is still perfecting it and making variations. She is lying on her right side on Level 1 platform. She has a hessian sack with her. She grasps the edge of the sack, rolls on her side to her left and in doing so automatically wraps the rough cloth around her. Then, sitting up, she makes a variation of this winding action, when, protected by her ischial callosities, she swivels around on her backside winding the sack around about her in the process.

Like other, related coiling actions we have seen her make the relationship she explores here is a *functionally dependent* one. She systematically tests rotation through both horizontal and vertical axes. Both result in the winding-up of the hessian.

The experienced learner

Helping little brother to spin (2)

11 April 2008 Video-recorded observation (Edited DVD) Singapore

A little less than a month later, Rah is hanging by her left arm from a sloping vine. Now an expert rotator, she performs a clockwise and anticlockwise wind or twist. The increased mobility of the chimpanzee skeleton makes possible a completely circular motion whilst hanging by one arm. Rah sees her little brother, Kimoni (7 months 28 days), nearby his mother, also dangling by one hand and arm from a vine above. Perhaps Rah thinks this looks promising for she tries to teach him how to rotate.

What follows is a continuation of an observation I introduced on page 153.

Attunement

> **11 April 2008 Video-recorded observation (Edited DVD) Singapore (observation continued)**
>
> First of all, to engage his attention, she makes a swaggering, side-to-side, pretend-display and then a couple of up-and-down jumps in front of him as he hangs suspended (though for etiquette's sake she is careful to include grumpy Ganga in her performance).
>
> Then she gives him some little pushes. She is trying to teach him how to wind and unwind on the vertical fire-hose – as her big brother did for her. We are witnessing a small part of the passing on of culture.
>
> However, Kimoni does not know how to respond to her promptings and he fails to rotate. Like herself, just over a year earlier (when she was 1 year 9 months 1 day old), he fails to see what the older chimpanzee is 'getting at'. After all, he has only just learnt to hang by one arm.

Ostensibly just a motor-act requiring only natural forces, the rotational movement, like many other of the primates' deceptively simple-looking acts, in fact has a cognitive aspect – exactly as such acts have for human babies.

It is worth considering Rah's teachings. This is the same morning she tries to play with Ainka who is just 16 days younger that Kimoni (see page 152). To Ainka, sitting at the far end of a short tunnel of branches, she suggests 'peepbo' whereas to Kimoni, suspended from one arm from a vine, she suggests spinning. Although Rah's actions are not perfectly adjusted to either infant, neither are they inappropriate or very far off, developmentally speaking. What each infant, Kimoni and Ainka, lacks at present is the emergent concept which will allow all the components of their muscular and skeletal structures to cooperate amongst themselves together with those of another conscious being according to the principles of a new play-assembly. This is a little beyond Kimoni's level of development which means that he cannot make use of Rah's teaching yet. Ganga knows that her son does not understand what is required of him and she takes him away from Rah just as Suzie rescued Ainka from Rah's premature 'peepbo' play.

> However, Rah quickly comprehends the problem. Unable to make him rotate, she adjusts her participation to a more simple hanging next to him, even performing a tangle-dangle with him upon the vine. This he does understand and is able to join in the fun, even contributing an upside-down tangle-dangle of his own.

Set spinning

It is nearly 6 months since Rah's first attempt to set Kimoni spinning. Kimoni is now ready to take some more advanced lessons in rotating around on his own axis whilst hanging from a branch or beam. It is his mother, Ganga, who knows just the right amount of help to offer:

Touch and go

10 October 2008 Video-recorded observation (Edited DVD) Singapore

Kimoni (1 year 2 months 3 days) is learning how to rotate and spin from one arm, his right arm, from the beam above. [As we noted above, the different skeletal structure of the chimpanzee arm will allow him to turn 360° upon it.] Ganga gives him a poke with her left index finger to assist his spin. First of all, he does not respond but shortly after, when she tries again, he gets the idea. With her right hand she gently prompts him, firstly with a push of her little finger and then her right index finger. He allows himself to be spun in a slow circle.

This is an example of very sensitive synergy and support. Ganga knows that Kimoni, who has already learnt how to dangle from one arm, is now ready to be introduced to the *spin*. She knows that all he requires is a gentle touch. This touch is almost a sign.

Holding hands

10 October 2008 (observation continued)

At the conclusion of his spin, as he comes to rest on the beam by his mum, he inserts his little fist into her big, open palm. This is perhaps what she was hoping he might do anyway – she wanted some reciprocated affection and physical kindness from her son. He briefly touches her big hand with his little one. He knows that his new act of spinning comes about because of a touch of his mother's hand.

Continuity and discontinuity

Continuous-rotation and rotation broken into steps

17 October 2008 Video-recorded observation (Edited DVD) Singapore

One week later, Ganga hangs her left hand down to her son (Kimoni is now 1 year 2 months 10 days old) and he takes it in his right. Using it as a supporting pivot, he rotates in a clockwise circle above the surface of the platform, stamping his steps a little as he makes his circular flight and, on completion of the circle, making a final stamp with his right palm-sole as he hovers above the deck.

This assembly is another example of how a route – a circular one this time – may be described in two different ways: (1) as a continuous path (in this case an elliptical orbit) and (2) as a series of discrete displacements in space (in the form of pacing out the same orbit).

The chimpanzees build upon the discovery of winding and unwinding in other ways too:

Wringing

06 November 2008 Video-recorded observation (Edited DVD) Singapore

00:04:00 This is the day I give the chimpanzees the magenta-coloured bag. [Remember that it gets wet and is reduced to a red-rag by the chimpanzees.]

Ganga (15 years 8 months 28 days) makes six pull strokes with the wet, red rag against the smooth wooden surface of Level 3 Platform and then pauses, looking over into my direction. 'Good girl, Ganga', I say to her, to encourage her. Her long, sensitive lips part and I am almost tempted to say a smile might be emerging on her usually disgruntled face.

She seems to consider my words. She makes two pull strokes before returning her attention to the pink rag in the transitional space formed between her parted knees and thighs. She picks up the cloth in her front-hands and squeezes it out very efficiently using both hands, each hand turning in a direction counter to the other, in a perfect *wringing* action.

06 November 2008 (observation continued)

Later on, Suzie wipes her face with the red cloth and then wrings it out in again a perfect wringing action, her two hands working against each other in opposite turning actions.

That the chimpanzees know something of wiping – in the sense of *cleaning* – has been testified by Jane Goodall (2000) and by my observations too. Earlier on in the day when the present observation is made, Ganga, with her left hand, wipes Bianca's left nipple with the red cloth. Two years seven months and twelve days earlier, when Ganga is 12 years 1 month 16 days old, she cleans her own ear with a piece of grass (24 March 2006, video-recorded observation, edited DVD, Singapore).

To return to our theme of rotation, here is another instance of two initially separate inventions being combined together.

Twisting and winding

26 May 2006 Video-recorded observation (Edited DVD) Singapore

00:28:07 Manzu (7 years 2 months 11 days) makes a Hanger which he twists around upon itself so that he 'winds it up'. Standing on a horizontal beam at Level 1, he hangs a strip of hessian sack over another horizontal beam just above so that both sides of the strip are roughly equal in length and their ends meet. Then, gripping both ends of the strip of hessian together in his hands he winds it up (storing up its kinetic energy). The winding-up is achieved by leaping at a calculated angle from beam to Hanger and setting himself spinning against the resistance of the gradually coiling hessian.

In other observations made the same day both Nkosi (5 years 10 months 29 days) and Makhulu (6 years 19 days) start to wind-up the sack but without the need to hurl themselves bodily onto it at an angle. They sit still, crossing and wrapping the two ends around each other, sometimes with front-hands alone; sometimes with the assistance of a foot-hand. The two chimpanzees understand the principles involved in winding and unwinding; that they are not confined to one way of achieving the same end.

Closed-shape

This is also an instance of the chimpanzee making a *closed-shape* – a Third Generation Structure of great structural and representational importance in human childhood development. You will remember that, in rope-workers' terminology, when a line or rope is folded so that the two ends meet but do not cross, this is called a *bight* but when they cross, so creating closure, this is called a *loop*. In their actions, the chimpanzees show that they differentiate between these two forms which, though physically closely allied, are structurally and conceptually far apart. This is because, the closed-shape (or *Jordan's curve, closed-curve* or *loop*) has a continuous, unbroken outline, without beginning or end, forming a closure which separates inside from outside in a way in which no other shape does. Hence, its appearance in human childhood development is an important milestone because a virtually endless variety of structural and representational possibilities proliferate from its discovery. Chimpanzees know how to form nearly closed-shapes in nest-building too, when they throw a long length of material over the back of their heads whilst retaining the two ends in their hands, so that the material (sack, thin branch, sheaf of leaves) forms a *near circle* around their behinds with the two ends meeting in front of them.

Later on in the current observation:

Nkosi forms another closure when he throws the sack strip over his head and neck to form a sort of 'scarf'; the two ends folded over each other at his throat. Right now, Manzu, still hanging onto the lower end of his wound-up hessian 'rope', lets it unwind, slowly at first then ever faster, spinning wildly around with it. Togo sits by his whirling son to his right and as the young male comes to a halt and sits down he looks up at his father as if for approval. After a pause, Manzu leaps from the beam and seizes the Hanger, starting himself spinning anticlockwise in order to wind the sack up again.

He sits down and briefly investigates the two ends of the Hanger. He touches its stringy ends to his lips. Then Togo moves off and Manzu, taking hold of one of the two ends, follows him, holding onto this one end so that the sack-strip is slowly dragged off and away from the overhead beam.

00:28:54 He resumes experiments with the hessian strip, hanging it this time over a fire-hose Slack Strap.

00:29:44 Later, Makhulu (6 years 19 days) has a go, bringing his particular stylistic variations to the performance. First of all, he dunks the sack in the pool at the Waterfall. Then he makes a Hanger from it by draping the sopping wet rag over the Long Slack Strap which runs over and parallel to Bridge Log. He spins around on this Hanger, wringing it out at the same time.

Mask

26 May 2006 (observation continued)

He swings from a Point A to a Point B, choosing a sawn-off tree stump as his Point B. Keeping the sack quasi-knotted over the fire-hose above him, he covers his face with the soaking hessian, spreading it out so that it is stretched across his face. He can probably see through it. [This is another example of how chimpanzees purposely alter their own perception plus, in this case, their appearance.] The wet sack follows the form of his face and skull like a mask. He swivels around on his ischial callosities to face me, so that I have a frontal view of his transfigured face. He may even take an in-drawn breath to seal the wet hessian more closely over his face. He uses the index finger of his left hand to push the sack right into his own mouth, so describing a slit-like mouth. His eye-sockets are dark caves. He looks at me through the veil of hessian for a few moments before peeling the wet sack from his face and staring me in the eyes. (See Figure 18.17.)

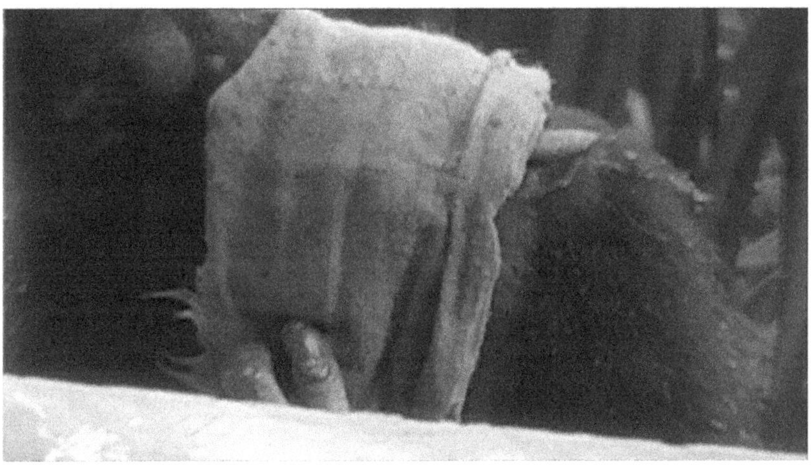

Figure 18.17 Makhulu (6 years 19 days) presses wet hessian over the bones of his face and with a fingertip pokes the cloth into his mouth to form an indentation.

19 Bottoms-up: Script-violation

'Well – I gave you every opportunity . . .!'

Max Wall, to the audience at Greenwich Theatre,
London (1973), after sticking out his backside at them.

The chimpanzees invent a number of ways with which to conclude a play-assembly. Here is another instance of rotation and sudden descent. This one ends (I believe) with a *joke*:

Punch-line

> **14 October 2005 (Part 5) Video-recorded observation (Edited DVD) Singapore**
>
> Nkosi (5 years 3 months 17 days) stands before me, on Platform Rock and sways exaggeratedly from side-to-side, shifting his weight from one leg to the other several times and then, most spectacularly of all, revolving at high speed around and around on the spot, often rotating on one foot as a single axis.
>
> He gives a striking display of recognition and greeting in the glance he gives me immediately preceding his performance. This glance toward me is again glimpsed at moments throughout the performance within the brief pauses which intersperse it, as if through the window of a madly revolving zoetrope.

Unlike some of his earlier displays, today's is not a belligerent demonstration of dominance but a friendly performance made specifically for me. The show he presents is one of great elegance. His high-velocity choreography on a central axis is extremely complex, involving rapid decision changes regarding its speed, directional variation, intensity, stress, cadence and amplitude.

He injects humour into it too, making a furious series of turns which end with his bending over and pointing his bum at me, whilst he looks at me, head upside-down, through his spread legs.

Depending on context and style, sticking your backside in someone's face is found sexy or funny ('peculiar' or 'ha ha') by many humans. However, though both erotic entertainers and clowns, from the anonymous child to a comic genius (for example, Max Wall, whom Nkosi now resembles), make use of this simple act, my observations strongly suggest that the chimpanzees' use of this posture is not imitated from humans.

For both human and chimpanzee, it may have an anal–genital origin but this would make up only one part of its history of meaning. When Nkosi sticks his arse at me and looks through his legs at me, I believe this is a creative 'script-violation' (Mitchell, 2002a); an intrusive insertion of incongruence into the congruent. It is natural comedy. Both human and chimpanzee society find this amusing but not because it has passed between human and chimpanzee society.

Another way in which Nkosi's performance disrupts reality is that, by looking through his opened legs with his head upside-down, he sees an inverted image of me (and I receive an inverted image of his face). This drastic alteration of vision may also be of interest to him in itself and is part of an array of strategies the chimpanzees have for transforming their vision. The masking of the face is one such (see Figures 16.2, 16.3 and 18.17) but there are others. That the chimpanzee finds entertainment in altering his or her vision directly contradicts Lenain's assertion that the chimpanzee has no awareness of, or interest in, a *field-of-view* (Lenain, 1997).

Here is another example:

26 January 2007 (Part 8) Video-recorded observation (Edited DVD) Singapore

Togo plays with Rah

There has been some noisy building work near the chimpanzees' enclosure which has made them all anxious. Togo (35 years), Ida (34 years) and Rah (2 years 1 month 19 days) (father, mother, daughter) are on Word Platform, Level 3. Ida lies on her back with one, left leg resting on right knee.

Baby Rah seems frightened and Togo plays with her. I say into the camera's microphone: 'He's trying to cheer the baby up'. He steps over her and she tries to grapple with his legs. The old Alpha male allows her to do this. He sits down with his back to me and handles and tickles the baby who rolls at his feet and wrestles with his hands. Togo then sticks his behind in the air and points it at Rah to amuse her and perhaps to show his upside-down face to her.

Bum-in-air is a spontaneously generated form which plays upon contrast and incongruity and is found subversive by both primate groups. This fits into Arthur Koestler's (1964) definition of the humorous in which two different planes of logic intersect.

There are many ways in which the chimpanzee subverts the norm. We return to Nkosi's act of 'mooning':

Clap

14 October 2005 (Part 5) Video-recorded observation (Edited DVD) Singapore (Observation continued)

After sticking his arse at me, he pauses, making an interval based on breath-recovery, and then he performs another rotation, starting it with a single clap of his hands and ending it with another. In this way he demarcates both the beginning and end of his dance.

After this, Nkosi makes some very clear and pronounced up-and-down nodding of his head at me. This additional act seems all-of-a-piece with the actions which preceded it – the high-speed rotation and clap.

Again, Nkosi is using an action which startles the viewer into taking *notice* in the sense that Dennis Atkinson (2002) means; it is an act which seems to shout 'I am here!' It is a non-accidental, dynamic-structure projected into the accidental world. It enables an observer (in this case, myself) to pick him out. As with all emergent signs, it makes use of the principle of *contrast.*

Nkosi's nodding is sometimes referred to by primatologists as 'head-bobbing' and is within the natural repertoire of chimpanzee behaviours. It is not derived from human–chimp interchange – at least, though humans may nod at the chimpanzees, chimpanzee nodding is not reliant on imitation of human culture.

The clap in slomo

A little later, still standing on the flat podium of Platform Rock, he makes a more expansive, theatrical-looking clapping, like a gesture of solemn applause, extremely slowed down, starting with arms stretched back horizontally and moving very slowly inward, symmetrically, to meet at the midline in front of his chest where he makes a pretended and silent clap.

A silent clap

Now, it is true that humans sometimes applaud chimpanzees, mockingly or for real, but if Nkosi's actions are derived from human clapping (and I don't believe they are), then his is not a simple process of imitation. This curious, theatrical clap, its defining sound subtracted from it, is like none he might have seen. Simply by subtracting from it one of its dimensions, that of speed, he has totally transformed it, making it an exaggerated parody of the act of clapping. His is a carefully controlled, slowed-down action with two arms moving through two quadrants of a circle, from the furthest stretched back position and slowly moving forward symmetrically to meet silently at the midline.

There are a number of ways to transform an action so that it may be seen in a new way. In this case, simply by altering its velocity Nkosi subverts the definition of the 'clap'. The techniques he uses to do this are *abbreviation* and *deceleration.* They are of a type employed as levers to separate an ordinary act from the here-and-now reality of adaptive struggle and shift it into a totally different psychological space; the analogue space of play.

Again, this act is another form of conscious variance from the norm. It is a conscious attempt to make something 'stick-out' (and not his bum, this time) to an observer in a more metaphorical sense as a marked moment in time. He uses the techniques of deceleration and abbreviation to transport this act into a conscious and (what tantalizingly resembles at least) *reflective* performance, for one cannot help thinking that this looks like he is commenting on the act of clapping.

It may be that we are looking at the dawning of representational skills. These strategies seem to me to be part of a cluster of interrelated actions which form the precursors of symbolic thought.

20 Rotation around an axis

Another developmental pathway the continuous-rotation takes is toward a *rotation around an axis*. The following example of this dynamic-structure is initiated with our now familiar vertical arc in the form of a single, downward thrash.

The pretend thrash

16 November 2007 Video-recorded observation (Edited DVD) Singapore (Tape 1)

00:10:27 Using a fully extended, vertical arc, Nkosi (7 years 4 months 19 days) thrashes Togo (his huge and fiercesome father of 34 years 10 months 16 days) with a stick to goad him into chasing him. Nkosi's action starts with his arm held straight up, vertical, before moving through a large sector of a circle down toward his father. Yet, in spite of its powerful-looking character, this is a pretend attack. He avoids Togo's head and gently taps the stick against his father's shoulder.

Compare Nkosi's overarm arc with that made by a 3-year-old human child:

March 2001 Video-recorded child-observation (Edited DVD) Singapore

00:08:23 A Chinese boy deliberates and delays over the downward thrusts of his handheld pencil. He modifies the vertical arc, allowing the pencil to hover momentarily above its target and then brings it down to hit the paper in series of overhand arcs. The vertical arc is controlled and modified. It is as though it has become an *idea* of itself.

This may also be true of the Nkosi's thrash at Togo which, even though it is a genuine whip-like action, is still nevertheless controlled and aimed so as not to hurt his father:

16 November 2007 (observation continued)

Rather like the Chinese boy, Nkosi begins his first 'hitting' action with a fully outstretched vertical arm delayed for a millisecond in an upright position, his stick held up over his head, whilst he looks and smiles directly into his father's face. The upstretched arm uses the principle of contrast to attract attention and act as a signal to Togo to 'tell' him that play is intended. Because he makes a clearly non-accidental structure and presents this to his father, this invites Togo to 'ask' what this action means. This all happens within a jiffy of time.

Then, Nkosi prudently places his next thrashes to left and right of the Alpha male, in the air, as if stopped by an imaginary surface; pretend-hits which avoid hurting or upsetting his Dad. Togo's anger can be awesome.

I observe Nkosi making such calculated, pretend, overarm strokes on other occasions. For example, when he is 6 years 8 months 5 days (02 March 2007, part 6, video-recorded observation, edited DVD, Singapore) he pretends to strike his mother Ida. She is sitting in a hollow of Twisted Tree at the end of Horizontal Log. Holding a curious, droopy stick (this is the funny stick I mentioned earlier), Nkosi firstly whacks it against the Horizontal Log, to get Ida's attention; then he thrashes at her twice more, but aimed to miss her, each blow to one side of her and neither of them hitting any actual surface but curtailed; stopping in mid-air.

There are interesting differences between the two episodes which depend on the exact circumstances of each situation. To announce start-of-play with his father, he makes a visual and kinetic signal not involving any physical blow, but just raising the stick high into the air. In this situation, his father is facing him, so the principle of visual contrast may be used. In the episode with his mother, she is not facing him so he whacks the Log near where she sits to get her attention with the sharp sound. In both situations however, the principles of *curtailment, abbreviation* and *contrast* (visual or auditory) are the techniques used to launch and pursue play.

Such acts of signalling and structuring are not instinctive but learnt. For example, Nkosi when aged 5 years 8 months 8 days (05 May 2006, video-recorded observation, edited DVD, camera 2, part 2, Singapore) introduces the idea to baby Rah, aged 1 year 4 months 28 days. She is facing him when he raises a sharp piece of coconut-shell high in the air as if about to strike her. His action is a truncated vertical arc. In response (00:27:55) she picks up a piece of shell and makes vertical arcs with it to her big brother but again, these are minimised actions, not actually touching him.

Rotation around an axis in chimpanzees

In the present play episode between Nkosi and his father, the pretend thrash signals a continuous-rotation around a vertical pole which starts with Togo pursuing Nkosi. Both Nkosi and Togo take turns in occasionally reversing the direction of the rotation. Nkosi makes some very good feigning movements and Tai Chi-like dodging of Togo's big, questing, arms and hands. Nkosi also swings out his arm into Togo's path, as if to delay him with an ephemeral 'gate'. Sometimes the notional barrier of the arm is swept aside by the huge chimpanzee; sometimes Togo uses it as a signal to change the direction of his circuit.

As Nkosi moves out of Togo's reach, so he speeds up the momentum of his rotation by hanging onto an overhead horizontal beam and hurling himself through a large sector of the circle, whilst Togo keeps up his steady, ponderous, yet ominous tread.

Togo is quite capable of faster speeds but is cooperating here by intentionally slowing down. Even when he reaches out as if to grab Nkosi, his actions are purposely slowed; he does not really want to capture Nkosi at all, for this would spoil or stop the play. [It is Dad being the Scary Monster for his young child.]

The two chimpanzees, father and son, make play grins at each other and also seem to smile to themselves. In addition to using the overhead bar to increase his distance from Togo, Nkosi daringly uses it as a fulcrum to lever himself suddenly face-to-face with his father, before quickly backing off, allowing Togo to chase him at the same speed again. [It is illuminating to contrast this play-situation with a *non*-play one, in which a younger male would not consider interrupting the locomotion of an Alpha male.]

In this lead-and-follow, hide-and-chase play, the structure is collaborative; the speed of the chaser calibrated to the speed of the chased. In an observation which follows shortly we will see that this is true of some children's play-chase rotation around an axis.

Togo does not often play with his son but today he is making the effort, like a tired dad on a Sunday afternoon. Though rare, I see other instances in which the old Alpha male supports his off-springs' play and sometimes takes part in it. Although the adult females usually do the main work of infant care, as we have seen, the father will, on occasion, play with his babies too and help them learn vital skills.

Rotation around an axis in children

We have looked at an example of rotation around an axis performed in play by chimpanzees, now we will look at human children's version of it:

14 October 2003 Video-recorded child-observation (Edited DVD) Singapore

00:16:06 We are observing a group of 4- to 5-year-old Chinese children in a Singapore kindergarten.

Joey, aged 5 years, happily operates a digital video-camera mounted on a tripod. [The children were introduced to video-cameras from age 3 years – see Matthews, 2006a.] Leonard and Vikki, also smiling widely, run anticlockwise circles around the video-camera and its operator. Zhou Quan, also 5 years old, jumps up-and-down in front of the camera's lens, laughing delightedly; his jumping maintained at a tempo which matches the running footsteps of his friends.

When my research assistant enters the space to see what is going on, Leonard and Vikki include her in their orbits, encircling her in addition to the camera and Joey. Zhou Quan continues to bounce up-and-down in front of the camera's lens, bouncing from his toes at a constant rhythm which seems to be in relation to the continuous-rotation Leonard and Vikki have set up.

Compare this with the following observation:

08 December 2006 Video-recorded observation (Edited DVD) Singapore

(Rotten Stump)

Makhulu and Nkosi (6 years 5 months and 6 years 3 months 11 days, respectively) chase each other around Rotten Stump. Rah (2 years) tries to join in, trying to catch the flying heel of one of her brothers as he disappears around the old tree stump. Unable to join-in (the young males are too fast and dangerous for her) she leaves the circle and takes shelter behind Ida.

Then, whilst Nkosi and Makhulu continue to whirl round and round the Rotten Stump, Rah jumps up-and-down and then hurls herself to the ground with glee.

Rah, unable to match the tempo of her big brothers' rotation and unable to understand its dynamic structure, may be trying to 'join-in' in another way, one of which is jumping up-and-down on the spot. This is an excited, histrionic performance but one which echoes the tempo of her brothers' play-assembly and may form a forerunner representation of it.

Rah's performance derives from her incomplete understanding of the dynamics of her brothers' play-assembly, making her unable to join in on their terms.

There is, however, one aspect of their play which she does understand and that is its *rhythm*. Her complete empathic engagement with its pace and tempo allows her to join in, in the only way she can, by jumping up-and-down on the spot.

This similar in essential respects to Zhou Quan's jumping up-and-down. Like Rah the chimpanzee, he also responds to the rhythmic aspect of the piece though, unlike Rah, he knows what he is doing in structural terms; he knows consciously what actions might 'go with' the actions of his friends and he places his up-and-down jumping rhythmically into the inherent beat of their rotation around an axis.

Maintaining the rhythm of his up-and-down jumping, Zhou Quan, positioned near the camera, bounces over to Joey's vantage point and looks into the Liquid Crystal Display screen of the camera. Here again is an example of an adaptational action being incorporated into the play-mode. He wants to look into the LCD screen but at the same time does not want to disrupt the rhythm of the play. As Vikki runs by him, he bounces on the spot, but this time whilst turning around gradually in an ellipse formed around the axis of his spine, so as to match and visually track her movement past him, as if she has swept him into a new orbit of his own. As she goes by he reaches out his hand to try to catch her but she whirls by to make another circuit around the camera and tripod. When, on her elliptical journey, she reaches him again, he tries to interrupt her path, by standing, smiling, with his arms and hands outstretched horizontally, as if he were forming a gate into which Vikki must collide. This does indeed take place, though Vikki does not pause but in her stride takes hold of his arm gently, as if it is a revolving door, and, by pushing it open (as it were), sets him spinning on the spot with increased speed. Although he did not know Vikki was going to do this, he instantly translates his own rotation into a spiralling rotation to the floor as she passes by. Vikki's impending impact is dissipated by its absorption into the flow and so her orbit remains unimpeded.

More laughter. Now Vikki, attracted to the central camera, adjusts her orbit too and moves in to take up a position near Joey at the machine on its tripod. Now Leonard spirals in toward the camera and, in a continuous movement, positions himself in front of the lens, while Vikki video-records his face. As he moves, she now pans the camera around to track his rotation. Leonard moves away, back amongst the other children playing on the floor and this clip finishes with Vikki moving in front of the camera, to see herself in the outward-facing, LCD screen.

No resistance is made to any impending impact; preserving the flow of the movement takes precedence in the minds of the players. This is true of the chimpanzees' rotation around an axis, too. The action-assemblies of both species are the product of perfect collaboration, in spontaneous, spatiotemporal events in which unplanned perturbations and variations are seamlessly incorporated into a fluid and flawless improvisation.

Although, for matters of clarity, we started off considering each dynamic-structure individually, their production occurs in a sequential and, I would argue, semantic relationship to other structures, forming patterns which seem to possess their own internal organisation based on a subtle balance between constancy and variation.

The children's and the chimpanzees' rotation share certain characteristics. Consider the conclusion of the rotational play-chase performed by Nkosi and Togo.

Rotation around axis in chimpanzees concluded

16 November 2007 Video-recorded observation (Edited DVD) Singapore (Tape 1)

After circling the tree several times, Nkosi finishes a rotation by performing a head-over-heels. The head-over-heels is another way the chimpanzees demarcate the end of a linear passage but in this instance, Nkosi only partially completes it. He does not fully roll over onto the back of his head and neck in the way in which this ensemble is usually finished but, instead, prematurely sits upright, so truncating the performance. Why is this? I think it is because he calculates that he has insufficient time to complete the head-over-heels in its true, canonical form before his father catches up with him, so he needs to shorten the ending. On the other hand, he does feel the need to emplace some sort of conclusion. For his part, Togo purposely slows up his pace in order that his son may fit in this important passage and punctuation point. Neither of them wants to disrupt and spoil the play-assembly. They want to preserve its fluidity and its structure.

Compare this situation with improvisations made during the children's rotation around the camera when, for instance, Zhou Quan smoothly fits in his act of looking into the LCD screen without spoiling the rhythm of the play; his easy absorption of Vikki's collision by translating it into a spin of his own body; or consider Rah's retrieval of a piece of coconut, an action inserted successfully into her rotation-and-descent assembly (pages 204–205).

Human children encode this action-assembly, rotation around an axis, onto the drawing surface.

Continuity and discontinuity encoded in drawing

04 December 1984 PAL VHS video-recorded child-observation plus written transcription, (Hannah) London

(Red raspberries; blue trails)

00:22:06 Hannah (2 years 2 months 28 days) makes vertical arcs in high, overarm, downward thrusts, with paintbrush loaded with red paint, onto a piece of cartridge paper on a table. She has a choice of colours in little pots. Her mother, Linda, sits next to her. Hannah's vertical arcs result in red blobs on the paper which follow a roughly elliptical course, a path which makes use of a central axis of a pot of paint standing in the middle of the paper. With each downward impact of the brush, the child synchronises a farting or 'raspberry' vocalisation in a one-to-one correspondence.

Between each downward stroke she brings the brush up above her shoulder and outside her field of view, poised for a moment, before crashing down again, just like Rara did between each whack of her knuckles against her clay lump, and like the way the Chinese boy raised his pencil out of his field of view between each of its successive descents.

In Hannah's painting a travelling sequence of red blobs explode into being; their rate of appearance accelerating as she beats down her brush; an increase in tempo matched by her mother's widening smile. At the end of this sequence, Hannah discovers a new type of marking-action when she drags the brush toward herself, making a short, vertical line. There now occurs one of those momentary pauses when we can see the infant has a new idea.

In a moment we will see that Hannah's new idea to make a new kind of mark with a new colour. This idea occurs to her when she slides the brush along the paper instead of bashing it. This is very like an observation of Rah (1 year 10 months 6 days), when she seems to change her mind about the type of action she is going to make and turns a slap into a tap.

Hannah selects a pot of blue paint instead of red. Hannah's removal of the paint-pot lid is assisted by her mother, Linda, but in a non-interfering way, allowing Hannah to accomplish a great deal of the task herself. On the video-recording there is an exquisitely orchestrated interplay of the two pairs of hands.

Cooperation and reciprocal support is supposed rare in non-human primates (Mitchell, 2002b) but my observations contradict this assumption. We have already read how older and more experienced chimpanzees show empathy with, and support for their young, comparable to that displayed by Linda to Hannah.

> With the lid now removed, Hannah now dips the brush into the paint-pot. Then, before making the first mark with this colour she suspends the brush from its end, so that it dangles over the paper vertically like a plumb-line.

Plumb-line

This use of a dangled stick, gradually becoming vertical as gravity pulls upon it, is seen in chimpanzee play and exploration too. For example, Rah, at age 1 year 8 months, dangles a stick from a limp hand, allowing it to become vertical before resting its point on the wooden surface beneath her (04 August 2006, part 3, video-recorded observation, edited DVD, Singapore).

When Rah places the point of her stick against the surface she immediately translates this into a trailing action and this is exactly what Hannah does now. She allows the brush to touch the paper, presses it down and then translates this initial contact into a trailing course which nearly completes a circuit around the central axis of the pot of blue paint. Just to reiterate: first of all, the point of contact is established and then the point is translated into motion. In contrast to violent downward thrustings of the red brush which were accompanied by explosive farting noises, she now trails a long elliptical course around the paper in blue paint whilst synchronising, in one-to-one correspondence, a soft, shushing sound.

In an episode like this, the kind of ephemeral rotation around an axis we have seen performed by chimpanzees around a tree is translated permanently into two dimensions whilst the children's rotation around a video-camera is recorded by the video-camera. [Depending on how it is introduced to children, this electronic medium can be as directly accessible as pencil and paper technologies. I am guessing that this might be so for chimpanzees too and Betsy Herrelko's (2010) work supports this notion.]

21 Rhythm

Throughout the book I have mentioned the rhythmic aspect of chimpanzees' play-actions. In fact, the presence of rhythm is one of the ways by which the play-mode can be identified. Rhythm does not appear in chimpanzees' adaptational, utilitarian actions in quite the same way as it does in their play (though rhythmically regulating one's heart-rate and breathing does assist in strenuous work). In this chapter I will pursue this aspect more deeply because I consider it a critical part of the formation of expressive, representational and symbolic thought.

Starting with the simplest examples first, we will now consider the origins of rhythm in chimpanzees' play.

Rhythm starts with repetition. An action is made and the chimpanzee or infant human finds he or she enjoys it. The sentient being then tries to duplicate the action which caused the pleasure. He or she makes a second action, then a third, and so on and, soon detects the rhythmical pattern within the series. The rhythm may not be perfect. In this book, for it to count as a rhythm, the chimpanzee or the young child needs to show the intention to make a rhythmic sequence (Young, 2000).

In both human and chimpanzee development, rhythm shows up clearly as an expressive component in the first of the First Generation Actions, the vertical arc.

14 October 2005 (Part 4) Video-recorded observation (Edited DVD) Singapore

00:05:31 This is the day Nkosi (5 years 3 months 17 days) finds the metal-tube or pipe. He takes it to Platform Rock by the Waterfall and uses two hands to bang it in vertical arcs against the rock. He looks at me for a moment. Banging the metal pipe against the rock, Nkosi becomes involved in beating out a clear, rhythmic sequence of sonorous, ringing sounds.

Another way a rhythmical sequence is started is by repeating a beat in order to attract the attention of another. By repeating a single, sharp sound (sometimes intentionally coupled with its visual and kinetic image), the primate makes a non-accidental structure which stands out from random, ambient sound. It tells the listener that this is a purposeful sound made by a thinking being. Again, when the action and sound is repeated more than once, a musical and rhythmic pattern sometimes emerges.

Whatever the circumstances of the rhythmic series, whether it is made as a signal to attract another, or simply to cause pleasure, the strike of the hand or other percussive object becomes a *beat* in time.

Sometimes, emergent signs are put together when an initial beat doubles-up or even triples-up in its meanings or its *denotational values* (Willats, 1997, 2005) first, as a way of attracting the attention of another; second, as an invitation to play and, third, as the initial, starting beat of rhythmic passage itself. This happens when empathy between the playing chimps is 100 per cent. More usually, the first strike is given as an introductory alert occupying a single beat in time, before the first real beat of the dance begins.

14 October 2005 (Part 4 continued)

00:31:33 Makhulu (5 years 5 months 7 days) comes up, wet from the stream, to join his brother Nkosi (5 years 3 months 17 days) and, raising high a fronthand brings it down in a vertical arc hard against a log, initiating mock conflict with a slap of his wet palm against the wood. Careful, frame-by-frame analysis reveals that, although there is but a single, audible slap, this is briefly preceded by Makhulu's highly visible rising hand, slightly decelerating as it ascends and which, like the rising baton of a conductor, makes a silent, preliminary beat which alerts the player(s) to get ready for the first real beat of the piece.

The next observation shows another way play is initiated. This time, the signal to prepare to play is given its own separate, kinaesthetic, tactile and auditory beat, followed by the first beat of the bar, also kinaesthetic, tactile and auditory.

08 December 2006 Video-recorded observation (Edited DVD) Singapore

It is 1 year, 1 month and 24 days later. Nkosi (6 years 5 months 11 days) and Makhulu (6 years 7 months) are poised to chase each other around Rotten Stump. Nkosi slaps Makhulu with two, full, very distinctive, vertical arcs, first with his right palm and then with his left. This sets off the ensuing play-chase around and around the crumbling old tree stump. Careful analysis reveals the changing semiotic values between each of Nkosi's two slaps. The first, preliminary slap is to alert his brother that the performance is about to begin; the second is the first beat of the passage itself.

Other interesting variations which occur within this same play-episode include direction changes, signalled by a slap on the bum of one brother by the other, to 'tell' the pursued that he is now to become the pursuer. Additionally, once play gets underway there is no further need to give introductory beats; players are sufficiently in tune with one another that they know when a note acts as a marker for an end of one passage or as a partition between that passing passage and the new one to come. It is also a bit like the way in which the same line in a drawing may stand for both the ending edge of one region and the beginning of another.

Human children also display this use of rhythm and sense of partitioning sequences in the organisation of their free-flow play.

11 November 2003 Video-recorded child-observation (Edited DVD) plus written transcription (Singapore)

Faith (5 years) taps her white-board marker, non-writing-end down, against the Formica-topped table at which she is drawing. This discrete beat signals and initiates a rhythmic sequence in which she enacts drinking from a can or bottle represented by a short, wooden column (from a set of wooden blocks). The 8-s long sequence ends just as neatly with a pat of her palm against the top of the wooden column to represent pressing closed the lid of the can or bottle (Matthews, 2006b).

Music and meaning

I hope you are seeing that one of the ways in which chimpanzee play might be fruitfully analysed is in musical terms. We see rhythm incorporated into First Generation Actions and it becomes stronger as play-assemblies are formed. It is gradually developed to a high degree in the more complex action-assemblies which follow.

If such collaborative play is to happen at all, then each participating chimpanzee needs to be sensitive to the structural and semantic possibilities involved. Furthermore, each chimp must act upon his or her understanding within the tiny window of time which opens when the first beat is heard.

After this first beat, and within the brief jiffy before the second beat is due, the partner must change his or her orientation to the first sound (which, milliseconds before, was just sharp signal for attention) and turn it into the first beat of a bar – to which he or she must respond – in time. Once into the rhythm, the chimpanzee will make the same kind of intuitive and rapid-fire decisions at subsequent key junctures.

The chase is not just a chase but a dance. This becomes a familiar action-pattern, becoming interiorised for its musical and rhythmic qualities; perhaps stored wholistically as a musical and configurative phrase in the chimpanzee's memory (Mithen, 2005).

It is not surprising then that this is not 'instinctive' but has to be gradually learnt. The chimpanzees start with the clearest, simplest and iconic exemplar, or canon, (say, Slap – Jump, or, Slap – Jump – Slap) and only gradually add variations. Here is a good example of a canonical rhythmic phrase.

03 January 2008 Video-recorded observation (Edited DVD) Singapore

00:05:09 Rah (3 years 26 days) is pendulum-swing-touching from a sloping vine to Lateral Log, reaching out with her right hand's fingertips to touch the Log at the point of apogee of her every inward swing. Then, on the final swing, she makes a trajectory-and-impact-type landing on the Log, ending with a follow-through, downward hunch. This phrase is punctuated by a final, downward knuckle-touch of her left hand against the surface of the log. This moment of contact is not functional in any sense; it is not used here (as it might well be on other occasions), simply to balance the body at its point-of-arrival. Rather, the use of her straight-down, left hand against the wood is as a marker, a punctuation-point only.

We could stop the observation right here at this convenient point, as Rah taps out the end of the sequence but this would be to miss out something else important. For this full-stop simultaneously signals the beginning of a new sentence. Marc Bekoff also detects in the play of dogs actions used as punctuation markers. Canid play is rhythmic too (Bekoff, 1995).

When Rah raps the log with her knuckles to mark the end of one assembly, the same downward stroke serves as a marker which initiates the next phrase which is to knuckle-walk slowly, head down, shoulders hunched, leftward along Lateral Log, in a pretend-stalking, as if toward an imaginary person. The Point B of this linear passage is a small patch of denuded trunk, the bark stripped off by chimpanzee teeth and showing up reddish, like the surface of Mars. As my camera zooms in, we see her also zoom in on her target, huddling over the dusty surface and inserting first her tongue and then her finger into a little orifice in the red wood.

My observational evidence suggests that the chimpanzees interiorise such rhythmical passages and carry them over to new situations, even after an extended, intervening passage of time.

Such rhythmic phraseology is based on the minute, natural time-scales of the limits of responsiveness of the primate system to a stimulus, involving breathing, heart rate, energy level and the orchestration of the muscular and skeletal frame into patterned sequences. The complex and collaborative play-assemblies of the chimpanzee are best thought of – not as a primitive version of human thinking – but primate thinking at a high level.

Kinaesthetic and visual mapping

16 December 2005 (Part 14) Video-recorded observation (Edited DVD) Singapore

(Greetings)

Baby Rah (1 year 9 days) is seen climbing onto Suzie's back. [Suzie, you will remember, is not Rah's mother. The females cooperate in the sharing of infant-care duties.]

00:04:19 'Hi, baby. Hi Suzie.' I say to the pair. Suzie stops and they both look at me.

00:04:27 'Hi baby.' Rah looks at me and then sits upright, looking directly at me. She makes a scratching movement against her cheek with her right hand and then brings this hand down in synchrony with her left to make a series of knuckles-down, double-vertical arcs on Suzie's back. As she makes this action she brings in movements of her foot-hands, which thrust downward in time with her hands.

This is a cross-modal association made between different bodily parts, the actions of which echo each other. They are a type of kinaesthetic and visual mapping, in which the movement of different body parts are linked or associated contiguously together (Guillaume, 1926/1971).

Also noticeable in Rah's action is the beginning of a rhythmical sequence. As we noted above, rhythm starts with a simple repetition of an enjoyable action; the action is duplicated because the chimpanzee has enjoyed it and wants it to happen again. Though this initial double-beat might be reflexive, its rhythmic aspect fortuitous, it may nevertheless suggest a genuine rhythm to the chimpanzee. A rhythmical sequence, being a non-accidental structure, is very noticeable to other conscious beings, including humans, and so acquires a communicative dimension.

Rah's emphatic double-punches are made whilst she stares at me. That the action of striking with hands is combined with looking is important, both to the being who does the looking and the one who looks back.

Rah is learning to make a 'display'-type of gesture. It is the start of her self-assertion but the vestigial element of aggression within her performance I take to be notional rather than serious. I believe this to be a mix of greeting and warning, a bit like the first greeting I received from 5 year 8-month-old Nkosi when he made his horizontal arc in the leaves (27 January 2006, video-recorded observation, edited DVD, Singapore, page 54) which was also tinged with a little warning.

Rah's action is a signal of communication similar to those I have seen given by human infants too (though I have not noticed the use of the knuckles-down strike in the human infant).

The essential, underlying beat is detected in more complex dynamic-structures too.

Continuous movement versus steps

Unbroken line versus separate, discrete displacements in time and space

04 August 2006 (Part 3) Video-recorded observation (Edited DVD) Singapore

With dangling arm and hand Rah (1 year 8 months), circles a stick (perhaps in time to the distant music coming from somewhere in the zoo), as she lays slumped, face-forward on Horizontal Beam at Level 2, with Ida.

Investing rhythmic structure into an indivisible circular movement is more challenging than detecting it in the one-to-one correspondence inherent in the repeated vertical arc. This is why it appears developmentally later on. We saw the first glimmerings of the circular motion being divided up into individual beats in the earlier observation of Kimoni (1 year 2 months 10 days, page 222) when he stamps his little feet as he spins on an axis. Shortly, we consider more fully these two ways of exploring movement; that of continuous flow and that of consecutive displacements in time and space.

Human infants also learn to imbue rotational movements with rhythm (which is why Nancy Smith calls what I call 'continuous-rotation', 'circular repetition' – Smith, 1983). We see such rhythmical rotation in painting and drawing and it can also be seen in any activities which elicit stirring, such as emergent cookery in early years' settings (H. Matthews, 3 December 2009, personal communication). I am interested to learn that Charles Darwin's daughter, Annie, as a little girl, also used to like stirring when making custard in an activity she called 'whirling round' (Keynes, 2001, p. 75).

To continue our observation . . .

Rah now dangle-swings her stick in horizontal arcs. Minutes later, she continues the same rhythm at Vertical Pole, Corner 3, where, hanging by one hand, she swings backward and forward against the pole, striking it with a foot-hand at the terminus of each swing. [This is a good example of an iconic trajectory and impact.]

In this case, she retains or 'carries' a rhythm between separate contexts. She is able to preserve the underlying tempo below and beyond its three different manifestations; first, in the close-up field-of-view-and-action play with the rotational movement of the stick; second, within its horizontal swingings; and third (and after a time interval), in whole-body play, when her entire person is involved in making the same rhythm.

Both Nkosi and Rah make a range of what I think of as a 'Silly Walks' (Monty Python, 1969). One of these is a 'stylised' walk in which each individual step seems 'measured'.

Taking steps

13 June 2008 Video-recorded observation (Edited DVD) Singapore

Nkosi (7 years 11 months 16 days) lopes along Lateral Log, seemingly doing this at a steady pace but with what I think of as a *stylistic* exaggeration, 1–2; 1–2; 1–2 . . . and so on, with stress placed on the Beat 2.

Rah also makes stylised walking, almost like a caricature of a threatening display walk:

29 August 2008 Video-recorded observation (raw footage on DVD) Singapore

Rah (3 years 8 months 17 days) makes a very clean trajectory and impact flight across Level 3 onto Mezzanine Level, ending with an audible thump. She sits down, looking back at her brother. She then creeps up on him in a stealthy, display style, spacing each step with deliberation in a fashion so nearly identical to one she made just previously that later, whilst reviewing the footage, I mistakenly think I have inadvertently repeated the same clip. [I have this experience many times when editing the material on my computer.]

This observation shows one of the circuitous, cyclic play-routes in which sequences are almost duplicated, with the chimpanzee making close copies of a passage but with subtle variations. In this example, Rah performs trajectories as both clean, continuous sweeps through the air, and also as discrete displacements in time and space, as 'steps'. The chimpanzees have an awareness that a linear journey may be expressed either as a continuous, fluid trajectory or it can be deconstructed into a series of steps or stops.

Human children also differentiate these two modes of locomotion:

23 January 2008 Child-observation, hand-written, Singapore

A boy (about 4 years of age) walks along a bridge and rhythmically slaps the wall to his left with left hand as he walks along, performing a repeating sequence which goes, two steps, one slap; two steps, one slap, and so on.

Both chimpanzee and child seems to encode onto flat surfaces these two different ways in which linear movement may be expressed.

01 August 2008 Video-recorded observation (Edited DVD) Singapore

00:09:15 Rah (3 years 7 months 14 days) has a piece of cloth in one of her front-hands. Issuing from the natural rhythm of her knuckle-walking, she makes controlled series of staged pushes of the cloth from right to left along Mezzanine Slope.

Again, this is a discontinuous series of what we might term 'dashes' (neither 'dots' nor a long, single line). The series of stops (or 'steps') of the cloth is based upon the natural, momentary pauses between each push.

This is almost a miniaturised version of this mode of locomotion and human children are very good at this too.

April (?) 2009 Child-observation, hand-written, Singapore

'Run, run, as fast as you can' sang the Gingerbread Man (Part 1)

A 4-year-old, white Australian, in a café, sitting at a table opposite his father, holds a Gingerbread Man biscuit in his right hand and, in a series of staccato stabbing motions which indicate individual footfalls, 'runs' it across his plate. Later, he moves the handheld biscuit through arcs in the air before him, in graceful, sweeping traces.

Round and round the garden

Children also make this distinction quite clearly in their drawings; continuous, flowing lines often representing continuous paths of travel, whilst the same journey may be made in a series of points which move, one-by-one, across the drawing surface. We saw the contrast between these different forms of motion in the painting by Hannah in the observation on pages 236–237. Here is more developed version from my study of Ben.

October 1973 Child-observation, hand-written, plus child's drawing, (London)

Ben (2 years 2 months) has been sung the following, musical passage accompanied by the appropriate verbal incantations: 'Round and Round the Garden, like a Teddy Bear . . . One Step, Two Steps, Tickly Under There!'

For readers unfamiliar with this nursery-rhyme cum performance, the script goes something like this: The adult takes hold of the child's upturned palm and with an index finger traces a continuous rotation on it in time to the first two stanzas: 'round and round the garden, like a Teddy Bear . . .'.

If you keep to the underlying rhythm, this makes seven times. Then, without breaking the rhythm, the adult proceeds to the next passage in which, in time to the three beats behind the words 'one step, two steps, tickly under there', the adult 'walks' his or her finger in three steps up the child's arm with the third and final beat corresponding to an abrupt tickle under the child's armpit.

This lengthy description is necessary for the reader to understand Ben's subsequent drawing about this. He draws a roughly circular shape for the 'garden', then three short strokes to represent the three 'steps'. He draws a tadpole figure 'standing' perpendicular on each of these short dashes. This is the 'Teddy' moving through three positions in space and time. The final mark he makes is an explosive squiggle to represent the 'tickle'.

This type of drawing is built upon such earlier ones as typified in Hannah's painting of red raspberries and blue trails (page 236). In a process of vertical decalage, the same underlying concept is revised and reworked at successive developmental levels. Whereas Hannah's painting was composed of the bare elements of rhythm and rotation plus contrast between continuity and discontinuity of line, Ben's drawing has further levels of meaning layered on top of this basic structure by way of language and the 'marshalling of complex graphic strategies' (MacKarell, 1988). Ben's drawing shows configurative and dynamic aspects of the same event; the continuous rotating line meaning both the boundary of the 'garden' and the teddy's movement around it, whilst the journey to the armpit is broken up into three stages and the final, moment of impact demarcated by a sudden squiggle.

Walking on water

15 December 2006 (Part 4) Video-recorded observation (Edited DVD) Singapore

Nkosi (6 years 5 months 18 days) crosses the mirror-calm, little lake above the Waterfall but makes no splashes. Instead of simply plonking his foot-hands into the water he places the sole-palm of each foot-hand carefully upon its surface, barely breaking its meniscus, and holds it there for a moment before allowing it to sink, and only then taking his next step which he makes in the same way. Perhaps he is impressed by the immaculate stillness of the surface of water. He likes the feeling of his palm-soles placed precisely on this surface. He is prolonging the sensation of walking on the water; delaying for as long as possible his steps across the stream but at the same time without wishing to break up the passage of movement completely.

The extraordinary act of walking on water is a script-violation which, along with a few other miraculous events, fascinates the human primate too. Trying to slow things down with the act of *prolongation* is another way we transform ordinary actions into the forerunners of representation or symbolisation. We saw an example of purposeful *deceleration* in an earlier observation of Nkosi's pretend clapping (page 228–229).

Perhaps, like humans, chimpanzees sometimes want an experience to last forever. As we all learn, this is not possible – at least not in this world. Perhaps this is the tragedy of all sentient primates. The desire to alter or even stop time is perhaps indicative of the feeling and thinking primate.

The experience of prolongation is achieved through the technique of *delay*. Like *deceleration*, it is used to take an ordinary action out of the here and now and help it on its way to become an expressive and representational vehicle. But symbolisation has limits. One can delay and prolong pleasurable experiences for only so long, after which they cease to be those experiences anymore and become something else – or nothing at all.

The placing of one foot carefully after another on a surface of water is an unusual example of breaking down a linear sequence into discrete displacements in time and space. As we have noted above, human children record this differentiation in their drawings. Perhaps it is not just chance then, that on the same day that Nkosi tries to walk on water he also makes his first marking actions. But more of this later.

22 A new kind of desire

So far then, we have been considering the background actions for representation, which are not representational in themselves but yet which show a new kind of interest; a new kind of desire born out of play (Vygotsky, 1966).

Breath-taking

Another way in which time is structured in chimpanzee play is by the use of *pauses*, *rests* and *holds*. These lulls in the action are inserted into play-assemblies at measured intervals. They are based upon acts of resting from effort and recovery of breath. The organisation of primate action, including that of human language, is based upon what you can do with a lungful of air. Interesting evidence comes from Marina Davila Ross and colleagues' (2009) studies in which humans tickle orang-utans suggesting that *laughter*, because its production requires special breath-control, maybe a precursor to speech. Her work suggests that we started laughing about 16 million years ago and have never stopped.

Pauses, rests and holds

When the chimpanzee pauses for a moment to recover breathing, these pauses, if extended, may become *rests* in the sense intended in written music. (I mentioned one such in the observation of Rah and Nkosi play-chasing – 07 September 2007, video-recorded observation, edited DVD, Singapore, pages 179–180.)

Once more, we see the developmental pattern in which a natural action becomes used for its structural possibilities. In this case, the inevitable consequence of exertion – the taking of a breath – is transformed by a switch in mental orientation and inserted as a timed interval within a pattern of action.

Pauses or rests can be made shorter or longer, their duration partly dependent on the energy expended but also strongly influenced by the chimpanzees' use and perception of the dynamic-structure in itself.

Still based upon respiratory rhythms, sometimes a rest is extended within a play-assembly over many seconds or even several minutes. This I learnt to differentiate from a pause or rest and I term a *hold*. Armstrong et al. (1995), when discussing the grammatical structure of sign-language, define a 'hold' as a brief period of time whilst everything is in stasis. When I first saw a 'hold' in chimpanzee play I simply assumed the play episode had finished. One player and then the other (or others) suddenly stopped. But the play was not yet over and just as suddenly it resumed, exactly where it left off, the players awakened as if from a sudden slumber.

The spell broken, the play is picked up again from where it left off. Sometimes the chimpanzees will just sit still while the play is put on-hold and seem to ignore each other completely. On other occasions they will start to engage in a completely different and non-play activity. A basic one is to stop to eat but there are many variations, including self-grooming or appearing to be engrossed in microscopic investigation of some small feature of the environment. This is all feigning. Even if the chimpanzee is really stopping for a meal-break, or even if he or she has indeed genuinely encountered something of interest, he or she will still have half a mind on the play and its anticipated continuance. Of relevance here is that although they are capable of sequential thought (as evidenced by the acquisition of sign language in a few of their kind) they are not stuck in this mode and appear to be able to think about several things at the same time (Fouts, 1997). Play-episodes are stories which may be stopped and resumed at various points in the action.

My studies are not detailed enough to tell whether such 'holds' might be maintained over a period of hours or even days but I suspect this might happen. After all, we have seen that the chimpanzee has a knowledge of time, a good memory and, if they do possess autonoetic consciousness, are able perhaps to distinguish between real events and imagined ones.

Of additional interest to us is that with experience and sensitivity, human interaction (in speech or action) may be inserted appropriately into briefer pauses within the structure of their action patterns and they, the chimpanzees, reciprocate. That, with regular meetings with a friendly human, other wild animals seem to do this too (for example, the English grey squirrel – sometimes referred to as 'Spanish squirrel' – and the London crow) perhaps points to a general condition across species as well as telling us something about predispositions toward, or, at least precursors of, language. In turn, this has important implications for our interaction with our own, human children.

Round and round the Rotten Stump

08 December 2006 Video-recorded observation (Edited DVD) Singapore

Makhulu (6 years 9 months) and Nkosi (6 years 3 months 11 days) play-chasing each other around Rotten Stump are joined by Manzu (6 years 9 months 8 days). The chimpanzees use symmetrically placed front-hands on the ground to propel themselves forward or into rotational movements after each other. Interjected into these breakneck chases are brief moments when they pause, before suddenly leaping into action again. The actions of more than one chimpanzee are often resumed in near-unison, making it difficult to see who (if indeed it is a single individual) initiates or breaks the 'rest'.

Sometimes, when they stop for a moment, they almost face each other at a distance of a metre or so. At these moments they usually do not look directly at the other's face but slightly to one side of it before springing into action again.

A strategically placed pause can be used to strengthen a sudden surprise as when, in the present observation, after a lull in proceedings, Nkosi unexpectedly pops up over the top of Rotten Stump (rather than the anticipated appearance around the occluding boundary of its trunk) to startle Makhulu.

Pauses also leave temporal gaps allowing their much younger sister Rara to insert a passage of her own. Starting with symmetrically placed, double-downward slaps of hands to the ground she performs a series of linked leaps forward. These emulate her brothers' movement. She stops for brief moments before resuming.

Sometimes the chimpanzees pause for breath and stop to sit and look at me, as if to check that I am catching all of this. In the present observation, a group of human visitors come by and stand quite still and silently, watching. It is significant that these human primates know what to do (they don't always). The chimpanzees now seem to time their pauses in response to the attentive humans, to whom they look each time they stop.

In such cases, pauses seem to serve the function of 'applause-breaks' (or even 'clap-traps'). However, while production of a pause might be spurred by human interest I do not think it is initiated by this or wholly reliant upon it or upon the human gaze either. I think these interludes arise from natural breaks from exertion and for the recovery of breath. Though the interaction of the watcher (human or non-human primate) plays an extremely important part, this is a later and additional feature to an originally self-generated dynamic-structure. The basis for linguistic interchange may have arisen in deep time from a pause for breath followed by a brief visual check of the other's whereabouts and his or her level of interest.

The ability to calibrate the insertion and timing of pauses or rests is especially important in collaborative play or else the whole thing breaks down. We can see this from the brothers' response to some of their little sister's efforts at joining-in. Sometimes, in the intervals they provide, she gets the rhythm almost right, as in the observation above, when her forward leaps are clearly a rough copy of theirs. At other times however she offers a very eccentric version of their actions when, for example, she responds to her brothers' wild but nevertheless highly-structured rotational play-chase with a bizarre interpretation of her own, consisting of frenetic whirls, spins and tumbles. When this happens, the brothers seem unable to respond and stop for a moment by their tree stump, 'stumped'.

Rah knows that some kind of script-violation is called for but she does not know quite what! This is really the use of the 'placeholder' again (page 104) and recalls one of the stages in which human children decode the structure of the 'joke' (Athey, 1977).

Apt humour is indicative of linguistic understanding and it may start in a process of acquisition similar to that of the baby chimpanzee, in which, through practice, the phraseology and timing of the play-assembly is formed.

Again we see how the younger, less experienced chimpanzee is gradually introduced to the structure of complex, expressive play-assemblies. First of all, the younger chimpanzee does not grasp the rhythmic structure or its integral signals. Like Poppy's attempts to play hopscotch with her older sister, Lily, joining-in (page 131) does not at first work because the apprentice does not understand the meaning of the sign. If some semblance of play does start, it quickly breaks apart for want of rhythmical structure. But the younger inexperienced chimpanzee will gradually learn the sign and the various ways it can be used. Whilst we cannot perhaps call it a language it is a semiotic sequence which is gradually made learnable to the novice by a more experienced, usually older, companion. The rhythmic and musical organisation of chimpanzee play is of extreme importance when we consider their representational and expressive thinking and the opportunities chimpanzee culture offers for its learning.

23 Deep structure

Insertion, going-into, going-through and right-angular attachment

In human childhood, the discoveries of certain key structures are important landmark events in development. These discoveries lead to the formation of what I think of as *structural principles*. One important structural achievement is being able to connect, join or link things together in some way. The making of *angular* or *near-right-angular joins* is one such powerful, key structure. This is also true for the chimpanzee and we have already encountered an example earlier, when Nkosi placed a straightish branch at right-angles on top of a horizontal beam making a *near right angular cross-over* (see page 197).

Another important dynamic structure is that of *going-through*. Some kinds of closure or bound-volume such as tube-like forms offer a passage of movement right through, by going into an entrance at one side (and, in some cases, coming out again through an exit at the other). This occurs on many scales. In both human and chimpanzee development, we see this dynamic-structure at both close-up, miniaturised scale, with handheld objects or toys being made to penetrate tubes and tunnels of various kinds and at whole-body scale, in which tubular volumes are bodily climbed through.

One example of the former is the observation of Joel and the coffee-grinder I described near the beginning of the book; other examples include the following developmental sequence of Rah.

> **27 October 2006 (Part 2) Video-recorded observation (Edited DVD) Singapore**
>
> Rah (1 year 10 months 20 days) is experimenting and playing with the action of *going-through*. She pushes a bendy, wand-like stick up-and-down into a drilled, circular hole in Level 3 Platform.

The next observation is made just over 4 months later:

16 March 2007 (Part 1) Video-recorded observation (Edited DVD) Singapore

Rara inserts a leaf into tube

Rah (2 years 3 months 6 days), sitting on Level 3 Platform, manipulates one of the flattened old fire-hoses and inspects its slit-like, opened end. She looks at me in response to my voice and then turns her head away shyly. She picks up a gold and yellow leaf with her left hand and with her right hand holding tube upright, she inserts the coloured leaf into the open end of the hose-pipe. I talk to her about this. Now she has inserted her entire hand, along with the leaf, into the pipe. With her right hand swallowed up in the open end of the pipe, she bends the pipe first one way and then another at a crease formed in the perished rubber. She looks into the tube in order to see the leaf inside.

Over the weeks, Rah becomes increasingly interested in the action of going-through:

04 July 2007 Video-recorded observation (Edited DVD) Singapore

At 2 years 6 months 28 days of age, Rah has hold of a transparent plastic, mineral-water bottle. She inverts the bottle, inserting it vertically, nozzle-first into the circular, human-drilled, little hole in the platform of Level 3. The bottle remains for a few seconds fixed vertically upright in the hole. Lying down on her stomach she leans over the edge of the platform and, with her face upside-down, peers at the bottle's nozzle as it pokes out ever-so-slightly from the hole's opposite entrance on the underside of the platform.

28 March 2008 Video-recorded observation (Edited DVD) Singapore

Some changes have been made to the Construction and Rah (3 years 3 months 21 days) investigates a new nut and bolt which connects two beams, one beam resting on and crossing over the other at right-angles. The bolt passes through a hole drilled vertically through the two beams. Rah clearly seeks out visual evidence to confirm that the bolt does indeed go vertically through the two beams. She looks at top and bottom of the nut and bolt and also in between the two beams to get a glimpse of the gleaming new metal of the bolt passing through the hole which runs congruently through the two beams.

08 August 2008 Video-recorded observation (Edited DVD) Singapore

Stick up backside

00:00:26 Four months 10 days after this Rah (3 years 8 months 1 day) is on Level 2, directly below Suzie and her baby Ainka (11 months 9 days) on Level 3. A circular, drilled hole penetrates a beam of the platform above. The hole is directly underneath Suzie's bottom. Rah inserts a long, sharp, palm leaf up through this hole to poke into Suzie's backside!

Suzie seems oblivious but it is noteworthy that Ainka watches her big sister do this and 42 days later is herself practising the same strategy on her own.

19 September 2008 (Part 2) Video-recorded observation (Edited DVD) Singapore

Going-through (observation continued)

00:04:00 Ainka (1 year 20 days) has nest-building materials on Level 3. Spinning slowly on her back, she catches sight of the line of drilled, circular holes in the beam just above her. Her mother is sitting on this beam directly over one of the holes. This gives Ainka an idea. She picks up a long stem and points the sharp end upward, toward one of the holes on the underside of the beam above her, above which rests her mother's big bottom. She manages to slant the stick at an angle so that she can manoeuvre it within the space between the two levels, insert it into the hole and into her mum's bum! Surely she learnt this from her big sister, Rah, who has also played this trick?

Aside from sticking something up someone's bottom, to be able to fix something permanently or even temporarily to something else is a crucial technological achievement in development and evolution – perhaps especially if you live in a tree. I do make one observation of Ganga succeeding in joining one thing to another in order to solve a problem (14 November 2008, video-recorded observation, edited DVD, Singapore). Ganga is carrying baby Kimoni (only 1 year 3 months old) at the top of Tall Tree, many metres from the ground. She also has hold of an interesting, burnt stick she does not want to abandon. This wouldn't normally be a problem but she additionally clutches to her chest an entire package of such sticks I have supplied the chimpanzees. (I will tell you more about these charcoal sticks later.) She knows that with her hands full she cannot descend safely with her baby but neither does she want to abandon her prize possessions for another chimpanzee to appropriate. What to do? She finds an elegant solution by driving the stick home vertically into a crack in the flat, sawn-off top of Tall Tree. Now she has a free hand and she can carry Kimoni safely downward and make a second trip later to retrieve the stick, safely jammed in the crack in the top of the tree.

Examples of whole-body *going-through* are also numerous but I will offer just one, a rather dramatic one, in the following observation of Nkosi.

Deep play

> **18 May 2007 Video-recorded observation (Edited DVD) Singapore**
>
> **Going-through the Electric-Web**
>
> Perhaps the most extreme version of playing with the idea of *going-through* is when Nkosi (6 years 10 months 21 days, after careful calculation, hurls himself through the narrow gap between criss-crossed, electrified wires (which I call the Electric-Web), high up in the fork of a tree.

This is an example of both *danger play* and *deep play*. Other than the thrill of risk-taking, there is nothing to achieve in playing it (and a great deal to lose). But aside from the danger involved, I consider it a form of deep play because it entails the chimpanzee becoming seriously engaged with dynamic-structure for its own sake, in this instance, that of *going-through*.

In the following observation we can see a synthesis of the dynamic-structures *insertion*, *going-into*, *going-through* and *right-angular attachment*.

> **21 September 2005 Video-recorded observation (Edited DVD) Singapore**
>
> 00:08:26 Nkosi (5 years 2 months 24 days) uses his left hand repeatedly and strenuously to fix a transparent plastic water bottle, by its nozzle, into a circular hole in a wooden, vertical post.
>
> The bottle's nozzle fits in okay but he is clearly not satisfied and wants a tighter jam between the shoulders of the bottle and the sides of the hole so that the bottle will stay fixed there, jutting out horizontally from the vertical pole. The problem is that the shoulder of the bottle is too wide and the bottle seems likely to pop out if he lets go of it.

Size and shape similarity: Comparison between positive and negative circular shapes

Makhulu (5 years 4 months 14 days) is hovering nearby with a leafy stem in his teeth. He watches Nkosi's activities with great interest. He points his stem vaguely in the direction of the hole.

Here is an instance of precursor *pointing* in the sense of directing attention toward a spot in space. Perhaps, influenced by his brother, he has the idea of inserting his stick into the hole.

Of great interest to us is that Nkosi tries to leave the bottle stuck in the hole. He sticks it in the hole and withdraws his hand from it, regarding it for a moment jutting out at right-angles from the vertical pillar. He has succeeded in jamming the shoulder of the bottle sufficiently into the hole so that the bottle stays fixed.

It remains there for a few seconds. Then it falls behind Nkosi's back and to the ground.

Nkosi clearly had the idea of fixing it there, in the hole. Other writers have commented on gravity-assisted structures or 'polypods' made by apes (Reynolds, 2002, p. 207) and we will encounter some of these later on in this book but Nkosi's present building is not assisted by gravity. He makes a fixture which, just for a few seconds, resists the gravitational pull of the earth.

The idea of permanence is important for us to consider, because an interesting question to ask is: Supposing he had succeeded in securing the bottle in the hole? Supposing it had remained fixed? Would this have led to further inventions, further developments? This leads one to wonder how the chimpanzees might develop ideas if the objects they use are not cleared away at the end of the day but left for them to return to, on subsequent occasions. In my child studies I found that children's play increased in inventiveness and creativity if the toys and other objects they used in a session were left exactly as they were (rather than being tidied away), so that the child could return to them the next day and develop his or her ideas. The thinking of both human children and chimpanzees is embedded in their actions with materials and objects. This means that in order for them to develop their thinking requires that they are able to pursue a continuous passage of activities, and therefore a continuous stream of thinking, with the same things and without interruption.

I also feel that my occasional spoken encouragement to the chimpanzees and my commentaries into the camera's microphone on what they are doing have the effect of supporting their activities, helping the chimps to prolong them. Exactly the same applies to interaction between human adult and human child – except that there is no moat and electric fence between them.

Sudden insight

Worthwhile research might involve introducing simple construction equipment and building blocks to the chimpanzees, based on their emergent concepts. At present, much ape research seems fixated on language acquisition. Why not build upon what the chimpanzees and other Great Apes actually do and are interested in doing?

A classic experiment by Wolfgang Kohler (1925) did just that. He put bananas just out of a chimpanzee's reach and then gave the chimpanzee rods which individually were too short to reach the food but which, if joined together, formed a tool long enough to reach the bananas. Kohler realised that the chimpanzee's success did not just depend on trial-and-error but involved sudden insight.

Such insights would be based on concepts. The concepts the chimpanzee most likely used in this case include *insertion* (one rod could be inserted into the end of another) and *alignment* or *co-linearity* (the rods had to be lined up with each other) and *congruence* (the ends which were to be joined had to be congruent with each other).

Nkosi also appreciates alignment and congruence whilst his insertion of the plastic-bottle into the hole can also be considered an example of *right-angular attachment* or the *right-angular join* which is an important Third Generation Structure in human childhood.

21 September 2005 (Part 2, Edit 2 continued)

00:19:24 Later, we see Nkosi swing to the ground to retrieve the plastic bottle. Oh! I suddenly notice that it has water inside it now! He has taken it down to the stream and filled it with water! He is now drinking from it, using it for the purpose for which it has been designed by humans.

This is another instance of the chimpanzee's appreciation of *inside and outside* relations.

Seven months and fourteen days later he does in fact invent the *cup*.

05 May 2006 Video-recorded observation (Edited DVD) Singapore

00:29:48 Nkosi (5 years 10 months and 8 days) has a piece of coconut shell in his hand. He runs down to the stream and, holding the shell in his left hand, cups water from the stream and drinks it. He repeats this action decisively and accurately, as a human might do. Of extra interest is that Rah is close by, also with a piece of shell, watching her big brother. Rah now scoops her 'shell-cup' into the water too, just as skilfully.

24 The image

Looking-through and line-of-sight (projective relationships)

I want to return to Nkosi's experiments with the plastic water-bottle because apart from trying to fix it into a hole and apart from filling it with water, he also does something else with it which is of great significance. He looks through it, along it, from its open neck end, in a similar fashion as one might look through a telescope. *Looking-through* is the next dynamic-structure to follow on, developmentally speaking, from going-through and we saw an inkling of this in the observations in the last chapter. In experiments about looking-through we glimpse the dawning of emergent concepts *line-of-sight* and *point-of-view*. This kind of knowledge is different from that which we have encountered so far because it implies the formation of ideas about visual relationships rather than purely physical ones. When Nkosi carefully arranges specific sightings of the bottle, from particular points of view, or station-points, he is forming *projective* understandings which have to do with how a visual array is projected to the eye.

Looking-into, looking-along and looking-through

21 September 2005 (Part 2, Edit 2 continued)
Video-recorded observation (Edited DVD) Singapore

00:23:08 I get a good shot of him adjusting the long axis of bottle so that he can look down it through the nozzle opening.

Looking along a long object is a preliminary action for the formation of a *line-of-sight*. A hollow, tubular object is ideal in this respect because one may look straight through it. Understanding lines-of-sight is a prerequisite for the building up of knowledge about *projective geometry*.

Human children's development of spatial understanding follows roughly the same sequence as that of the chimpanzees. First of all a *topological* geometry dominates, concerned with inside and outside, closure, boundary, hollowness and things which *touch*. This is coupled with *dynamic* knowledge derived from movement made through space and time, and the shapes one can make by simply moving.

Later, a *Euclidean* geometrical understanding emerges in which the shapes of objects are noticed. Later on still, the relationship in space between objects is noted in relation to the viewpoint of the observer. This last type of geometrical understanding is *projective*, in that it captures relationships in three-dimensional space as if projected to the eye of a viewer.

Children's drawing shows this sequence of development clearly (Matthews, 1984, 1999, 2003, 2010; Willats 1997, 2005). I have not as yet discerned projective relationships in chimpanzee drawing and (though I could be wrong), I suspect it is not present. Of course, it would be fascinating if someone could contradict this.

Even so, in the present study, the first part of a developmental sequence we see in human childhood, from topological, haptic and dynamic thinking to the first primitives of Euclidean geometry can be seen in the actions the chimpanzees perform upon objects. In human childhood, an awareness of projective relationships is the last of a series of geometrical understandings to appear, when children begin to grasp the significance of view-specific information.

These emergent concepts – first, *going-through*, then *looking-through* – are extremely powerful in terms of the cognitive events to which they lead. Glimmerings of this type of thinking are present in the 4-year-olds' use of a video-camera (page 233), which of course depends upon an understanding of line-of-sight and point-of-view (Matthews, 2006a). (See also Figures 24.1–24.4, later in this chapter.)

Another pertinent example is offered by Betsy Herrelko's fascinating (2010) experiment in which she gives the chimpanzees a video-camera in a rugged box. The videographic results she obtains seem to be more the product of the chimpanzees' random handling of the box than of any awareness that they are video-recording or even just viewing a videographic representation. When I give video-cameras to 2- to 5-year-old children (Matthews, 2006a), over the course of weeks and months they learn naturally, with no overt instruction necessary, to use the video-cameras because they base its use on their emergent concepts of going-through and looking-through. They come to realise that they can look-through ('virtually') this electronic device in a fashion which emulates the way they can look through tubular volumes.

In contrast, in Dr Herrelko's experiment, the chimpanzees almost certainly understand the 'chimpcam' first and foremost as a *box* and are probably preoccupied with trying to break it open to see what is inside (despite Herrelko's care to insert the camera securely so that there is no tell-tale rattle). Chimpanzees are tool-users and if you give someone a tool, hoping he or she will use it then the function of that tool must be apparent or at least discoverable. I predict that, if the chimpanzees are given a more tube-like device, they will, like children, quickly realise that it can be (virtually) *looked-through* and then we will see our first genuine videographic representations made by a non-human person.

As we have seen, in their everyday actions, chimpanzees show an excellent understanding of line-of-sight and point-of-view, being able to hide from each other, both for real and in play and also sometimes (when they are in the mood) from me, in what looks like a 'peepbo' type of play. As well as presenting objects clearly within their own field of view (using usually their front-hands) they also know something about another's point-of-view and, if they want to be noticed, will intentionally insert themselves, or an attractive object, into another's range of vision. One example, is when a young chimp will purposely run across another's field-of-vision in order to provoke a chase. After 2 years 3 months of observations I am intrigued to find that some chimpanzees have started to insert themselves purposely into the range of vision of my video-camera.

Glass-eye

The chimpanzees look at the video-camera

In fact, I become aware that the chimpanzees are gaining some insights into my use of the video-camera only 2 weeks into my study. They are of course inured to having cameras pointed at them but I notice a marked development of their insights after 2 years 3 months 8 days of observation. Their insights are based upon understandings of the act of looking and of line-of-sight. I begin to notice that I sometimes unintentionally initiate a 'peepbo' play with them simply by my act of standing behind the camera and looking through its viewfinder. It dawns on me that they think I am hiding myself behind the camera. Sometimes this seems to worry them and so I desist. After all, in the wild, a partially concealed animal could be a potential threat.

In any case, I realise more and more that if I want to talk to them I will have to move slightly away from the camera and face them properly. This means that I will have to allow myself to miss some good shots and leave the camera on its tripod, pointing only approximately in their direction whilst I physically face them. It is worth it however, because I find they are more likely to interact with me in this way.

Later on, their ideas about the video-camera and my use of it seem to change again. Frans de Waal (2000, 2005) notes that chimpanzees do not like to be continuously looked at. My chimpanzees seem to start to realise that my use of the camera is a special part of my looking at them. The younger chimpanzees notice that I am following their movements with movements of the camera when I pan or track it after them. They apparently treat the tracking lens of the camera as a quasi-eye.

I learn to give them a break from my gaze, both naked-eye observation and the electronically assisted.

Coupled with the chimpanzees' understanding of line-of-sight and point-of-view is the dawning and related understanding of what is, in reality, *light projection.*

One indication that chimpanzees know something about the projected image is their interest in shadows and reflections. Dating from fascinating experiments by Gordon Gallup (1970) about chimpanzees and their use of mirrors there have been interesting studies which show how chimpanzees and a few other sentient beings become aware of their own self-image. First of all, chimpanzees respond to their reflected image as if it is another animal but, after some experience they learn that it is a picture of themselves. Only a few other animals, including magpies, humans and dolphins (see Reiss, 2010; Reiss & Marino, 2001) seem to show self-recognition (and therefore have an idea about visual representation) and even babies do not seem to realise for several months that the image they see in a mirror is a picture of them.

18 February 2007 Video-recorded child-observation (Edited DVD) plus written transcription, (Chloe) Singapore

I am video-recording my granddaughter Chloe (7 months) and I open out the LCD screen toward her so that she catches sight of her face in it. She gives a little start and reaches her left hand toward the screen.

00:03:00 She had been starting to cry but she stops suddenly and I say to her: 'Oh can you see your hand there?' To which she immediately responds 'Yeeh!' as her crying gives way to a cautious enjoyment and interest in what she is seeing.

She sees the relationship between her now controlled and purposeful movements of the hand and their visual consequences on the screen. At the end of the sequence it looks like she might be having some thoughts about the relationship of the lens with the LCD screen, for she looks into the lens and then back toward the LCD screen. (See Figures 24.1 to 24.4.)

My chimpanzees look at their own images reflected in water and I think they must see themselves vaguely in the reflective, painted surfaces of the wooden

Figure 24.1 Chloe (7 months) catches sight of herself in the outward-facing LCD screen of my video-camera (the LCD can be imagined just out of frame to the bottom left). Chloe's father, Ben, is partially visible to the left of the screen. Chloe tests the device by purposely rotating her left hand in front of the LCD screen.

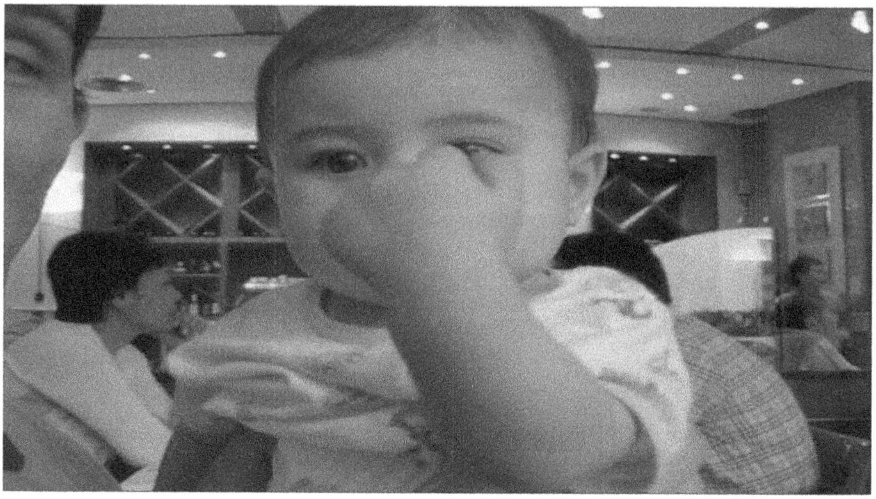

Figure 24.2 She closes her hand into a fist . . .

Figure 24.3 . . . and then opens it again, spreading out her fingers. She sees this in the LCD monitor.

platforms. I also start to become aware that they have noticed their images reflected in the lens of my camera. Other researchers report wild gorillas also intrigued by their own reflections in camera lenses (de Glanville, 2009).

Another way in which they find out about the projected or cast image is in their awareness of their own shadows.

Figure 24.4 She looks right into the lens of the camera and then back to the LCD screen. She may be thinking about the relationship between these two.

06 October 2005 (Part 2) Video-recorded observation (Edited DVD) Singapore

Nkosi leaps onto a plastic vine and, holding on with both hands whilst stretching his body as straight as possible, makes long pendulum-swings, seeming to look down at his own shadow changing shape like a pool of ink on the dazzling white rocks.

Suzie seems to be watching with curiosity the way in which I pan the camera around as I keep it trained on Nkosi's movements. It dawns on me that she has noticed that the panning movements of the camera immediately follow and keep track with the locomotion of the juvenile male. She is starting to be aware of the connection between my visual attention and my use of the camera. She seems to realise that the camera's movements are indices of my interest and watchfulness. She looks toward Nkosi, presently my videographic subject, and then pointedly stares at me.

An example like this suggests that Suzie appreciates the geometrical basis for different viewpoints (and so may form part of the necessary background for the making and sharing of the 'pointing' gesture, which occupied us earlier).

Ida, at her favourite Corner 4 on Level 2, is also staring at me. It is an eerie moment because it is as if these clever females are thinking: 'What are you up to?'

Chimps in liquid crystal

22 June 2007 (1) Video-recorded observation (Edited DVD) Singapore

It is not until 1 year 8 months 16 days later however that I make the decision to show them their own images in the Liquid Crystal Display screen of the video-camera. The LCD is a screen of 8 cm × 5 cm. Like many video-cameras of this period, the little screen can be hinged outward and turned in a variety of positions, including that of facing toward the video-recorded subject. It is to this latter position that I adjust the LCD screen today, so that it is facing outward to the apes.

Togo comes over to his favourite Look-Out place at Lever Log, apparently to check-out the outward-turned LCD screen of my camera. Nkosi looks toward me and he may also be looking at the LCD screen. 'Yes, Nkosi – can you see this?' I tap the screen with my left index finger and, sure enough, Nkosi fixes his eyes intently on the screen, stepping slowly, steadily, on all-fours, closer, to investigate the image he sees there. There is a look of wary but intense curiosity on his face. He does his typical, left forearm fling; hand swinging up to his eyebrows (an avoidance and displacement action? Or to test the relationship between his voluntary action and the image on the screen?). 'Can you see that?' I am asking.

Togo also rolls his head toward me, to look at me and at the LCD screen.

'Can you see it too, Togo?' I call out, but he wants to avoid face-to-face contact with me and rolls his head away from me. Nkosi echoes this facial aversion as if on cue from his dad. Togo turns his head away still further to look, questing, into the face of his son. Nkosi turns away and holds onto an overhead branch with his left hand.

00:26:54 Then Togo does something startling. He lifts into the camera's range of vision a bunch of twigs and leaves which he gently rotates in front of the lens. I see this event clearly in the centre of the screen when I review the footage. What makes this extra interesting is that I think he does this for his son to see. Sure enough, Nkosi notices and looks with great interest at the image of the moving twigs in the LCD screen along with his own portrait.

Nkosi walks up and along the projecting Lever Log, ever closer to me and the LCD screen, as close as is possible for any chimpanzee to get to a human on the other side of the moat. Perched on the end of the log, he looks into the screen and at his own portrait he finds there. Does he find the image on the LCD screen overwhelming? I had hoped he would stop for longer but he moves to the left along Horizontal Log as if in avoidance.

My granddaughter Poppy (2 years) also shrinks away nervously from her first sight of her videographic representation on a television monitor (Matthews, 2005). It is worth noting that though Poppy starts with fear, her much younger niece, Chloe, only 7 months old and in a similar situation, has no such reaction. This is interesting because it may mean that Chloe's sense of self and its representation is not as developed as Poppy's so it cannot be threatened by any uncanny appearance of its image. Perhaps it is the case that, because Poppy, at 2 years of age, is just starting to be aware of being a separate self that novel representations of it are disturbing in a way they are not for Chloe. Is this also the case for my chimpanzees?

25 April 2008 Video-recorded observation (Edited DVD) Singapore
(Edit 1)

It is about 10 months later. Today is the third time they will see their here-and-now portraits on the LCD screen of my video-camera. I intuit that the chimpanzees are now aware that these are images on the LCD screen are of themselves. I sense that they do not behave toward the images as if these are those of other chimpanzees.

Not surprisingly, their line-of-sight remains fixed upon the outwardly-turned screen rather than the lens (which, of course, is where their optical images are being received). Again unsurprisingly, any actions they make to test the causal relationship between their own actions and the effects on the screen are directed at the screen (and not at the lens).

Some 3- and 4-year-old children also behave in this way when they encounter the video-camera for the first time. In an attempt to video-record a favourite toy, one child holds it against the outwardly turned LCD rather than to the lens. He persists in doing this despite advice from an adult and – even more telling – despite the fact that the image of his toy in the screen disappears as he presses the toy against the screen, so dominant is his physical and topological understanding of connectedness. Meanwhile, other children in the room of the same age are ready to relinquish understandings based upon physical contact and replace these with the understanding that the camera's videographic relationship with the world is a purely optical one, based upon the line-of-sight from its lens (Matthews, 2006a). I predict that chimpanzees will also be able to move through a similar developmental sequence with videographic lens media – provided the use of the tool is transparent.

25 April 2008 Video-recorded observation (Edited DVD) Singapore (Edit 1) observation continued

Meanwhile, Ida gazes into the LCD screen as I gently talk to her. 'Hi Ida. Can you see it? Can you? Good girl.' She responds to my voice with little adjustments of her gaze and tiny movements of her face. She stares, rapt, into the screen. She looks away for a moment, to her right.

This is like the facial aversion chimpanzees practice when interaction with a human becomes overwhelming. The human baby also responds like this when meeting a stranger. We call it 'shyness'.

At this moment, Rara (3 years 4 months 18 days) finally becomes aware that her mother is looking at something in the distance. Ida is fixated on the forward-facing LCD screen. Rah has at last picked up the little cues her mother is giving. Looking at something fixedly may be a way that a chimp can make another chimp look in the same direction. In many circumstances there is no need of a pointing gesture. Rah crouches down by her mother and stares intensely at the LCD screen. Her eyes are wide and there is a little smile on her face. She is concentrating totally. She then moves upward and away and out of the visual field of the camera, to Sloping Vine above her. Then she looks directly at the LCD screen again. She huddles low on the vine and with increasing interest she stares at the image of herself on the screen. She clutches the vine more firmly with both front-hands plus her left foot-hand, the right foot-hand being let dangle below her. I zoom the camera in closely to her face, so she receives a big close-up of her own face. She scratches her arms, perhaps nervously, unsure of her response. She touches the fingers of her left hand to her lips. She stares into the screen. I am sure she knows that it is an image of herself but she cannot understand how this can be. She climbs down again to the horizontal support beam which is perched nearest me and leans as far as she can toward the camera from the sawn-off end of the beam, staring with evermore interest. I zoom even more tightly into her face, so she receives a very good, big, clear close-up of her own, staring face. Her lips compress slightly together, suggesting intense interest. She has a curious smile on her face, as if she knows I am playing a new kind of joke with her.

She purses her lips a little, and moves her hands, arms and head around a little, surely purposely, to test the image of herself against her own kinaesthetic sense. She now stills her body and face as I zoom into an extreme close-up. She suddenly backs away and turns her head and line-of-sight toward her left, as if suddenly a little frightened.

This is remarkably like the observation I made of my granddaughter Poppy (2 years) when she turns her face away from the image of her face on the television monitor. I think that chimp and child are both on the verge of an extraordinary realisation but one which is more than a little frightening (Matthews, 2005).

25 Shapes of things to come

Earlier on, when Nkosi managed to fix, temporarily, the bottle into a hole in the timber, he formed a *near right-angular attachment* or *join*. I mentioned that right-angular structure, along with the closed-shape are important Third Generation Structures in human childhood development and that we can also see these structures emerging in chimpanzee investigation and play. When Nkosi fixes the plastic bottle into a hole at right-angles to a beam he is essentially making a *Euclidean* geometrical relationship. The bottle does not merely touch the beam, as in a topological relationship, it is *joined* (albeit for a moment only). Moreover, its fixture into the hole by its bottle neck necessitates this rigid, perpendicular connection with the vertical post.

What other kinds of angular shapes possessing measurable, Euclidean properties can the chimpanzees make?

27 October 2006 (Part 1) Video-recorded observation (Edited DVD) Singapore

Rah makes connections between things

Rah (1 year 10 months 18 days) has found a tool-stick made by one of her seniors.

Other researchers have noted that chimpanzees will make use of a discarded tool (Sanz et al., Morgan 2009 – though whether this qualifies as a 're-cycled' tool, as the authors claim, is debatable). As we noted earlier, they seem to recognise when an object is designed and made by another primate, hairy or naked.

The tool she picks up now has a headdress of leaves left, by its maker, on its unsharpened end. [As I noted earlier, this is an intentional design feature made to prevent loss of the tool through a hole.]

Rah inserts the stick down a hole in the corner of Level 3 Platform and practises bending it back and forth. She balances the stick on her chin like a juggler. She allows the stick to drop through the hole in the platform and lets go of it, knowing that the remaining cluster of leaves at its end will prevent it dropping through the hole entirely. The stick penetrates the hole vertically and, sure enough, just hangs there, held in place by the bunch of leaves at its top which cannot pass through the hole. She climbs underneath the platform to look at the stick poking out from the other side of the hole. She bends the stick from underneath the Word Platform, using the lip of the hole as a fulcrum. In doing so, she forms both elliptical and angular shapes.

Minutes later she has another idea of using the long, thin stick to bridge the gap between the platform and an adjacent post about a metre away. She tries to fix one end of the stick into the hole in the platform and the other end into a hole in the nearby post. She retains one end of the stick inserted in the first hole whilst bending the stick over horizontally so that its other end can be inserted into the opposite hole. She almost succeeds in her attempt to attach the stick at both ends, using the technique of inserting each end into a hole but the stick jerks out of the hole in the adjacent post after staying firm only for a moment. She looks underneath the platform at one end of the stick which has remained in its hole, perhaps considering the problem. She tries again, again using the under rim of the hole as the fulcrum against which to bend the rod over to the other post. The stick bends into a near right angle. This time, instead of poking the other end of the stick into the hole in this nearby post she wedges it just above the hole, between the top of the post and a beam which crosses over it. The springiness of the stick supplies the tension required to keep it firmly in place and she looks at her work with some satisfaction. Here she is making use of tensile strength to help her achieve her end. [It occurs to me that certain human-made construction kits work on this principle in which rods, like Rah's stick, neatly 'ping' into place.]

The chimpanzees get good at gauging tensile strength from their many experiments of bending, plucking and twanging thin, flexible stems, plastic vines, rubber fire-hoses and wire cables. Nkosi has watched Rah's activities with interest and comes along to be with her, to involve himself in the same play and construction.

27 October 2006 Video-recorded observation (Edited DVD) Singapore

Of great interest to us is that Nkosi does not disrupt her approach to the play by introducing totally different ideas from hers. Indeed, by bending the stick about with her but without dislodging it from its hole, Nkosi, because he understands her project, adds to it. He knows what it is she is trying to do and feels he can help. Rah looks underneath the Platform to see the results of her big brother's technique.

This is real, deep 'scaffolding' or, better, 'synergy' – depressingly a rarity in much contemporary, institutionalised human education.

One of the techniques Rah uses in the above observation is that of levering a stick through successive *moments-of-turn* or *points of the compass*. This will also help her make shapes which have angles or corners. This is essentially the beginning of the Euclidean geometry which shows up clearly in children's drawing.

This is the kind of drawing in which the child feels it no longer sufficient to compose representations of objects as undifferentiated topological closures which capture *inside*, *outside*, *hollowness*, *adjacency*, *boundary* and *connectedness* but yet fail to show true shapes of things and spatial relations between them. Now the child wants to make carefully configured shapes; shapes which have measurable, straight lines, controlled curves and line-junctions with carefully calibrated moment-of-turn. This is important if you want to draw the corners of windows and doors and other so-called 'carpentered' objects like cars, spacecraft and houses. Such drawings are often composed of proto-Euclidean shapes, including near rectangles, triangles, circles, and diamonds and so on.

Waving and bending, twisting and turning

Angular-variation and moment-of-turn

Linked with new knowledge of turning, twisting and rotating, the chimpanzees also find out about *angular-variation* and *moment-of-turn*.

23 January 2009 Video-recorded observation (Edited DVD) Singapore

Kimoni

Kimoni (1 year 5 months 10 days) is trying to pull a long palm-stem up onto Level 2 Platform. He leans over the edge of the platform and pulls the bouquet of green leaves up towards him. Bending the green stem gradually over and onto the platform, he visually tracks the fanning action of the long, blade-like leaves as they pass through their individual moments-of-turn like dancers in a chorus line.

Seesaw

03 January 2008 Video-recorded observation (raw footage on DVD) Singapore
Tape 1 (1 of 2 tapes)

(Seesaw)

Rah (3 years 26 days) comes to acknowledge me immediately when I arrive back at the zoo after my two week Christmas break back in London. She recognises me. Suzie gives me a glance.

Rah uses the Seesaw Log. It is a relatively new installation, having been here for 2 weeks. This is the first time I have seen this being used by any of them with any enthusiasm or even appreciation of its affordances. Its use by Rah shows, firstly, that she is now strong enough and heavy enough to lever the log downward from its resting position which is at an angle of 50 degrees. Two or three weeks back she would not have been able to do this. It is a heavy log of about 20 cm in diameter by 4 m long. It hangs obliquely at an angle of about 30° from the vertical column to which it is attached. This Seesaw Log is attached by a tight coil of thick, plastic vine wound around both the worn groove at its middle and around the vertical support post. This sort-of figure-of-eight knot forms a squashy seat on which the big log sits. Upon this seat, the heavy, sloping beam can be levered or seesawed up and down – if you can exert enough pressure down upon it, that is.

Rah sits on the top end of this log and causes the log to swing down to the opposite direction of axis, at about the same angle of 30°. The continued observation shows me that she is able to sit at varying positions along the log so that it can be turned downwards at differing speeds. She can exert less or more strength on it, just as she desires, by hanging from it by her front-hands and arms, again at varying positions. Though she does not need to go to the extreme top end to make it turn through a downward axis, she nevertheless likes to do this, using one hand to grip the log's higher end in order to exert maximum force upon it and bring it down quickly. When it has moved down as far as it can go, she leaps off it, allowing this same end to spring back up with a jarring shock. She may also be doing this a little for me. Her speed and tempo seems to be activated by my return (after my break of 2 weeks), for she follows through her final seesaw with an additional embellishment; a lively zigzagging descent through aerobic chicanes which terminates, with a little punctuating bounce, at Yellow Puzzle-Box. I have a strong feeling that this performance is for my benefit.

During this session she seems to favour letting go the log so that it springs suddenly upright on its own. Next week I see she has learnt how to ascend whilst seated on the beam so that she travels up, very quickly, aboard the log, as if on an elevator from lower levels to higher ones. If she gets very close to the right-hand end of Lever Log she will always bring the beam down easily and quickly but she has learnt that by varying the distance of her sitting position in relation the fulcrum point of the seesaw, she can choose her own rate of ascent and descent, faster or slower as she wishes. She is operating a set of scales or a balance.

Moment-of-turn and angular-variation are also enacted with the chimpanzee's own body with few, if any, props. Some of these performances seem to have representational and expressive aspects.

23 January 2009 Video-recorded observation (Edited DVD) Singapore

Clowning, Ainka (1 year 4 months 24 days) seems to pretend to fall down sideways whilst looking at me. She lifts an arm up in the air (perhaps imitating me, for I am holding up my arm and hand in a greeting to her) and slowly pretends to fall sideways, through an increasingly oblique angle until she sprawls against her mother. I laugh at this and she repeats it.

'Falling down', if done skilfully, is found humorous by humans and has been performed in human theatre for hundreds of years. Again, the roots of this curious emotion, humour, seem to grow from a primaeval juxtaposition of fearfulness and relief from fearfulness. Unintentional falling is, in real life, dangerous and feels bad. Its intentional representation, on the other hand, feels good because we escape without injury. Its enactment can express sadness too, but both happiness and sadness can be made to feel good in primate-play.

Human children also play with this idea of turning through points of the compass. A person, falling from the vertical, 'standing up' position to a horizontal, 'fallen down' position is a favourite.

April 2009 Child-observation, hand-written, Singapore

'Run, run, as fast as you can, you can't catch me I'm the Gingerbread Man' (Part 2)
(Cafe)

Standing a Gingerbread Man biscuit upright on the plate and maintaining contact between the biscuit-figure's 'feet' and the plate, the 4-year-old Australian boy moves the Gingerbread Man slowly from this pivotal point, through 90° from its initial vertical position, as if the Gingerbread Man is slowly falling down, until the biscuit is lying horizontal on the plate. Later, he grasps his father's chin with one hand and his upper-lip with the other and speaks of an 'opening door', as he opens and closes his brilliantly cooperative dad's mouth.

Here, the boy is building up knowledge of *moment-of-turn* and *angular-variation*, the scenario held together by subject matter which includes moment-of-turn in both the act of falling and in the movement of opening and closing too, when the Gingerbread Man is finally devoured by a munching mouth. The father's imaginative support of his son's play, including allowing his son to move his jaw through moment-of-turn in pretend mastication, is crucial to its development.

Whether chimpanzees can layer play-episodes with such rich meaning is controversial but they do seem to supply the substrate for such symbolism in their investigation of moment-of-turn.

26 May 2006 (Part 3) Video-recorded observation (Edited DVD) Singapore

00:42:47 Nkosi, high up in the Construction, briefly looks toward me and, with the wet hessian mask over half his head but with his right profile presented clearly to me, makes a pretend and comic fall to the side, almost falling off his high perch but saving himself at the last moment. As he falls, he opens his mouth in a wide yawn of mock terror and flails his arms and legs around futilely in empty space.

Polygonal polypods

As in human childhood, the chimpanzees' investigation of moment-of-turn and angular-variation may lead to the construction of a variety of shapes which have *corners*. Sometimes, even gravity-assisted *polygonal* structures or *polypods* (Reynolds, 2002, p. 207) are made by the chimpanzees. In my observations, I do not see any chimpanzees *drawing* these shapes, as children do. Whilst chimpanzees enjoy rapid direction changes and bisection of lines, I do not see clear evidence in chimpanzee drawing in which the principle of angular-variation in linear junctions is abstracted from its origin in the act of abrupt cornering and used to make polygonal shapes. When children notice the effect of sudden direction change in a line, or when they find they can join two or more lines together at angular-junctions, they immediately appreciate the potential of this structural discovery.

Again, I could be wrong but it may be that the kind of psychological space the chimpanzees inhabit is predominantly a topological and dynamic one.

However, in their manipulation of other materials we do see the possible dawning of some aspects of Euclidean space. These include germinal parallelism, division and subdivision and acute direction change in line.

My observations in Singapore Zoo show that the chimpanzees are aware of, and use, the geometrical properties of the human-carpentered Construction and in the occasional human artefact they discover. In their play and navigation they use the *straight* sides and the *angles* or *corners* of rigid, geometrical shapes of platforms and beams.

Straight edges and straight lines

10 October 2008 Video-recorded observation (Edited DVD) Singapore

Kimoni (1 year 1 month 27 days) slides a white-coloured, forked stick back-and-forth along the straight edge of a platform. The stick neatly hooks over the edge by its forked branch and this junction guides its movement in a straight line. Kimoni stops to inspect this fork in the stick, clearly interested in it.

Like a template, the wooden edge engages and guides this newly formed 'tool' along a straight path. He regards with great interest the white stick's acutely angled junction and the way this fits neatly over the right-angle of the platform edge.

In my observation of human infants when painting, straight lines are frequently created in similar ways; the brush (for instance) encountering a straight edge (the edge of the table on which the painting surface rests, for example, or the edge of the paper itself, or the corners and crevices of an easel). Given more freedom, the human infant may use other features of the carpentered environment as templates which cause straight lines to appear:

08 May 1973 35 mm colour-slide film-recorded child-observation with handwritten transcription, (Ben) London

Ben paints a straight line

When Ben is 1 year 9 months, he paints upon a window-pane in my studio. His brush slides easily along the glass into the bottom left corner of the window-frame and as he moves the brush about, so the straight edge of the wooden frame acts as the template for painting actions. As with one hand he paints up-and-down, up-and-down, with his other hand, he simultaneously fans the window open and closed, open and closed, by pushing and pulling upon its handle; an action which seems to reiterate his painting and also a good example of functionally dependent moment-of-turn.

Nowadays, children will sometimes slide a stylus against a virtual edge or into a virtual corner on the glass screen of a computer. In all these instances, the natural arcing tendencies of the moving primate limb have been modified by a template's straight edge to become a *straight line.*

Tracing and co-linearity

17 November 2006 (Part 5) Video-recorded observation (Edited DVD) Singapore

Nkosi (6 years 4 months 20 days) trails a branch along the side of Horizontal Beam Level 3 from one end to another as a lure to baby Rah (1 year 10 months 20 days).

The Construction has many right-angled or near right-angled joins within it (although there is an hotchpotch of other angled junctions too). The chimpanzees are seen frequently to trace courses along the edges and junctions of the rough and ready carpentry, even occasionally tracing with fingertips the chiselled, wood-relief of block capitals on the underside of Word Platform. This platform, like all the others, is rectangular but this is to change after I have made about 14 months of observation.

21 March 2008 Video-recorded observation (Edited DVD) Singapore

When I come into the zoo today I see that the humans have made a significant change to the chimpanzees' Construction, replacing old timbers with new ones and two of the rectangular platforms with triangular ones.

00:03:02 Rah is on one of these new platforms (Level 2) and, with a loop of thread over head and neck, she drags a hessian sack in a straight line from one corner to another and back again. This is a familiar A to B route but then, as if a new thought strikes her, she takes the end of the thread to the third corner of the triangle, a new end-point, Point C.

From habit it seems, she starts a familiar run along an A to B route before realising that she can now use a third corner, Point C, to which she now runs. Rah transforms a linear (A–B) course to an A–B–C sequence; from a two-point, single straight *line* to a three-point course defining a triangular shape. Piaget said that shape is the product of action (Piaget & Inhelder, 1956). Perhaps we are witnessing the evolutionary dimension of this process.

In addition to discerning and using the geometrical properties of human-made objects the chimpanzees also make polygonal forms of their own with natural materials.

The first ways in which angular-variation and moment-of-turn are investigated is in the performance of early dynamic-structures described above. Repeated enactments of many different kinds of waving, thrashing and spinning all garner knowledge about angular-momentum and moment-of-turn. Sometimes a spray of stems is compressed together in the hands to form a straight bundle and then allowed to splay out again into a bouquet; the same object but shown in two quite different states (26 January 2007, part 1, video-recorded observation, edited DVD, Nkosi 6 years 6 months 29 days)

Other ways in which angular-variation is investigated is with bending and folding long objects, like palm-stems or long, bamboo poles.

05 May 2006 Video-recorded observation (Edited DVD) Singapore

Nkosi

00:19:51 Nkosi (5 years 10 months and 8 days) experiments with folding and bending stems (in order?) to make polygonal shapes. He brings a bamboo pole of about 3 m in length up close to the group. The pole is bent, broken nearly in half, so that when he holds it upright the upper portion flails around, describing a half-circle or thereabouts.

He sticks the pole onto the ground and holds it there vertically and then looks up to view the broken portion swinging slowly around above him, like a boom on a sail-boat. The other chimpanzees also watch it, though rather more warily.

Human children find out about angular-variation and moment-of-turn from similar experiments with any object which can be made to turn through, or show, points of the compass, a revolving chair, the hands on an analogue clock and any number of dial, segment-timer, thermostat and the like; the hinged LCD screen of a digital camera, a steering wheel or a bicycle wheel or just an ordinary door which can be opened and stepped through.

Doorways

19 January 2007 Video-recorded observation (Edited DVD) Singapore

Sitting side-by-side, Nkosi (6 years 6 months 22 days) and Rah (2 years 1 month 12 days) are intensely occupied in bending, folding and snapping the long stems of some new nest-building material provided by the Keepers.

This is a kind of bamboo in long lengths of about 4 m. In earlier observations we noticed that some stems and branches are pliable and bend more or less evenly to form curves, whilst others are more brittle and snap sharply into angles but without breaking completely. The stems now in the hands of the two chimpanzees are of the latter variety.

> Nkosi folds a very long stem into a large, roughly rectangular shape, with long sides about metre long and roughly right-angular corners. He lifts the whole thing up to a vertical position and, holding it upright like a door-frame, walks through it.

To achieve this he has used a modified nest-building manoeuvre which entails passing a long stem over the head and the back in order to bend it. The only difference here is that he omits the final, concluding part of the process.

> Instead of completing the building sequence by dropping the green frame over and around his back, he curtails the manoeuvre once he has raised it over his head and it has reached its vertical position.

The technique of *omission* (not quite the same as *curtailment*) is another way in which a usually routine action might be significantly modified for another purpose. In this case, by snipping off the last part of a nest-building process, the action is effectively stripped of its meaning and becomes a vehicle for expression alone. Simply omitting the concluding passage of this action-assembly enables the act of bending and shaping to be performed for its own sake, as an activity of interest in itself.

Nkosi uses a combination of familiar strategies here but adapts and alters them, freeing them from their normal connection to the goal of nest-building. It is the play-mode which allows him to liberate these strategies from their usual functional end-point and reassemble them for another purpose altogether, quite removed from survival or subsistence needs. If he has a goal at all it is the making of a form which pleases him plus the enjoyment of the actions involved in making this form, in this case a polygon and near-rectangle.

> Nkosi, still holding the green frame vertically, now gives a kick against its side whilst exerting downward pressure upon it too. There is a sharp snap and a rectangle instantly becomes a triangle!
>
> 'Well done!' I say to him. He shows appreciation, as he has before, to my words of praise. I talk to him about his construction as if he is one of my art-students. 'I thought earlier on that you were trying to balance it – when it was in the vertical position.'

Looking on from the visitors' path are two humans, a mother and baby. The baby is looking fixedly at Nkosi and at the green frame he has made and asking: 'Wazzat? Wazzat?'

I think that the human baby to be asking, not what the animal *is*, but what the animal is *doing*. Humans and chimpanzees seem to share an interest in the same forms of constructional activity. Even this little human notices that the chimpanzee has made something unusual and interesting.

Togo, who had been lying in the grass nearby also becomes interested and comes over to help strip bark off the long stems and to eat with Nkosi. Nkosi smiles at Togo and Togo smiles back. (They are reconciled over an earlier argument.)

One year and seven months ten days later, Nkosi is still interested in using the long, green stems to make polygonal shapes:

29 August 2008 Video-recorded observation (Edited DVD) Singapore

00:17:57 Nkosi (now 8 years 2 months 3 days) holds a palm branch vertically and stands it on its thick base whilst holding it upright in his left hand. Above him, the remainder of the palm bends sharply over at three points under its own weight, so forming an angular bower around him. He sits like this for a few seconds, holding the rod vertically before him like a King with a Mace, enthroned within a green proscenium arch; the rest of the stem kinked into a series of sudden angles around him.

Then he picks up a leafless, second piece of stem and momentarily touches one end of this at an approximate right-angle to the upright section in his left hand. He crosses this upright rod with his laterally aligned rod, a bit like a cellist putting a bow to a cello, these two sticks forming for an instant a cruciform shape.

25 October 2009 Child-observation, hand-written, London

A Chinese child (about 3 years old) is with his mother and father in Greenwich Park, London. His parents are searching the ground for sweet chestnuts. The little boy watches me as I play Tai Chi nearby. I think his dad must tell the child something about this Chinese art-form, for a few moments later the boy has hold of two sticks, each about 35 cm long, both held between the fingers of one fist so that they cross each other roughly at their middle points and at near perfect right-angles. He swivels this cruciform around in front of him making martial arts types of high-kicks. Later, he points it like a bow and says 'Hands up!' to his dad, who obligingly surrenders.

Later on still, the boy contrives to release one stick as if it is an arrow from a bow. This act is of special interest because the boy, knowing full well that he cannot actually propel the stick from a string (there being none) has to make do with *pretending* to shoot the arrow.

Such pretences arise from 'making-do' and this is one way in which representation starts and is practised by the chimpanzees too. Some of these instances we have seen already (for example, pages 103, 104, 105) and some are still to come.

Meanwhile, back in our observation in Singapore Zoo:

Nkosi, who for a moment seemed to toy with the idea of cruciform structure, now lets go the whole contraption and it falls to the ground. I think that Nkosi just had the dawning of an idea – but for a moment only. At the same moment he lets go his upright stem he lets go, along with it, his idea.

'Nkosi had some sort of idea then about building a structure', I note for the videotape, but by now the chimpanzee is stripping down the palm stem in his usual way. For a moment, I had thought he was pondering the right-angular cross-over of two sticks in his hands, attracted to this structure as are human children, but the moment passed.

Though perhaps not quite. He again picks up the main branch – the one with which he just now formed a polygonal frame – and stands it against the deck, almost vertically before him, with an attitude reminiscent of that of a moment ago. I wonder if he is going to resume his building and this prompts me to ask him: 'Have you got some sort of idea there?' He holds the staff upward to his face and surveys its end.

Then he starts stripping pieces off the stem and this is the moment when he finally lets go any last residual traces of the idea he might have had moments ago. He simply reverts to a chimpanzee's usual way of stripping the palm-stem down. Ending up with a small, butter-coloured rod, he chews its soft core, sometimes holding it before his eyes to look at it.

As has happened so many times before, a young chimpanzee seems tantalisingly on the brink of a discovery but is unable to sustain the new operations required and reverts to a more routine system.

Moments before, he seemed to be on the verge of constructing a polygonal shape. Just as had happened before. Just over 19 months ago. Nkosi became interested in the way a rigid, polygonal frame could be vertically aligned, resting on the ground, and that he could pass through it as if it were a window or doorway. But also as before, he finds no way to support either the physical structure or the idea embodied within it, and he lets the whole thing go, palm-stem, idea and all. Perhaps, lacking language, he had no way to sustain and store the idea for long enough to work upon it.

Storing information about shape

Drawing is a good way of storing information about shapes. In human childhood the discovery of the angular-join, the cruciform and the 'corner' are momentous discoveries made in the act of drawing (and only later in language). The child achieves and sustains much conceptualisation about space and time through drawing. Long before the child knows the geometrical terms 'right-angle', 'rectangle', 'polygon', 'parallelism', 'symmetry' etc., he or she has developed the concepts for all of these through drawing. (Note well that by the term 'drawing' I mean spontaneous, *free* drawing. I differentiate this from much of the brain-deadening drill which passes as 'art education' today).

The chimpanzees learn about angular-variation and moment-of-turn but they cannot practise or ponder this in the hypothetical space generated on a drawing surface – at least, not in natural or semi-natural conditions – at least, not as far as we know. Whilst the child, like the ape, may not be able to join or fix two parts together in the three-dimensional, physical world, the child – unlike the ape – may, in two dimensions (without nails, glue or knots), form – and so be able to think about – the structure of the *join*.

Nkosi momentarily glimpsed a right-angular association between two separate, physical objects but had no way of 'fixing' it, either physically or metaphorically.

The Chinese boy was able to form a temporary cruciform shape by holding two sticks crosswise between the fingers of one hand. This is not as easy as it sounds, by the way, but I have observed chimpanzees hold sticks in similar ways (for example, Rah, age 2 years 6 months 15 days, on 22 June 2007, raw footage on DVD at 00:28:51), but Nkosi does not do so here. Is this because, whilst Nkosi has the manipulative skills to do this, the structural concept, so clearly the driving force in the boy's play, is not strong enough to sustain Nkosi's?

For the child, the concept of the *join* is revolutionised with the growing ability to vary and control the angle of attachment in the two-dimensional world of drawing. This understanding converges with the new-found strategy – also formed in drawing – of making polygonal shapes. The child is now moving on from an essentially topologically characterised space, to one in which the exact shapes of configurations, and the precise degree of moment-of-turn in any corners of polygonal shapes, are increasingly perceived as important. This is the beginning of an appreciation of the uses of *Euclidean* geometry.

The chimpanzee also seems to glimpse these geometries in experiments with structure which, to me, seem no less intense and thoughtful than those of the human child. What the chimpanzees do not possess is some means to store these ideas 'for the time being', as it were, and return to them. Even the Keepers' daily tidying up of the enclosure militates against any sustained development of an idea.

As we noted earlier, if the materials were left as they are, just as the chimpanzees abandon them at the end of the day, perhaps they might be able re-work the ideas embodied within them.

Surprisingly complex structures have been developed by very young children in this way, especially when given the supporting interest of a watchful and sympathetic human observer. If drawing and construction materials were made available, along with a constant and sympathetic human companion, might not captive apes be able to develop their thinking too? Surely this is a valuable way future research could go.

26 The dynamic knot

In this chapter we consider other ways in which the chimpanzees show the glimmerings of ideas about how to join things together (in order?) to make new structure.

Hanging-over and going-over

We observed earlier that repetition of the same type of action across different media results in permutations of the same class of dynamic-structure. For example, Rah, from around 2 years 6 months of age notices the similarity between the different instances of bending a saliently long object (branch, long leaf, long blade of grass), over a fixed, stable object. Despite differences in its manifestation, she knows that this dynamic relationship remains essentially the same and she wants to investigate the different forms it may take. We have seen such systematic investigation of structure many times so far, in the various ways things may go up-and-down, go-through, go-over and go-around.

In my observations of children investigating these exact same dynamic-structures, they too pass the same idea through a similar series of transformations, noting that which changes and that which remains unchanged.

The line

The young chimpanzees start to take note of objects which are saliently long and thin. They sometimes seem intent on making a *line*; a 'line' in the sense of a thread, a string or a rope. This would be in contrast to making a rigid type of tool. In tool-making activities, perseverance in the act of stripping down twigs, sticks and grasses sometimes results in an extremely pared-down tool, which, though it retains some of its self-supporting resilience, is no longer like a rigid stick but more like a thread or fishing line.

On some occasions, the chimpanzees will thread such a line through one of the holes in a Fake Termite Tower, presumably because with a line (which retains some tensile strength but is not rigid like a stick) they will be able to negotiate the little kinks in the tunnels which penetrate the tower. The process is more like *threading* than just poking. Here's an observation of Manzu, when he is 7 years old, selecting and using such a line.

03 November 2006 Video-recorded observation (Edited DVD) Singapore

Manzu, on his way to Tower 2, picks up the unusually hard, straight stick he was investigating minutes before [and which we discussed earlier on page 48]. He considers it briefly before discarding it in favour of a smaller, wand-like stem. Taking its leading-end, or what rope-workers refer to as its 'working-end' (Budsworth, 2004) in a pincer-grip, he threads this very long, thin stick into a hole in the tower.

Such strategies are made possible because of the emergent concept of *going-through*.

Chimpanzee tool-making is not a clumsy, crude activity as some might assume. Though all the work is done with the hands, fingers, nails and teeth, the tools they make are, if not precision instruments, then skilfully crafted artefacts in their own way. Chimpanzee dexterity allows numerous variations on nest-building and tool-making behaviours. This may lead to the emergence of new objects and the new ideas which go with them. In the present example, the continued paring down of a stick may draw the chimpanzees' attention to the *linear*.

Consider the following observation of Nkosi in which he is using a length of rubber fire-hose like a rope for tying.

The knot

17 March 2006 Video-recorded observation (Edited DVD) Singapore

Nkosi (5 years 8 months 20 days) takes the fire-hose around turns of the uppermost branch of Faraway Tree by throwing himself over the branch and carrying length of the hose with him. Such leaps might be thought of as the first movement in knot-tying but instead of just the fingers alone carrying the working-end of the line, the knot's configuration is described by the entire body moving through space, the working-end carried in the hands.

Sometimes he makes turns of rubber around the uppermost branch which form 'loops' he can tighten or loosen as he wishes, so making small loops or big, long ones. He can also hold the working-end of his rubber line up simply touching against a section of the curve of hose in his other hand and so arrive (by another route, as it were) at a slightly different type of *closed-shape*. Rah, only 1 year 3 months 10 days old at the time, comes along, watches him and she jumps onto the fire-hose as if trying to emulate what her big brother is doing but then, as if appreciating that he is performing skills still beyond her understanding, makes way for him to take over.

It is 1 year 3 months and 5 days later and Rah, now 2 years 6 months 15 days of age has learnt a great deal from Nkosi about coiling rope-like lengths over beams and branches. Branches themselves, especially very long, thin branches, may also be thought of as 'lines' – lines in the sense of threads, strings or ropes.

22 June 2007 (1) Video-recorded observation (Edited DVD) Singapore

00:02:00 High on Level 3, Rah is pulling at a long, leafy, whippy stem. I realise after a while that this stem is a limb of the newly installed Hanging Branch. [A 'Hanging Branch', you will recall, is typically a big, leafy branch hung upside-down by the Keepers from Level 3 by a chain and padlock (at Padlock Point) and on which the chimpanzees can swing and climb. (It looks a little like an upside-down tree).]

In this most recent version, some of the branches are thin and flexible and still bear leaves; her favourite kind, spear-shaped and brightly coloured, this type green. Rah is trying to pull one of these long branches from its tip – the end furthest from the point at which it joins a bigger branch which is about 2 m below the stump of the main branch, secured to the Horizontal Beam at Level 3 by the chain at Padlock Point. She is trying to pull this thin branch from its extremity, its working-end, back up to Word Platform and over the other side of Horizontal Beam Level 3 (that is, its further side from my vantage point); a very difficult project and one she has tried several times before. It is difficult because it entails bending the branch back upon itself,

Figures 26.1–26.4 The dynamic knot. Rah tries to coil a slender branch around a horizontal beam.

Figure 26.2

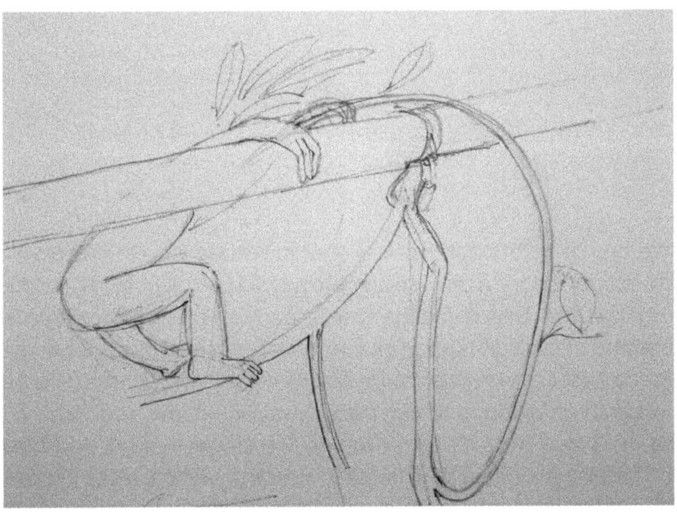

Figure 26.3

against its natural inclination to droop downward, directly away from the beam. Additionally, she has to contend with the weight of the rest of the Hanging Branch to which her thin stem is joined (see Figures 26.1 to 26.4).

Rah has to maintain a firm grip on its end, because the further up she tugs it, so the more its tensile strength increases, working against her, tending to make the thin branch spring away from her. She succeeds in bending the

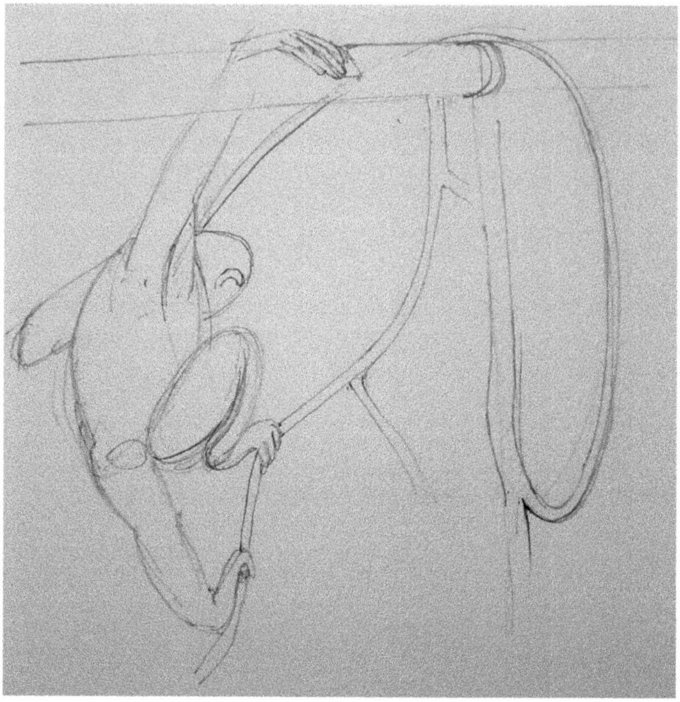

Figure 26.4

flexible little branch right back, up and over in a huge loop to the Word Platform. With extraordinary agility she improvises strategies to overcome the branch's springiness. She holds the stem steady by biting on a single leaf at the working-end of her thin branch or 'line' so freeing her hands for a moment to allow her to reposition herself more solidly on Word Platform. The branch is all the while threatening to spring back down again and, inevitably it does so, tearing itself free from its flimsy tether, the single leaf gripped between her teeth. The branch whips downward to become vertical again, joining the rest of the upside-down tree; its leaves once more hanging downward. With torn leaf still in her teeth, Rah seems to abandon the task. She appears to get engaged in other activities.

But she cannot put her plan out of her mind. A little later she is once more on Hanging Branch trying again to bend the same, long, leafy stem upward from its junction with the thicker branch. Once more, she is successful in pulling its end right up to Word Platform.

As ingenious as her previous strategy was of biting on a leaf at the end of the stem in order to leave her hands free for climbing, the fragility of the leaf meant that it was bound to fail. The problem is that, if she grips the branch firmly the same vexing problem simply returns; that is, as she climbs, so the tensile strength of the branch increases proportionately, threatening to make it whiplash back down again.

She solves the problem by loosely cupping her hand around the branch, rather than gripping it tightly, allowing the branch to pass through the 'collar' or 'sleeve' formed by her coiled fingers as she herself travels up along its length back to the platform. In this way she reaches the Horizontal Beam whilst not losing the branch but without increasing its tension either, which would unbalance her. Once stable on the platform, she can of course resume her tight grip on the branch's extremity.

Implicated in this achievement is the emergent concept of *going-through.*

Sadly, in her manoeuvrings on her precarious perch she loses hold of her bendy branch which swishes away from her and back down again, leaving her with only a single leaf from its tip. She does something interesting with this leaf though. She slides it along the top of Horizontal Beam Level 3 just above her. The leaf is a longish one and she runs it crosswise over the Horizontal Beam keeping its surface pressed flat to that of the beam so that it follows its circular form.

I think she is performing a miniaturised rehearsal, or 'model', of her real project which is to pass the entire bendy branch (of which this single green leaf is but a member), over the beam. Rah knows that the chain passes over the beam and now she practises this *going-over* dynamic in a miniaturised form, at a more manageable scale, with one of the bendy branch's green leaves.

Now Rah climbs down to Hanging Branch again. Here, she takes hold of the same, supple branch and tries again to take it up with her to overhead Horizontal Beam.

00:03:30 Hanging on with her left hand to the overhead beam, she places her right palm-sole on the base of the pliant stem at a position just a few centimetres from the place where it grows out of a bigger bough of the Hanging Branch. She presses down with this foot-hand on the bendy branch as a fulcrum for leverage whilst at the same time pulling the branch from a position further up its length with her right hand.

She ascends to Level 3 and tries to pass the long, supple stem through the groove in Horizontal Beam through which runs the chain securing the Hanging Branch. She is not successful and, losing her grip on the stem, it whiplashes away from her and down again.

The long branch has got itself lodged on Swinging Platform and Rah reaches down with her right front-hand to retrieve it.

00:03:43 She hauls the springy stem back up again to Padlock Point and tries again to slot it into the channel through which the chain runs. She has difficulty because of her precarious position plus the tensile strength of the bendy branch. She wrestles with the stem for a second or so as if fighting with a wriggling serpent which nearly succeeds in dislodging her from her perch. She stands up bipedal and, whilst clenching the branch in her front-hands, hauls it along Horizontal Beam. She adjusts her position so that she can walk backward, still in bipedal position, pulling on the branch. This gives her better leverage but also makes her prone to stumble. She is trying to coil the branch over the beam but this time at a position a metre or so away from the groove at Padlock Point. It may be that she is hoping that from this position the springy stem may just slip back into the groove.

The next part is interesting because it looks like she appears to lose her footing and fall off the beam – over the far side of the beam from my position.

I think that she was planning something like this anyway; to jump over the beam carrying the stem with her but, instead of the precise pitch she had planned, she tumbles prematurely.

She manages however to retain hold of her long, whippy stem and carry it with her as she lands on her palm-soles on a bough of Hanging Branch. Though her fall is somewhat accidental she succeeds in pulling the stem right over the beam by a good 2 m or so! The long whippy branch now forms a graceful bow, several metres long, curving over the Horizontal Beam. Now, tugging it a little more she even guides the stem into groove in the timber! She continues to pull the thin stem even further out with her left hand, leaning over to her left as far as she can possibly go. She does not seem satisfied however; perhaps she does not feel she has sufficient purchase on the stem; perhaps she would like to pull it out even further and cannot do so from this position. So she allows the stem to re-run part of its course back along the groove in the beam and, after adjusting her own position, pulls it once more outward and down the far side of the beam.

Rah clearly tries to encircle the beam with the pliable branch. She pulls the stem tightly downward so that it runs precisely into the deep channel in the beam like a cable through the wheel of a pulley. She hauls the stem up and around the other side of the beam – the opposite side from the one down which it descended. She is trying to coil it around the beam in the same way as the chain is coiled around it.

Frustratingly – for her and for me – as she attempts the climb back up to the beam with the stem in her hands it slithers from her grip and springs, whip-like, away from her. It would have resumed its upside-down position along with the rest of Hanging Branch had it not got trapped again on Swinging Platform.

00:04:08 Tired and perhaps dispirited after her efforts she rests, slumped face-down over the Horizontal Beam, arms and legs flopping limply. Moments later, whilst hanging onto the Horizontal Beam with both front-hands, she has recovered and she takes hold of the stem with her left foot-hand and pulls it toward her. Letting go of the beam with her right hand, she leans way over to her right and transfers the stem from her left foot-hand to her right front-hand and carries it up to Horizontal Beam. Once there, she sits down upon it, thus preventing it from escaping, the foliage at its tip bunched in front of her and some of the leaves pressed into the groove in the wood through which the chain runs. She looks down closely at this groove and, with her right thumb, presses one of the leaves into the channel, as if, with the use of a small-scale, demonstration model, confirming the viability of the project.

00:04:32 Carrying the tip of the stem – the working-end of her line – she walks a step or two along the Horizontal Beam and throws herself off and over the other side of the beam, carrying the long thin branch with her. This time her leap is made purposely and with precision.

Frame-by-frame analysis reveals that she skilfully combines the pivot formed by her left arm and hand hooked over the smooth beam with the tensile strength of the branch itself to translate her headlong dive into a near-circular course over the beam and underneath it, partway around again, so that she ends up almost directly below her point of departure – where she started her leap.

She alights on a strong bough of Hanging Branch still holding onto her long-fought-for stem all the while. In this, her latest attempt to coil the long, whippy branch around the beam by carrying it with her on her leap into space, she has almost formed a complete *closed-shape*.

Such leaps might be thought of as the first movement in knot-tying but rather than the fingers alone carrying the working-end of the line, the knot's configuration is described by the entire body moving through space and carrying the working-end with it in the hands.

Sadly, after all this effort the pliant stem breaks free from her grip once again and springs back down. She looks back at it where it once more snaps to a standstill, on top of Swinging Platform.

She bounds along Horizontal Beam to Corner Post 4 and tucks herself down into a small, black ball on Word Platform. She picks up her remaining leaf and chews it.

00:05:17 Rah looks at me again and I respond with a 'Hi, Rah'. At this, she leaps to the base of Hanging Branch, just near the place where it is attached by the chain at Padlock Point and, without pausing, leaps back to Corner Post 4 around which she swings, pole-dancer style, on an elliptical course, propelling herself once more along Side 3 of Word Platform.

This is a variation of a previous circuit and again I get the impression that it is a reiteration in another form of her efforts to encircle the Horizontal Beam with her long branch. Her orbit concludes with her heading toward her mother but stopping short of her and sitting down in the vicinity of Corner 3. She is still chewing her last leaf and she once more looks into my watching face.

00:06:33 Just when I think she is going to grasp the longed-for branch once more it she seems to change her mind and makes what looks like a last despairing pass at it with an outward fling of her right front-hand before finally abandoning the task.

Her perseverance in her work is remarkable. Even when she takes a rest from it she surely retains it as an *idea* in her mind and eventually resumes work on the problem. Her activities with the bendy branch involve the emergent concepts of wrapping, coiling and bending. These are present in her techniques of nest-building but her present activity is a long way from nest-building. Her project is a bit like a version of the 'hanger' too but it is more advanced than this device involving as it does a complete encircling of the post rather than a simple hanging or hooking over; it involves a *loop* rather than just a *turn* or a *bight*.

Since the time that the young ones achieve self-locomotion, I have seen them many times stop to inspect the silver lock and chain used to secure the Hanging Branch from the platform. I named it 'Padlock Point' because it seemed so salient a landmark for the chimpanzees on their circuitous play-routes around the Construction. Rah has an inkling of how and why the chain which carries Hanging Branch is associated with this channel and she is surely trying to learn from this exemplar in her attempts at dynamic knot-tying. Just over 6 months later, she will show interest in a similar groove cut for a similar purpose by the humans into a vertical post on which the Seesaw Log is to hang (03 January 2008, 'Seesaw', tape 1, video-recorded observation, raw footage on DVD, Singapore).

Such grooves or channels fascinate human children too. They figure in some of their popular imagery. A famous example is featured in the George Lucas film, *Star Wars*. This is the circular channel which runs around the 'Death Star' and through which spacecraft loop themselves in lines in space.

Later on in the day I notice Rah experimentally wrap a long, stringy, straggly, grass blade over the sawn-off top of a post; another example of the dynamic-structures, going-over and going-around and surely the fading memory of her ambitious knot-tying project.

Connectivity and attachment

Rah appreciates that the chain secures the Hanging Branch in some way she cannot understand. This means that she has glimmering concepts of *attachment* or *connectivity*. These are two overlapping concepts we met earlier when, for instance, Rah (page 253) and her brother Nkosi (page 255) both attempted, in their different ways, to fix a plastic bottle into a hole and when Ganga (page 255) jammed a stick into a crevice. In human childhood the emergence of these concepts is associated with the onset of Third Generation Structure, when (for example) children and become interested in forms of connectivity in the environment (a dangerous interest in electric cables and plugs is one example), and when they start to attach lines one to another in drawing.

The knowledge Rara needs is beyond her zone of proximal development. She needs to understand how to tie a *knot*. Is this really beyond the understanding of any of her specie?

Snagging a line: Untangling the mystery of the knot

I have never seen a chimpanzee tie a true knot but in this observation Rah comes very close to doing so. A knot is a sort of ordered 'tangle' and new ones are discovered from time to time in human rope-workers' craft (Budsworth, 2004). Rah knows a lot about tangles and how things get stuck or 'snagged'. For example, she notices such occasions when her bendy branch gets trapped and tangled. The same cluster of concepts involved in Rah's activities with the long, slender branch (coiling, wrapping, going-over), is also employed by very young human children when they attempt the primitive precursor of a knot. As with Rah, these concepts emerge from random tangling:

For example, Ben (3 years 3 months) wraps string haphazardly around toys in attempts to tie them together. Serendipitous knots are sometimes produced in this way. In his play scenarios, the physical links he makes between things often unconsciously symbolise emotional ties, connections and knots. He wants things and people tied together so that they cannot come apart, part company and get lost (Winnicott, 1971). Whether the tangling play of the chimpanzees has similar deeper levels of meaning we just do not know. Chimpanzees are complex, social and emotional beings so it is not out of the question.

In the observation above, when Rah wraps the thin branch around the beam, she is acting in a more developmentally advanced way than Ben when, hopefully, he winds string around toys. She is busy untangling the mystery of the 'knot'. That Rah even considers the need to join or attach something together with something else is surely of extreme interest. Perhaps future research might study whether apes could learn the mechanics of the knot. Piaget writes about the concepts needed to tie one and Bruner suggests that in order to do so, a kind of representation called *enactive representation* is necessary. This internal, mental model captures the movements required to configure a knot or other complex form in time and space. For example, it is an enactive representation humans use when they tie a shoe-lace (Piaget & Inhelder, 1956; Bruner, 1964, 1966).

Rather than try to get an ape to master unnatural tasks (as they are often required to do) it may be more realistic and fruitful to study how far along a developmental route an ape can travel toward formation of enactive representation. The sophistication and complexity of their navigational skills in three-dimensions of space and time suggests that they are close to forming such understandings. Rah's manoeuvrings of the thin branch are probably based on *enactive representation.*

Note that, once more, we see a chimpanzee taking advantage of what was initially an accident; when she firstly tumbled off the beam whilst retaining hold of the tip of her branch. The intelligent primate takes note of any useful *fortuitous* structure which might transpire. Really, to be more precise, we need another term rather than 'accident'. Like the human baby, the chimpanzee sets out in play to purposely randomise events, so to create a rich pool of possibilities. This process of randomisation is interpenetrated with conscious gleaning of any structural principles which may be present in fortuitous configurations. Rah's first fall, for example, is probably not wholly unintentional. She is probably operating at the edge of control. Children take the dual process of randomisation and careful investigation further in their use of graphic representational media. I anticipate that chimpanzees would be able to do this too and there is surely a suggestion here for a powerful approach to research on primate intelligence. Although Rah does not possess graphic technologies of image-making with which to test her ideas it may be that she employs another kind of 'analogue space' (Wolf & Fucigna, 1983) when she thoughtfully passes a single leaf over the curvature of the horizontal, wooden pole, or when she tries to tuck a leaf into the groove cut into it at Padlock Point. This is similar to how a human sculptor makes a small maquette in preparation for a larger sculpture.

Three months, twenty-seven days later, Rah is still developing understandings about encircling something with a line. The following episode is built upon her experience described above, in which she jumped over the edge of the platform with the whippy stem held in her hands like a thin *line*, apparently trying to encircle a timber. I suggested that this action, made with her whole body in space, is the first linear trajectory involved in the tying of a knot. Now follows some confirmation that this is the case.

The logic of the Knot

19 October 2007 Video-recorded observation (raw footage on DVD) Singapore (Tape 1)

Rah is now 2 years 10 months 12 days old. She is pendulum-swinging from the fire-hose which hangs from the top of Faraway Tree and intercepting Suzie each time the older female comes along Bridge Log toward her.

The next part of the observation shows Rah still holding onto the fire-hose but now whirling around the top of Faraway Tree whilst carrying the old, flattened hose-pipe with her as she goes. She is winding it around and around the top of Faraway Tree by moving her whole body up and around the top of the tree whilst holding onto the strap and taking it with her.

She was introduced to this technique 1 year 7 months 12 days ago by Nkosi (17 March 2006, video-recorded observation edited DVD, Singapore, page 283) with this very same length of hose at Faraway Tree. She was only 1 year 3 months 10 days old and at that time she knew only to jump up and down excitedly on the rubber-strap. Now she understands the principle behind her brother's actions. As she swings up and around the top fork of the tree so she coils the strap around it.

Rah repeats this action several times, sometimes completing the circuit and so succeeding in coiling the strap around the fork formed between uppermost branch and top of tree. Soon, there are several coils of strap wrapped around the forked branch.

Another important consequence of her whole-body winding technique is that the remainder of the strap, dangling vertically downward, becomes gradually shorter with each successive wind she makes. She is 'winding-up' the rubber strap. The relationship between her winding action and the gradual shortening of the hanging length of strap is another *functionally dependent* one (Athey, 1990/2007). The apparent 'shortening' of the strap is, in actuality, the gradual upward movement, by small increments, of the working-end of the strap.

She swings up again, again holding the strap in a foot-hand but does not complete the circuit of the branch this time and descends to the grass.

Rara later sets free these sorts of understandings in an exquisite linear adventure made when she is 4 years 5 months 6 days old:

13 May 2009 (Part 2b) Video-recorded observation (Edited DVD) Singapore

Loose-thread

00:10:53 From the frayed corner of a hessian sack, Rah has torn loose a very long, white thread, perhaps 10 m in all – it just keeps unravelling – which she carries through the Construction, trailing it behind her, clearly interested in the way it wraps, winds, twists and tangles around things as she moves but yet still follows her. She uses her teeth and sometimes her hand to pull the string.

She brachiates along Slack Vine, the long length of thread flowing out behind her. She carries it by its end down to Twisted Tree and then on to the middle of the parallel pair of Slack Vines which run from Twisted Tree to Faraway Tree. She stops at about the midway point. She knows how to find the middle of the vine by using one those measuring techniques we considered earlier. She can tell she's at the middle of the Slack Vine between the trees because her weight drags it down into the characteristic 'V' shaped configuration at that precise point.

She trails the white thread all the way along the bridge to Faraway Tree, where she hunkers down; the end of the thread in her teeth. She pulls it tautly down through her teeth with her left hand, 'dental-floss' fashion, and then she pulls it out tight to her right side with an outstretched right hand and arm. When her right arm is fully extended and she cannot pull any more thread through her teeth, she adjusts the position of her hand on the string, pulling at it from a fresh starting point close to her teeth and slowly stretching her right arm right out to the right once more, thus pulling another arm's-length of thread through her teeth. She measures out three arm's-lengths in turn this way; methodically, one length at a time. It's not unlike how a human builder might unroll a length of rope or cable in measured lengths from a spindle formed by upper-arm from elbow to hand.

Holding the string in her left hand, she reaches up as far as she can into Faraway Tree. She stays for a few seconds and then brachiates off at a brisk tempo, through a series of arcing loops along the double Slack Vine, the sinuous line of white thread flowing behind her in the unusual breeze.

Slow-motion analysis of her passage shows that, despite the speed of her flight, she still manages to manoeuvre the string into different positions as she moves, using her free foot-hands and sometimes a front-hand to make the light thread billow in delicate curves in the air. Whilst we noted the prominent role played by the front-hands (like us humans) in some activities the foot-hands 'come in handy'.

Requiring high-speed, in-flight calculations she makes skilful adjustments to the way the white thread trails behind her in order to watch the effects, which she looks back upon even whilst yet flying forward. She 'pays-out' the line from one foot-hand to another, using the chimpanzee's unique pincer formation of the foot-hand and, as each front-hand is alternately freed from its role in brachiating, she uses it to adjust the flight of the white thread. As she approaches the outskirts of the Construction, in midflight and with a cupped, left hand she sweeps up a long, lacy curve of line, gazing at it. Her facial expression is one of joy.

Eventually, Rah's line links chimpanzees Ganga and Kimoni like black points on topological map, both of whom are intrigued by the geometrical shapes formed by Rah's movement. Finally, Rah ensnares the entire wooden Construction, like a fishing line around an ancient shipwreck in the depths of the sea.

27 Drawing and writing

An important part of human evolution and development has been the recording of ideas about space, location and movement onto flat surfaces of various kinds. I have hinted at such a possibility in the chimpanzees' behaviour and it is now finally time to look at chimpanzee mark-making and drawing more deeply and thoroughly.

In human children, the beginnings of drawing show up in important precursor movements which, first of all, do not look anything like drawing at all. Even before any mark or trace is made, actions of the hand coordinated with visual attention make 'descriptions' in empty space. In the first instance, the hand is empty. Later on, a handheld object is driven or flown around, perhaps as an extension of the self. Later on still, a mark-making instrument of some kind is held and actions which were previously invisible leave a trace of their passing. They are recorded in marks and shapes on surfaces. Children quickly become aware of and very interested in these trace-making effects (Michotte, 1963). The consequences for human evolution have been enormous.

Chimpanzees and other apes also show abilities in drawing and painting. During the 1950s and 1960s there was a flurry of excitement about this. But then serious questions were asked. Are these non-human primates really interested in drawing? Are they really aware of the significance of their efforts?

As soon as human children start to make marks they become interested in them. Perhaps not with the very first unconscious dashes they might make as babies but soon after this, when they are aware of what has transpired. They quickly exploit the expressive characteristics of mark-making and it is not long before they realise its potential for representation.

When they make their first marks they enjoy an extraordinary process through which, and through their own agency, something appears out of nothing. In itself, this phenomenon seems striking to the child. Even during a phase of development which was, until quite recently, disregarded as mere motoric marking (the so-called 'stage' of so-called 'scribbling'), they are already thinking about the making of meaning.

But is any of this true of apes? For many human children, when they make marks and they know that they are doing so, they are almost instantly captivated by this activity. The action of leaving a trace is interesting enough to want it repeated. It is not long before they become interested in how to organise shapes and colours and think about how their marking actions and the marks themselves might refer to events and objects quite external to the act of drawing and the drawing surface.

Children's play seems similarly layered in meaning. Given that in this book I have described the chimpanzees producing strikingly similar actions in their play and investigation, the same question arises as to whether these actions are as loaded with meaning for the chimpanzees as such actions are for children.

The physical act of mark-making is a visible event whereas the earliest expression of meaning occur in actions which leave no trace. In the beginning, meaning can only be inferred. We make this inference about children's actions because we know the teleology of their development. In some of the baby's earliest actions, though we suspect the presence of communicative meaning we have no way of knowing for sure. Even so, we do know that eventually, in very similar actions, there will be meaning.

Even before they start talking there are many reasons why we know the human infant's early vocalisation bears meaning. When they do start talking this confirms our earliest impressions. Of course, there is no comparable teleology with the chimpanzee. Or is there?

With human children, because they eventually produce speech, we can be more certain about the meanings of their earliest marks but even so, much of our understanding of children's understanding is inferred too. Indeed, in recent comparisons of chimpanzees' and children's pretence and imagination it has been suggested that perhaps we have inferred too much conscious awareness on the part of the playing child and, contrariwise, underestimated the play and imagination of apes (Mitchell, 2002b).

For me, the attitude of the child in play offers complex assemblies of signs and signals which tell me that the child is playing and imagining. As Dennie Wolf et al. (1984) comment, its very repleteness is a strong indicator of play's rich symbolic nature. The beginnings of children's play cannot be relegated to a category of routinised patterns of behaviour.

Likewise, when chimpanzees switch into the play-mode the change in kind and quality of their body movements, coupled with distinct changes in their attentional behaviours and cognitive engagement is convincingly indicative of acts of pretence and imagination too.

So, if the actions which make up ape drawing and painting are very similar to actions of the human infant and if, in the case of the human infant, they possess meaning and organisation, then, the intriguing question arises as to whether the rather similar actions which go into ape painting and drawing are similarly endowed with cognition and imagination.

Here, I am challenging the assumption that ape art is really only a contrivance of a close and unusual relationship with a human keeper. It is accepted that, in atypical, non-natural relationships with human companions, when all the distractions of the natural environment have been taken away, then, and only then, might an ape become interested in drawing and painting. But many people think that the kind of painting the apes produce in these situations is really only an unnatural contrivance devised by humans. The most extreme version of this notion is the idea that animal art is really only an anomalous kind of human art arising out of the combination of the cleverness of both human and ape, each animal perhaps deluding the other in a curious reciprocity. Perhaps, in such situations, there occurs between the two primates a subtle, two-way exploitation – the one wanting to 'magic' art out of an animal; the other wanting to gain food and good treatment.

Setting aside for the moment the marked similarity between this arrangement and that between certain contemporary artists and the art market, in fact, Desmond Morris, for one, was careful not to reward his chimpanzee friend Congo with food and, though Congo surely enjoyed the companionship of this special human, Desmond Morris's unique (1962) study shows convincingly that the chimpanzee painted and drew for the sheer pleasure of these activities. Nevertheless, this has not stalled the general assumption that drawing and painting arise from interests which are solely human. Up until now, the verdict has been that ape art does not occur naturally in the wild.

But is this so?

Field-of-view and field-of-action

The precursors of drawing in human children start with the establishment of a field-of-view and with presentations of events especially produced within a theatre set up before the eyes and before the questing hands and fingers.

Here is an example from my child studies.

18 February 2007 Video-recorded child-observation (DVD) plus written transcription, Singapore

Chloe (7 months) catches sight of her left hand as it happens to pass into her field of vision. As soon as she sees it, she decelerates its motion and focuses her eyes upon it; first, on the hand in its entirety and then, a millisecond later, on the index finger, which she extends away from its neighbouring, coiled fingers. Visually tracking this finger, she traces a graceful arc across her field of vision. With her other hand, she intercepts the flight of the finger and smiles when she catches hold of it.

Early structure and expressive movements involve the child's entire body. Active, moving limbs, sometimes including the legs and feet but especially the arms and hands, become selected targets for visual attention. They become the first 'toys'; even the first expressive art objects and events. This appears to be the case with some non-human primates too.

One of the reasons Thierry Lenain (1997) claims that chimpanzee painting is not really like human art is because he thinks the chimpanzee does not appreciate a field-of-view as a human artist does. Yet, in the present study we have seen over and over again that the establishment of field-of-view is very important to chimpanzees too. As many of my observations suggest, the sophisticated field-of-view demonstrated in artists' work (the iconic example perhaps being the very notion of the 'picture' itself) is really only the developed form of a field-of-view established by the primate in his or her natural interactions with the ambulant and ambulatory environment.

16 November 2007 Video-recorded observation (Edited DVD) Singapore

00:03:33 Kimoni (6 months 3 days), holding onto his mother's hair whilst leaning down from the sawn-off top of a tree, looks at his own leg and foot-hand which hang against the bark of the tree. He fingers the hairs and skin of his ankle.

As with human infants, his awareness of his own vision as an optical sample, captured from a single station-point, is testified by experiments he makes with this selected range of vision:

He brings first his leg and then his little pink hand out of his field-of-vision. He looks at the scene alone, without the visual presence of arm or leg. It seems that he purposely withdraws these limbs out of his visual field, as if to avoid unnecessary visual distractions.

His experiments in testing the range of his field-of-vision are very like those of human baby Chloe at 7 months described in the observation above. Kimoni, like the infant human-primate, is selecting a visual sample and noting that he can penetrate this scene when and where he wants with his own arm or leg, or else he can withdraw these limbs from the visual sample. Being able to know where your arms and legs are when you cannot see them is made possible by a kind of perception called *proprioception*. Tiny sensors in your joints tell you where your hands and arms or other parts of your body are.

After using empty hands to enter into and exit out of your field-of-view, the next step is to extend the empty-handed gesture with a handheld object and enhance its expressiveness. We have already seen the chimpanzees use sticks, stones, leaves and occasionally the illicitly thrown-in plastic mineral-water bottle. Some of these objects become favourites and are used like toys.

27 October 2006 (Part 1) Video-recorded observation (Edited DVD) Singapore

Rah (1 year 10 months 18 days) moves a handheld palm stem of about a metre in length through curves in space. Nkosi sits nearby on a platform.

Rah takes the stem to him for him to cast his eye over it. It is as if she is asking for his 'advice'. He looks at it briefly and hands it back to her. This act of giving, or at least returning, an object is interestingly clear for, on this occasion, he does not merely allow her to take it out of his hands – he actually hands it back to her.

She plays with the long stick in one hand whilst with her other she makes synergistic pushing, pulling and arcing movements on the surface of the wooden platform.

The same activities occur in human childhood. Very young children use First Generation Actions to push, pull and sweep a handheld object or toy around on a surface whilst visually tracking the event. As the child grows older, he or she seems to replicate that which started off as whole-body movements but on a smaller scale, in a sort of *miniaturised* space created within the close-up field-of-view. Whether or not the chimpanzees also make this revolution in the way they view what they are doing with their hands remains unclear. It certainly looks that way but as we noted above, some writers think that to do so involves a 'dual representation' (DeLoach, 2000, p. 329) or 'dual orientation' (Boysen and Kuhlmeier, 2002, p. 211) to the scene which chimpanzees find difficult.

In childhood, in this new representational play, movements through space of the whole-body reappear but are transformed by a process of physical miniaturisation coupled with a change in psychological orientation. In human infancy this usually commences with a smaller body-part (usually the hand or the fingers) and later with a handheld object or toy acting as the pivotal lever around which emergent representation revolves.

We saw the precursors of this miniaturised play performed by Chloe (7 months) in close-up field-of-view actions in which, holding her hands before her face, she moves her fingers across her field-of-vision and visually tracks them.

These actions are not in themselves representational but may form a substrate for later representational acts. Full-blown representational play is built upon the kind of actions of the early months in which hand and finger movements are coordinated with visual attention.

08 July 1977 35 mm colour-slide film child-observation and written transcription, (Joel) London

Joel is now 3 years 2 months old. He is holding a small 'Batman' figure in his fingertips. He moves it gracefully through rising and falling arcs in space whilst making vocalisations which also move through ascending and descending scales. As he moves the figure through these orbits, analysis reveals the toy-figure being rotated on its own axis in Joel's deft fingers.

His slowing down of a figure serves to give him time to enjoy viewing the figure. Prolongation is the strategy used here, in which deceleration and replay make a pleasurable experience last longer. We have seen earlier on in the book that the chimpanzees also use techniques of deceleration and repetition so as to delay the passing of pleasure. Nkosi's standing a footstep still on a surface of water is a good example (page 246).

Sometimes Joel will dexterously rotate the figure in his fingertips and, with the toy held close to his eyes, regard the changing views he obtains from it. Such viewing actions strengthen the idea that this is a radically revised psychological space. I have found the terminology of film-making useful in both discerning and describing this spatiotemporal theatre. Joel replays or 're-shoots' certain actions. He makes digressions or 'cut-aways' and he 'intercuts' details in 'close-up'; he 'edits-out' interludes of stage-management and he tries alternate versions of a sequence. His deceleration of an ascending handheld object sometimes results in the object being held still at the moment of apogee just before it is caused to fall. Possibly with the influence of slow-motion cinematography on TV and at the cinema, such actions eventually go on to represent the object's moment of weightlessness at its point of apogee. But though cinematography and videography might later support such a notion it would be a mistake to assume that this kind of visual praxis initially derives from movies he sees (Wartofsky, 1980). Embodied within it are the elements of art forms used in adult, human culture but this type of play does not start life as a copy of these things but is a natural visual practice which predates film-making.

It is built upon earlier manipulation of hands, fingers and objects in small-scale actions initially made small simply that they be encompassed within the close-up, field-of-view.

Chimpanzees also perform similar-looking small-scale play with their front-hands before their eyes but whether this contains levels of meaning comparable to children's play we just do not know. I have already described some of the techniques used to construct this close-up field-of-view and field-of-action, including *curtailment, abbreviation, deceleration, miniaturisation* and *prolongation.* All of these are present in the following example:

26 October 2007 Video-recorded observation (Edited DVD) Singapore

(Smile)

Rah (2 years 10 months 19 days) lies on her back whilst, gripped in a pincer between the finger and thumb of one hand, a coloured leaf is caused to fly through her field-of-vision as she watches it with great interest.

For me, this recalls Joel flying his hand-held 'Batman' figure. In Joel's life, the next phase of this process occurs when these movements in space are recorded in some way. The flight of 'Batman' or other imaginary beings will be later represented in physical pigment or in electronic pixels. Only a few chimpanzees get their hands on artists' materials or computers but it seems to me that everything in the child's life leading up to the great event of mark-making is echoed quite clearly in the chimpanzee's life.

Drawing the action

The transformational process of *miniaturisation* is one of several possible routes to representation. As we have noted, even though the chimpanzees have effectively four hands, they, like humans, favour the front-hands for manipulating an object in play or investigation whilst the foot-hands usually play a supporting role. Also of importance is the finding that nearly all the chimpanzees' natural signing is performed with the front-hands. In my observations, the foot-hands and toes (foot-fingers?) are rarely, if ever, used indexically or iconically.

Precursor drawing movements appear early on in chimpanzee infancy in ways similar to those in which they appear in human infancy. Before the age of 2 years, Rah produces actions suggestive of drawing.

29 September 2006 Video-recorded observation (Edited DVD) Singapore
(Tape 2)

Driving a stick along a flat plane

Rah, at 1 year 9 months 22 days of age, is climbing up and down independently now. She is practising other self-locomotive skills too. In a couple of weeks she will start dangling upside-down from her foot-hands. She will learn how to swivel around on her behind. Today, I observe her learning how to slide her body along smooth surfaces; by her hands, arms and legs along the old, rubber fire-hoses and, with either her whole-body or a part of it, along the wooden platforms.

Now, with a small stick in her hand, she is engaged in a solitary, imaginative-looking, stalking play. Hunched over, she moves along, step-by-step, the platform at Level 2, as if creeping up upon an imaginary prey.

Then she presses her stick at a near-right-angle to the platform and pushes it along its wooden surface, describing a line of about a metre in length. As she makes this action she focuses her eyesight on the moving point of the stick as it slides along the flat surface.

The stick's direction of travel takes it on a course to a small hole in the platform through which she inserts it and lets it go. The stick drops through and down to the ground, several metres below.

Instead of climbing down to retrieve the stick, she climbs onto a horizontal beam made shiny and slippery by constant contact with chimpanzee hair. Prostrating the entire length of her body along it, she slides herself, face-down and head-first along the beam.

Most likely, what has happened here is that, because she has dropped her stick down the hole and does not want to be bothered fetching it, she emulates the forward-sliding of the stick with the sliding of her whole person along the beam. We have seen comparable examples before when, for instance, failing to get one of her brothers to play chase with her, Rah instead chases a stick (see page 118). We considered whether such acts were ones of substitution or the forerunners of representation. Certainly, the kinship between these different examples is seen in the speed and convenience of the act of substitution; in the earlier example, one kind of chase for another kind of chase; in the present example, one kind of sliding for another kind of sliding. In both cases the transfer is made as economically as possible. This economy may be one of the hallmarks of the beginning of representation.

It could be that it is her own newly discovered act of sliding which she represents a moment later with her little stick sliding along a wooden surface. This would be a common enough transformation in the equivalent play actions of a human child. In the case of the chimpanzee we just don't know.

28 Description

29 September 2006 Video-recorded observation (Edited DVD) Singapore (continued)

Precursor drawing actions – Rah aged 1 year 9 months 22 days

Rah's sliding of her stick, held perpendicular to the wooden surface in conjunction with her close, visual tracking of it, recalls a small child's 'driving' of a handheld toy car, or perhaps a pencil, across a surface. Whether she had planned, in advance, the stick's journey straight to the hole and down it, or whether this was a fortuitous event of which she took immediate advantage, is not known.

Situations very similar take place in Joel's and Hannah's drawing when both are around age three:

The Hole: Looking at it from both sides

24 April 1997 Child-observation, hand-written, (Joel) Lake District, England

Joel is 2 years 11 months and 16 days old when he places a rubber eraser in the centre of a sheet of paper. With a fine, felt-tip pen, he draws a continuous-rotation around the eraser, using it as a central axis, to describe, as he explains, climbers going around and around a mountain. Then he draws a line which travels a little way up the paper and down again, to show the climbers ascending and descending the mountain. He uses the eraser in an attempt to rub out a mark in the middle of drawing but he rubs so hard that he makes a hole in the paper. Then he enacts the journey of the climbers *going-through* the mountain by physically poking his pencil through the hole in the paper!

Compare this with the following observation of Hannah:

12 February 1985 Child-observation, hand-written, (Hannah) London

Hannah (3 years 5 months 6 days), whilst drawing says: 'I danced through the hole and fell through . . .' as she pokes her pencil through a little hole she has made in the paper and then, as she turns the paper over and looks at it from the other side, 'It has a hole in the other side . . .' She pauses for a moment, reflecting on her last statement and, seeing the humour of it, exclaims 'It has to have!'

It is as if, along with the moving point of her pencil she runs, like Alice, into an interesting philosophical issue about the definition of a 'hole'. Chimpanzees are interested in the 'hole' too and in their investigations of going-through we have seen that they too consider it from both sides (as it were).

All three of them, Rah, Joel and Hannah are, at this moment in their development, interested in the concepts of *going-up*, *going-down*, *going-around* and *going-through.*

The representation of shape and action

For human children, however, they are able to reflect upon, and develop their ideas using a range of semiotic systems. The next step for them in drawing is to use a very special stick (pencil, charcoal, computer-mouse or stylus) to record the trace-making effect in pigment, physical or virtual. Again, although plenty of mystery remains as to the exact denotational values children at these young ages attribute to lines and shapes, their speech allows us to be reasonably certain that representational meanings are being formed in the act of mark-making.

09 September 1973, 35 mm colour-slide film child-observation plus written transcript, (Ben) London

'There's a car there . . . it's going around the corner.'

Ben (2 years 1 month) is painting. He makes a spot of blue paint on the paper with his brush and says: 'There's a car there'. Possibly he is referring to the shape and location of the car simultaneously; both these properties being embodied in the splodge of pigment. Next, he considers the car's movement, starting it off by dragging the brush away from the patch of blue paint and moving it in a clockwise circle, as he says: 'It's going around the corner . . .' and, as this elliptical line becomes subsumed and invisible in other blurs of blue trails, Ben goes on to say: 'It's gone now . . .'.

The idea of the single, stationary point versus its passage of movement is not only at the heart of young human art but is also a strong presence in an extremely wide range of adult art around the world, both western and eastern. The *point* and the *moving point* form the basis of Islamic art.

Shape, location and movement are the subject matter of human children's earliest painting, play and exploration and such interests are possibly present at the dawning of chimpanzee symbolisation too. Knowing where things are and where they go helps to uncover what they are. These are questions both primates need answers for. These are the issues which shape the organisation and character of both human and chimpanzee enquiry.

At a further level of description, concerns about shape, location and movement turn out to be the forms in which deeper ontological issues manifest themselves; issues about the appearance and disappearance of loved persons. Consider, for example, the 'emergency drills' practised by Ganga and Suzie, in which they both separate and hide themselves briefly from their babies (see pages 137 and 138).

Binocular and monocular vision

1994 Hi 8 Videotape-recorded child-observation plus written transcription, Singapore

Two 4-year-old Singaporean Chinese children move handheld toy spacecraft through orbits in space. The spaceships move through ascending, descending and rotating flights. As they make these movements, the children visually track the course of their spacecraft by continuously making adjustments of the head, shoulders and hips so that their eye-gaze remains locked-on target at all times. One of the children, a girl, rotates on the spot in a complete circle whilst she moves her spacecraft around her, maintaining a face-on view of the toy. At one point, she even clamps a hand over one of her eyes so as to look at the spacecraft through one eye alone. Using this strategy (almost certainly not taught to her by an adult) she obtains a single-image or monocular view of the changing shapes presented by the moving spacecraft (Matthews, 1994).

Sometimes one needs to differentiate the *shape* of the object from the *object itself* (Piaget & Inhelder, 1956; Willats, 1997, 2005). This is what this child is doing. Ordinarily, looking at an object with both eyes presents the viewer with two overlapping images which help us obtain a stereoscopic view of the object. This is very useful in day-to-day life because it helps us see depth relations and three dimensions. Occasions arise however in which it is necessary to differentiate, from its overall form, the viewed shape of an object as if seen from one fixed position in space. This might happen when it is important to be sure of the true shape of one aspect of an object when, for example, gauging if your settee can be carried up the stairs, or when threading a needle, or should you need to insert a USB drive correctly into a port in your computer or a mineral-water bottle into a hole in a pole.

To make these judgements does not necessarily require the physical closing of one eye, though there are occasions when this is useful. When looking at an object very close up with two eyes, a confusing blur of two images arises because the image from one eye overlaps with the slightly different image from the other eye. Try it yourself and see.

A special occasion when knowledge of the projected, monocular image is useful is when drawing from observation. In this case, the draughtsperson has to map a three-dimensional scene as seen from a fixed position or 'station-point' onto a two-dimensional surface. This is why human artists, when drawing a scene before them, may, from time to time, close one eye to obtain a clear, single image of the scene. *Linear perspective* is essentially a one-eyed vision of the world. The young Chinese girl (in the observation '1994 Hi 8, videotape-recorded, child-observation, Singapore' above) is not drawing her spacecraft but she is using the artist's strategy of monocular vision to obtain a clear, single 'picture' of its changing shapes, almost as if she is making a pretend film of it. Of additional interest is that she does this only for visual pleasure.

I do not see chimpanzees physically cover one eye with a hand whilst leaving the other open, though I do see them cover an eye with other objects, including branches and the occluding boundaries of tree trunks as well as their beloved hessian sacks. This is sometimes done with the intention of partial concealment but of necessity it results in the obtaining of a monocular image. A fascinating movie by Crickette Sanz and David Morgan (in Foer & Nichols, 2010) in which a young, wild chimpanzee, noticing the lens of the researchers' otherwise hidden, remote camera investigates it by probing the lens with a finger and, at one moment, looking at it with one eye alone whilst the other eye is occluded from the camera's range of vision. If my interpretation is correct, then this is very important – this chimpanzee has never seen a human or a human artefact before, suggesting that chimpanzees in the wild know how to use monocular vision in order to clarify novel phenomena.

I suspect that other chimpanzees have occasion to check out something with monocular vision and it would be valuable to find out if I am correct, for this would be a clear sign that they are aware of their own optical sampling and know how to obtain a monocular image. Again, this would mean that they appreciate a field-of-view, a concept crucial for visual representation.

They certainly experiment with changing their vision; looking through translucent materials – brightly coloured leaves; scraps of polythene; the transparent, plastic, mineral-water bottle; the magenta-coloured, cotton bag; the mesh of a hessian sack. They will even blindfold themselves (with a sack or hairy forearm) to experience darkness and to make their gymnastics even more difficult and dangerous.

Foreshortened view

03 November 2006 Video-recorded observation (Edited DVD) Singapore

Manzu (7 years 7 months) sits upon one of the flattish, Platform Rocks near the moat. This is the day he finds the straight, cylindrical stick of hard wood measuring about 30 cm in length. Holding it in first one hand and then the other he hammers it in a vertical arc and then holds the banged end close to his eye to inspect it.

Not only does he looks at the end of the stick to see what effects he has made upon it, he also seems to perform another important action whilst the stick is positioned in this fashion, with the end near to his eye. He seems to *look along* the stick from this end. In this way he may obtain a monocular, foreshortened view of the stick.

He then moves the end of the stick further away from his face to rest it on his lips whilst he continues to look down the rod toward this same, opposite end. Tasting is one of a series of tests chimpanzees routinely run on any new object they find but in this case it is important to note that he looks whilst he sucks. From this viewing position he produces, if not a true monocular image, then a narrow convergence of eye-focus resulting in an accurate display of the foreshortened stick. An important feature of this technique is that by resting the other end of the stick on a part of his face (his lips) he will always be able to accurately reproduce this viewing-angle and focal-length, should he need to. His station-point in relation to the object he wishes to view is, so to speak, 'measured'.

We have earlier seen other instances of measurement as embodied in proportions and actions of the primate body. Rather like techniques used by human builders, he makes use of an easily reproducible gauge for the taking of a *sighting*.

As we observed earlier, my chimpanzees have an interest in the unusual straightness of a rod like this. I believe that, like a human carpenter, Manzu might be using the technique of looking along the length of the long rod from one of its ends to the other in order to confirm its straightness (since any minor kinks will show up in this foreshortened view). No other tool is needed to perform this; it can be done quickly with hand and eye. Ordinary builders knew about perspective long before Brunelleschi and Alberti wrote down the rules in fifteenth-century Florence. If the behaviour of the chimpanzees is anything to go by, perhaps we discovered it long, long before this, perhaps over 6 million years ago, before we became human.

Strategies utilising projection of a monocular image and the construction of a foreshortened view form part of the natural basis of a perspectival understanding.

Observation continued

Manzu then pulls back his head and rests the banged end of the stick on his lips, balancing the stick vertically for a moment.

As we have seen, my chimpanzees learn to balance long objects on their heads and faces when they are very young. The chimpanzees do this for fun but it is another simple technique for finding out if the stick is truly straight. Straight sticks balance differently from bent sticks.

These techniques are part of an entire and systematic repertoire of tests chimpanzees perform upon a novel object, including tasting it, sniffing it and looking at it from different angles. This, in fact, is the general sequence in which chimpanzees conduct their sampling (often interspersed with bashing it or bashing something else with it), though the order and type of tests to which a novel object is subjected varies, depending on the age and experience of the chimpanzee and the type of object being investigated. (Of interest is that I do not see young Rah using such clear techniques for obtaining foreshortened views, even at the end of my 3 years with the chimpanzees, suggesting its use has a developmental and learning dimension.)

Experienced chimpanzees recover two, key views of a straight stick of great value when navigating through an arboreal environment. These two aspects of a long object are very different from each other. Looking at a stick laterally, or 'side-on' to its length, reveals its saliently long axis which looks like a straight line, whilst looking at the stick from its end ('end-on') offers the very different image of a little spot or circle.

Other forms of object offer up very different optical shapes. When Rah manipulates her lump of clay, for example, the shape it presents to the eye remains roughly the same, whichever way she looks at it, as a form extended roughly equally in all dimensions as John Willats (1997, 2005) would say, or, in a word, a 'lump'.

Rah's lump of clay, can be tasted, sniffed and looked at but it is harder to wave around and, unlike a stick, *cannot be pointed.*

Lumping-together versus pointing-out

This may be another reason Betsy Herrelko's chimpanzees never use their box-like 'chimpcam' unequivocally as a device for pointing and *looking-through.* For this strategy to be induced, the chimpcam may have to be constructed so that it retains its salient length and its highly discernible axes of *going-through* and *looking-through.*

As we noted earlier, chimpanzees do not seem to point at things and I think this is an intriguing puzzle, since 'pointing' as a physical act is very easy. However, they do show actions which suggest they are on the edge of realising this strategy. To be able to point out something to another requires some understanding of another's point of view but the use of pointy things (as opposed to lumpy things) might suggest (or 'point out') such an idea to the chimpanzee.

In the case of the unusually straight wooden rod, Manzu adds to the typical repertoire of trials a couple of other tests especially appropriate to its long form.

One test, as we have seen, is to balance it on his lips. Another test is to attempt to insert a smaller stick down its centre. At the centre of many sticks and stems the chimpanzees discern a softer spine or core of pithy material which they like to dig out and chew. This is what Manzu is trying to do here, though with no success, for this rod is pretty much solid. Note in passing though, Manzu's knowledge of alignment, co-linearity and congruence, understandings involved in even attempting this act and also related to line-of-sight.

Manzu makes another addition to the chimpanzee standard-test procedure which is to spin the rod lengthways on the flat table of rock on which he sits. This is another affordance specific to this class of long, straight, rigid object. (You could of course spin a lump but it would look very different from the stick in motion and it would not reveal any significantly different axis, remaining, even in motion, generally 'lumpen'.)

The projected image

He goes off and offers the stick to Togo for his superior inspection. Here, a junior-rank male 'asks' 'advice' from a senior, his father, who is also Alpha male of the group. Like his son, Togo holds the end of the stick up close to one eye and looks along it. Surely he must see a single, optical image from a foreshortened view? He makes a similar sequence of viewing, smelling and tasting strategies as did his son; placing the end of the stick close to one eye, then redirecting this end of the stick to his mouth and from there to his nose. Again, being able to look-along a straight stick and note that it is indeed straight and to also be able to look through a metal tube suggests that he could learn to use the virtual tube of a video-camera (see Herrelko, 2010).

Apart from the acts of sniffing and bashing, similar strategies are employed by humans for obtaining understandings about a novel object (and we are even known to sniff and bash things too) whilst more or less identical-seeming strategies are used to obtain a monocular view.

Later, Manzu is again looking at the stick and further investigating its 'affordances' (Gibson, 1979). He hooks it over a Level 1 Horizontal Beam so that it rests on the beam positioned at right-angles to it, cruciform fashion. Then he hangs from it, his front-hands gripping the two ends of the wooden rod. This is the *Hanger* invention which we have already noted, used by both chimpanzees and orang-utans.

Both species use a variety of materials for making a Hanger, including fabric (hessian), sticks, grasses and palms. Manzu's use of such a rigid, straight rod is interesting because it more or less obliges his selection of *right-angular cross-over* when positioning the rod over the beam. We have noticed the geo metry of this relationship before when, Nkosi hooked a branch over a horizontal wooden beam (page 197).

Nkosi's branch was slightly crooked but with such a rigid, perfectly straight stick as Manzu's wooden rod, and with his hands grasping each end of it as it crosses over the beam at a near right-angle, the symmetry of this transverse relationship is especially striking.

Manzu experiments with other ways of hanging from it, including hanging upside-down. The functionally dependent relation between 'balance' (and imbalance) and 'symmetry' (and asymmetry) shows up in all his variations. This relationship is embedded in all our (primate) interactions with the world.

29 The scratch

Manzu performs one more experiment with the stick using an action for which it is admirably suited. He jumps down to the ground again, retaining his stick in his hand. He holds its end against the soil and pulls it toward himself, making a longitudinal scratch line in the leaves and mud. He looks briefly at the scratch he has made. He looks to the left and walks away in that direction, carrying the special stick with him.

With human infants and chimpanzees, the making of marks seems to be, if not an inevitable part of the techniques used to test a novel object, then an important member of a family or ensemble of such actions.

Chimpanzee drawing and writing

After consultation with zoo-director, scientists and Keepers, I give sticks of charcoal to the chimpanzees. These are of two kinds. One kind is made from stout sticks, ranging from about 25–30 cm in length and about 5–6 cm in diameter, made from fallen tree branches, each with one end roughly sharpened (though not dangerously so) and then burnt into a charcoal tip. The other kind is shop-bought, artists' charcoal-sticks which measure roughly 10 cm in length by 2 cm in width.

Sam, the chief keeper of primates, agrees that the charcoal will not hurt the chimpanzees should they eat it and in fact will be good for their digestion! [Orang-utans in the wild purposely seek out and chew found charcoal pieces to aid digestion – Schuster et al. 2007.] Other apes, including chimpanzees, know something of the medicinal properties of certain plants – see Goodall (2000).

I have supplied some black sticks of artists' charcoal to the chimpanzees but it is the interesting white branch I described earlier (page 274) which Kimoni (1 year 1 month 27 days) first encounters on Level 2 of the Construction.

10 October 2008 Video-recorded observation (Edited DVD) Singapore

The white stick and the black stick

00:06:44 Pale-coloured sticks are rare in the chimpanzees' environment and perhaps this is why Kimoni finds it attractive. Inevitably (so it seems to me),

Ganga takes the branch from him. She puts her right foot-hand on the branch, trapping it against the floor and there ensues a 'tug-of-war' play between mother and son for possession of the stick. Ganga, resolutely, implacably, seals her right foot-hand over the white branch from her end, ensuring that her son cannot take it.

Later, Kimoni picks up a piece of charcoal and places it in his teeth. Then, taking it back in his hand, he makes a single marking action against the deck.

The action is a little like a classic horizontal arc except that it starts from an accurately placed, piercing stab of the charcoal against the wooden floor and then moves about 15 cm to his right where, meeting some resistance, it skids off the surface and into the air. I cannot see the surface of the platform, but there must be a line drawn on the wood, for the charcoal-stick is pressed against the surface so forcefully – the final resistance it encounters surely testifies to this.

This kind of drawing action, in which friction causes the drawing instrument to skid off into space, occurs in human childhood too. When, for example Hannah (2 years 2 months 28 days) paints her red blobs and her trailing blue line (page 236) there is a moment when her brush of sticky paint meets resistance and though, just like Kimoni with his charcoal, she retains hold of it, its working-end flies up into the air. In both cases, this happens because chimp and child are in the process of modifying a three-dimensional action in relation to a two-dimensional plane. Also in both cases, after the accidental skid, the marking instrument is immediately re-directed back on target by the young artists.

Intriguingly, these two observations, the one about a chimpanzee, the other about a human, resemble each other in another way too. This is in the interaction taking place between adult companion and infant. The interaction between Kimoni and me recalls that between Hannah and me and between Hannah and her mother. Before starting her painting, Hannah looks into my eyes and into my wife's eyes and then looks down at the white paper. Then she starts painting. Her painting episode ends with a final clonk of the metal ferrule of her brush against the paper on the table. Then she looks up at Linda and then at me.

Kimoni too gives a glance in my direction as he ends his one and only charcoal stroke, though whether the gaze of his amber eyes includes me is hard to say.

30 Agon

The fight with oneself

10 October 2008 Video-recorded observation (Edited DVD) Singapore observation continued

He bites at the end of the charcoal-stick, then withdraws it from his mouth and investigates it by visually focusing on, and fingering, with his right hand, the chewed end of the stick.

He then puts the stick on the floor and presses down upon it with the palm-sole of his left foot-hand and with his right front-hand he now tries to pull the charcoal-stick from under his foot-hand. He is pressing very hard downward upon the charcoal-stick with this foot-hand, so purposely making it difficult for himself to withdraw the stick with his right hand. He may be emulating the strategy his mother used just moments earlier when, with her right foot-hand, she trapped the white branch in which he was interested and which he was unable to pull away from her.

As with all play-assemblies, the present study shows that the 'tug-of-war' has a developmental trajectory all of its own, entwined with those of the other ensembles. Although I have not the space here to tell you about this development, it again involves interpersonal companionship between the young, apprentice chimpanzee and his or her older and more experienced friends and relations, who provide and support learning opportunities for the learner. Perhaps now, in a fight with himself with this black stick, Kimoni represents the tug-of-war of moments ago, with his mother, over a white stick.

Finally, he pulls the charcoal out from under his own palm-sole. He now purposely traps the stick under his ischial callosities and plays the same play of tug-of-war with it, trying to pull it from under his backside yet at the same time pressing down upon it with the weight of his body.

Teeth-marks in the charcoal are clearly visible as he again raises it, held between his two hands laterally this time, to his lips. A moment later he has the stick in his left hand and he tucks the charcoal-stick this time under his right armpit, whilst gazing upward, probably toward his mother.

Again, this entrapment of the charcoal-stick is done in order that he can pretend to struggle to withdraw it. I am becoming more convinced that he is experimenting with his mother's strategy to which she introduced him when she trapped the white branch under her foot-hand, so making it difficult for him to remove. What is happening now is that the little chimpanzee has seen further possibilities of pressing the charcoal-stick against a part of his own body in order to test his own strength in removing it. He plays out an imaginary conflict, as if with an imaginary opponent, over possession of the charcoal-stick, the stick held down by a part of his own body.

The idea is to produce a sensation of being locked in battle until suddenly relenting and releasing the charcoal, as if succeeding in wresting the stick away from an enemy.

Finally, this self-contrived fight between oppositional forces results in the charcoal coming loose from his mouth and, gripped by his right front-hand alone, Kimoni again presses the charcoal against the wooden floor, this time using his right front-hand but with assistance from his left foot-hand.

These latest mark-making actions, involving front-hands and foot-hands coordinated skilfully together are unlike nearly all of the drawing actions of the human baby. I write 'nearly all' because there occurs a small passage of developmental time in which the human baby will use his or her feet to touch or attempt to manipulate an object before the hands soon take over. In Kimoni's actions too, the front-hands are starting to become the star players in object manipulation.

He stops doing what he is doing with the charcoal-stick and replaces it cigar-wise between his lips, exchanging hands to hang onto the beam above him.

He swings from his right hand and whilst doing so, bends over, withdraws the charcoal-stick from his mouth and, holding it about 20 cm from his eyes, surveys it with interest. He returns it to his mouth and sucks it, still swinging from the overhead beam.

This is the beginning of *agonistic* play, which is to do with conflict, struggle and eventual overcoming. I consider it central to primate symbolisation. It is the background for an imagined war between what some humans term good and bad. We can see its full development in the following excerpt (see Chapter 31) from an observation of Rah when she is 3 years 9 months and 19 days old.

31 Dying slave

26 September 2008 (Part 2b Edit 1) Video-recorded observation (Edited DVD) Singapore

Rara's rain-dance

00:03:40 It is pouring with rain so fiercely that the chimpanzees are reduced to grey spectres in the mist. Struggling to keep my video-camera dry under a rain-beaten umbrella, this is the day I observe Rah performing, perhaps for the first time, a *rain-dance* (see Figure 31.1).

Figure 31.1 Rara's rain-dance (*Dying Slave*).

Rain-dances were first observed and described by Jane Goodall (2000). Sometimes, a downpour of rain will elicit from a chimpanzee, often a high-ranking male, a spectacular dance. No-one really knows what these dances mean but they are performed with extraordinary passion. Do the chimpanzees love a sudden rain-storm or do they hate it? Or is it a balance between both these emotions? Jane Goodall was impressed by the almost religious awe which seemed to consume the rain-dancing chimpanzee (Goodall, 2000).

In Singapore Zoo, I have seen both Togo and Nkosi perform rain-dances but today these males are locked up in the Den. Perhaps because these tough guys are not around this is why Rah feels free to perform her version of the dance today. It starts with a quick dash of her hands and arms over the wet, black hair of her head and shoulders followed by a scampering dash from one end of Lateral Log to the other, where, with the knuckles of her left hand, she raps the sawn-off top of the post there, marking the end of her linear journey. She sits there for a moment, huddled up into a black blob against the rain. She shakes her head and arches up her opened mouth skyward to catch rain-drops.

Her dance resumes in earnest when she runs from one end of the log to the other, punctuating each pause with a series of frantic wrigglings and writhings of her wet body; movements which, though they paraphrase each other, are never completely replicated either, but always incorporate some additional detail or inflexion; an up-stretched hand and forearm flung up and around the girth of the big tree; her back, head and neck arched back along its rain-blackened bark, her eyes rolling closed as if in hypnagogic trance.

00:12:05 Rah makes a pretend-fall allowing herself to topple backwards in slow-motion off Lateral Log but at the last moment saving herself by hanging onto sloping vine.

Leaning back, she falls through her final moments of turn, her lips compressed, her right arm crooked over her head, her eyes rolling closed, mouth stretched open in a wide yawn of ecstasy. Rara the chimpanzee, wet through, reminds me of a child in dramatic and agonistic play, the tragic heroine, the wounded soldier, the slain cowboy, even Michelangelo's *Dying Slave* in his death throes; an exquisite balance of pleasure and pain; as she just enjoys being alive in the rain.

32 Sack-cloth and ashes

31 October 2008 Video-recorded observation (Edited DVD) Singapore

Twenty-one days later, Kimoni (1 year 2 months 17 days) is sitting within the transitional arena formed between the knees of his seated mother, Ganga. After checking with my colleagues at the zoo I supply the chimpanzees with rectangular, plywood boards. Each board is about 28 cm × 22 cm and 4 mm thick and has rounded-off corners and bevelled edges. The boards, wrapped in a hessian sack, are left unpainted and I anticipate that upon their smooth, pale, tawny surfaces the chimpanzees might be able to produce lines of greater contrast than the fainter ones they make upon the dark timbers of the Construction and the rough stones around it.

Kimoni now draws with a stick of artist's charcoal on one of the boards. Holding the charcoal in a delicate palmar grip in his left hand, he firstly raises it high above his head and shoulders, outside of his field-of-action-and-field-of-vision. Then he gently commences its dive downward, its drawing end aimed at the wooden board beneath him. He does not visually track the descending flight of the charcoal but looks toward its impending point-of-arrival. It is a graceful and controlled descent, not a wild stabbing. Retaining the charcoal in a near vertical position, he touches its tip against the deck. After its gentle moment of arrival (not a fierce impact), he pushes the charcoal-stick along the surface and at the end of this straightish line, skids it a few centimetres to the right, making a little inflection at the 'higher' end of the line, like a near right-angular 'hook'. Rah (3 years 10 months 17 days) is watching carefully nearby and she nuzzles down even closer to him to be a part of the event in some way.

This is a good example of early, non-human primate drawing. The charcoal was first introduced 14 months and 17 days ago, on the day that Kimoni was born, yet he has already superseded the first generation vertical arc, adapting this type of stabbing motion into a carefully focused descent and point-of-arrival followed by a continuous-contact charcoal line.

First Generation Structure

14 November 2008 Video-recorded observation (Edited DVD) Singapore

Today, Ganga (14 years 9 months 6 days) has taken possession of a hessian sack in which are some sticks of artists' charcoal together with four of the plywood boards.

Sitting with the package on Level 2, and with her 1 year 3-month-old son, Kimoni, by her side, she takes out a stick of charcoal from the bag but leaves the remaining boards inside. [She and the other chimpanzees know what the bag contains from their previous experiences with such packages.] Ganga mouths her charcoal-stick whilst allowing Kimoni to handle the sack. Rah (3 years 11 months 7 days) comes up to have a look and Kimoni raises the sack up like a screen between the pair of them. Then he repeatedly stamps his left foot-hand on the sack. He reaches up with his left hand and grasps the charcoal stick Ganga has in her left hand. Ganga allows him to withdraw it from her fingers which are only loosely cupped around the charcoal-stick. Gripping the black stick palmar fashion, Kimoni makes some gentle marking actions with it against the hessian. The crumpled sack is now lying on the deck, temporarily in his possession, with Ganga sitting at one side of it and Rah, watching closely, sitting on the other, bending her head down into the arena of her little brother's marking activity. She seems fascinated by it, her eyes only a few centimetres from any charcoal marks which might be made. Kimoni stands up bipedal for a moment and makes an experimental marking action on the horizontal beam just above his head. Rah continues to study at close range the sack where he was just now marking, Realising that if she is going to get a turn at this interesting activity she will have to obtain a stick of her own, Rah sets off down to Hollow Rock, stopping for a moment to enlist the support of her mother, apparently not wishing to go near the novel objects on her own. Rah takes hold of Ida by the scruff of the old grey female's neck and coaxes her toward Hollow Rock. Rah jumps down the little cliff and finds a stick of charcoal in the grass by the moat and then makes her way back to the Construction.

Kimoni is in Ganga's lap-space double-punching the hessian sack at his feet with his knuckles. There are some pieces of charcoal on the sacking and, because of his actions, these bounce about before him. It may be that he is trying to hook up a stick of charcoal into the air so as to catch it. Whilst Kimoni punches the sack-cloth Ganga makes a series of downward brushes with first her left and then her right hand against her son's back, as if trying to keep in time with his jabbing of the hessian sack. Actually, her action is only an interference to him.

It may sound paradoxical (and you certainly will not find this in today's repressive and dreary curriculum rhetoric), but some genuine 'scaffolding' is not even intended to be helpful! Interaction between infant and parent and the way that love and affection are translated between them is extremely complex and cannot be reduced to a simplistic set of steps (well-meaning or otherwise) imposed by the 'teacher' onto the 'learner' whose activities are naturally free-flowing and ultimately unpredictable.

The charcoal jumps and leaps out of Kimoni's grasp, which though interesting to him is not, I surmise, his main intention. His mother's actions do not seem to support these intentions, which are to investigate cloth and charcoal and exploit their expressive potentialities.

Slightly misguided parental or over-teacherly support of the young learner is commonplace amongst humans – but that is not to say that its effects are always bad. (It is a tragic irony that human education tends to go badly wrong when governments try to iron-out mythical errors.)

Perhaps in response to his mother's unhelpfulness he stands up bipedal to get out of her way. Reaching up with his left hand for the support of an overhead, horizontal beam, Kimoni makes a stunning series of six, very dedicated, vertical arcs with his charcoal-stick. His first downward stab partially functions as a crutch to assist his move upward to a bipedal position – he presses the charcoal for a moment against the floor like a diminutive, black, walking-stick – but it is this move, seemingly extraneous to the act of drawing, which suggests to him his second strike, the impact of which is so forceful that a small charcoal fragment flies off into the air. Three more, rapidly executed, vertical arcs are made, the charcoal hitting the wooden platform in a succession of fierce impacts.

Synergistically accompanying these downward plunges of the charcoal-stick, Kimoni makes a series of four downward stamps of the sole-palm of his left foot-hand; the first stamp occurring in a one-to-one correspondence with his first vertical arc; the second one missing a half-beat and so impacting in-between beats; the third one back in time again and striking the deck in synchrony with his charcoal-stick, whilst his fourth stamp is maintained in time with his fourth vertical arc. The analogy he makes between actions of his drawing arm and non-drawing arm seem based upon the memory of the earlier synchrony formed between his mother's softened blows to his back and his own double-thumping of the hessian sheet.

Note that his strong, vertical arcs do not stop as singularities on the wooden surface but are extended and continued as *pulling* movements which, though I cannot see them from my position, must be scoring the wooden surface with saliently long, black lines. This means that the little chimp has, in minutes, progressed from a dynamic and tactile enjoyment of exploration of the charcoal to a purposeful making of lines – presumably because he finds this enjoyable too.

He slows down on the fifth stroke, the charcoal brushing the hessian sack his mother is handling. She is holding a stick of charcoal between the thumb and first finger of her right hand. There follows a pause as Kimoni looks around to eye the sack which his mother is holding up from one corner to look at. In picking up the cloth in this way she makes a screen between her and her son. I think she may be helping him draw. (See Figures 32.1 to 32.8.)

Figure 32.1 Kimoni (1 year 3 months) is about to make some marks on a hessian cloth with the charcoal he is thoughtfully sucking. Big sister, Rah (3 years 11 months 7 days), looks with interest at what he is doing.

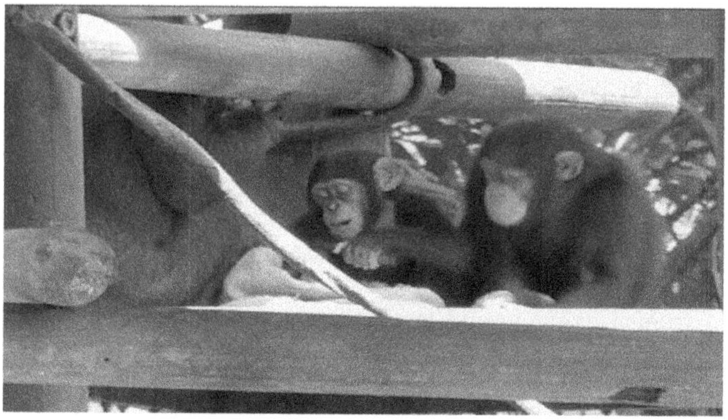

Figure 32.2 Kimoni makes marks on the hessian whilst Rah watches with interest.

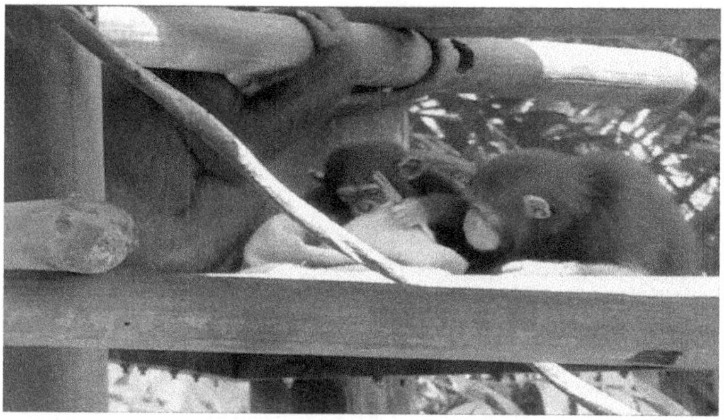

Figure 32.3 Rah and Kimoni both look closely at the marks he (Kimoni) has made.

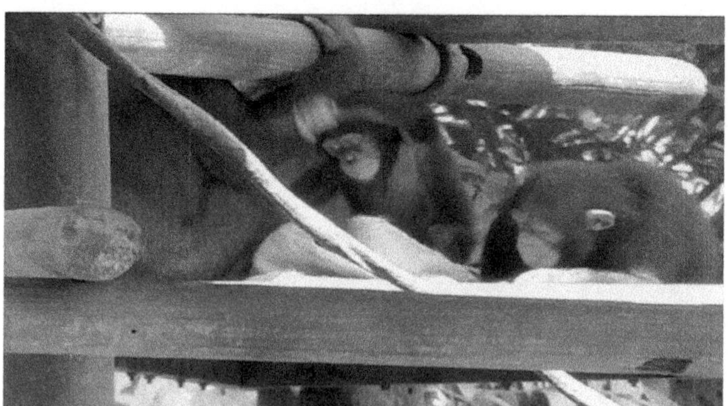

Figure 32.4 Kimoni begins a fast series of vertical arcs, dashing the charcoal against the sack-cloth whilst Rah watches closely.

Figure 32.5 Rah continues to look at the cloth while Kimoni pauses for a moment.

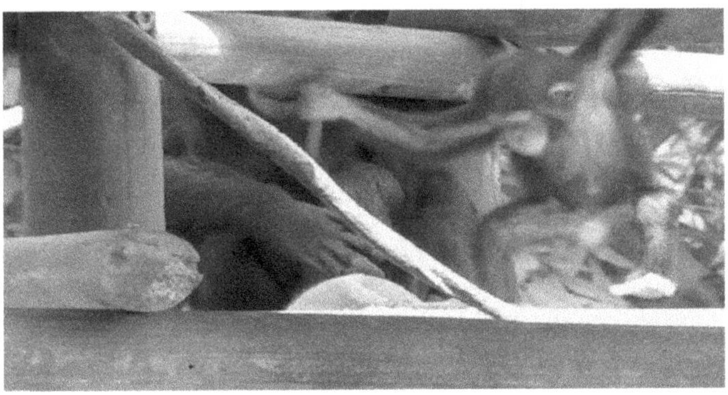

Figure 32.6 Kimoni makes a series of powerful vertical arcs, stabbing the hessian sacking with the charcoal.

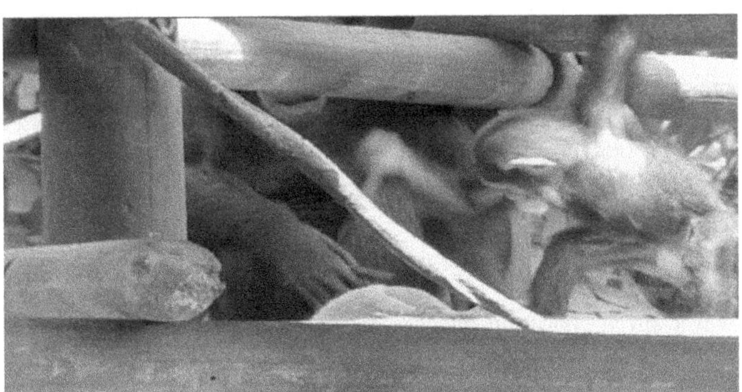

Figure 32.7 His strike is powerful and accurate. He keeps his eyes on the target. He bends closer to see the effects of his action.

Figure 32.8 Ganga (14 years 9 months 6 days) holds the sack up and Kimoni pokes it carefully with his charcoal.

14 November 2008 (observation continued)

Kimoni pauses, looking carefully at the roughly straight, vertical plane of brown hessian hanging before him. He aims the charcoal at it in a sixth and final stroke which, though derived from the class or family of vertical arc, is no longer, strictly speaking, one of these but more a conscious and deliberate point-and-touch at a selected location. It is more like a poke than a stab. As he touches his charcoal-stick gently against the sack it leaves a faint, black smudge.

Kimoni bends close to the sack to look at this black mark, the effect of his action. Then he looks back to the marks he has made on the wooden deck.

Rah is also interested and comes closer to watch. Kimoni sees her coming, crouches down and, using all four fingers of his left hand makes a series of four *pull* movements across the charcoal traces he has already made, surely making another, more extended, smudging effect which crosses over his original longitudinal stroke at right-angles to it. I write 'surely' because again (pathetically, crucially), I cannot see the marks from my station-point.

Targeted-marking

In beautifully written observations, Thierry Lenain (1997) also describes this type of mark-making action in which a previous mark is crossed-over with a later one but unfortunately he interprets these actions in a negative way, focusing attention on the obliterative aspect of what I would term in very young children's drawing, *targeted-marking*. The use of a salient target at which to aim the marker has been noted by Paul Schiller in an important study of the chimpanzee 'Alpha' made in 1951. Professor Lenain misunderstands them as destructive in character rather than compositional. I think that here he makes a significant error because such *cross-overs* in human infants' drawing signal the emergence of *right-angular join* and *cruciform*.

Note that the first pull is a long, slow one, starting off as a haptically and kinaesthetically perceived pulling of all four fingers as he looks up to check Rah's approach. As the stroke progresses he looks down and away from his big sister and toward the mark-making event in order to visually track the remaining pull lines.

He then sits on the cloth, bounces about on it and slaps it with his hands but Ganga grasps the infant by the back of his neck with her huge left hand, a black leather gauntlet so big that Kimoni's entire cranium fits into it too. She does not hurt him; she means only to restrain him. It is as if she is saying, 'Okay, that's quite enough of that'. As she lowers her hand to hold him companionably around his left shoulder so he, with charcoal-stick held by his right hand in his mouth, cranes his neck back to look up at his mother as if for guidance. (See Figure 32.9.)

Figure 32.9 Kimoni looks up at his mother for guidance whilst she puts her hand around his shoulder.

Kimoni does not want to stop however and, with charcoal-stick in his left front-hand, wriggles free of his mother and knuckle-walks in my direction, looking me straight in the eyes for a moment. He looks away from me, bends down and makes a slight graze of the charcoal point against the wooden platform. As he turns around to look at his mother, so the charcoal describes a partial circle on the wooden deck as it is spun around in his left hand.

Very young human infants also describe circles in this way when, sitting on the floor and equipped with a marker they trace a circular course upon it when they spin on their hips (for example, 35 mm colour-slide recording and written transcription of Ben, aged 1 year 9 months, May 1973).

Returning to Ganga's lap-space, Kimoni wants to continue to investigate and play with the charcoal but she grasps his skull again, this time in her right hand, to restrain him, perhaps even to re-direct his line-of-sight.

Rah now returns to the platform along with Suzie and baby Ainka. Rah has two charcoal-sticks in her left hand and, with this hand leading an outstretched left arm she makes a sort of swimmer's side-stroke lunge along the platform, upon the left side of her body, whilst at the same time sliding one of the charcoal points under and along the edge of the package, laying flat on the floor, containing the plywood boards. At the end of this gliding movement, she hooks the point of the charcoal-stick further under the edge of the flat package and levers it up through gradual moments-of-turn so that it rests on its opposite, bottom edge forming a roughly vertical barrier or screen between her and Kimoni. The hessian is stretched over the boards like canvas over a stretcher. Rara ducks down as low as she can behind this screen, trying to conceal herself from Kimoni. He happily slaps the stiff, flat package, his lips smiling. Jostling behind this screen, Rah makes a mark on the cloth, an acutely angled line, which both she and Kimoni (peering over the top edge of the screen), look at with interest.

The screen

The use of the *screen* is particularly interesting here. The screen or barrier has been mentioned several times in the book. That a chimpanzee will hold up a screen for another to draw upon is of vital significance and we will look briefly at a few moments of its developmental history. We'll start with an observation made only 20 days before.

24 October 2008 Video-recorded observation (raw footage on DVD) Singapore

Rah (3 years 10 months 17 days) presses a small plywood-board against her face as a screen behind which she can hide from Ainka (1 year 1 month 25 days).

The above examples show the use of a physical screen or barrier and have a potential developmental direction toward ideas about line-of-sight, occlusion and concealment. The physical barrier also serves as a *divider* which may help along emergent ideas about what lays *in-between* two ends of a path.

However, this physical barrier is soon minimised in play to become a token, even a sign. Consider for example, the rotational play-chase between Togo and Nkosi (16 November 2007, video-recorded observation, edited DVD, tape 1, Singapore, page 232), in which Nkosi raises an arm to form a horizontal barrier in his father's path, or in the similar rotation by the Chinese children (video-recorded child-observation, edited DVD, Singapore, page 234) in which Zhou Quan raises his arms horizontally like a kind of 'gate'.

In such cases, the physical barrier is moving in a developmental direction toward a future existence as an idea represented by a sign. This is the sense in which it is used in the following example too.

23 February 2007 Video-recorded observation (Edited DVD) Singapore

It is one of those rare occasions when Nkosi (6 years 7 months 26 days) wants to look at me and listen to what I am saying but Rah (2 years 2 months 16 days) is sitting next to him and wants to gain his attention too. Fed up with being interrupted by his little sister, Nkosi deters her approaches by raising the palm-sole of his right foot-hand to her face, as if shutting her out of the conversation. If he could speak he might be saying: 'Sssh! I want to listen to this!'

There is a further sense in which the screen operates. The following is from an observation we considered earlier with regard to a pair of chimpanzees working together to form a Hanger but in the part of the play-episode we'll look at now it is a barrier or screen they cooperate in erecting.

07 September 2007 Video-recorded observation (Edited DVD) Singapore

Stars of the screen

00:16:08 Rah (2 years 9 months) and Nkosi (7 years 3 months 10 days) are wrestling and rolling on Word Platform, Level 3, taking turns to attack each other. Nkosi holds the sack up like a screen between them and she pummels this curtain in a tattoo of vertical arcs. Rah dives head-first into the hessian screen and he wraps it over her.

When, in some observations chimpanzees erect screens which can be drawn upon – *and not only for themselves, but for another* – this seems to be a critical evidence that the chimpanzees have an understanding of a basic principle in art-praxis; that concerning the use of the physical support-surface required for drawing

Figure 32.10 Two oblong pieces of plywood remain in the corner of the hessian sack. This makes it into a good surface on which to draw. Rah and Kimoni situate themselves either side of the dividing screen they have made. Rah makes an acutely angled line and both she and Kimoni study it closely, Kimoni leaning right over the top of the cloth-covered board to see.

actions and the accompanying allusion this use makes to the idea of the *visual-sample*. At its deepest level of description, a painting or a drawing, whether by Rara or Rembrandt, is a surface made to reflect light in particular ways. As drawing develops (in human society) so this physical screen simultaneously alludes to an analogue space with some attendant and profound implications about field-of-view, visual sample and even to an imaginary 'window'. The beginning of this extraordinary process is being enacted by Rah, when she lifts the hessian surface, stretched over the plywood like a canvas on a stretcher, into a near vertical position and on which she draws, and over the top of which Kimoni peers, that he too might see what she is drawing. (See Figure 32.10.)

14 November 2008 Video-recorded observation (Edited DVD) Singapore (continued)

Marking the spot

Shortly afterwards, Ganga appropriates the package and takes it up to Level 3 where she further investigates its contents by opening up the sack and looking into it. Meanwhile Rah is luring Kimoni into chasing her. She does this by inserting herself into his field-of-view and scurrying past him vertically up to Level 3. Kimoni follows quickly and peeps out over the top just in time to see her disappear over the edge of the opposite corner. Kimoni hurries to this corner and, with his charcoal-stick held palmar-fashion in his left hand, makes with it a very precise, over-arm arc aimed just a few centimetres from the edge over which Rah vanished.

Although within the category of vertical arc, Kimoni's action is more a careful, downward placing of the charcoal point against a selected position on the wooden platform, its highly salient corner. He has greatly reduced the stabbing quality of the action so that the charcoal point arrives gently at its target. *Deceleration*, which in previous observations transformed violent blows to graceful trajectories is again the technique used here to translate the vertical arc into a controlled drawing action. For a jiffy of time, the charcoal tip rests at its point of arrival at this important corner and is then translated into a pull line which travels roughly 15 cm toward his body.

Although the physical action is important to him, I feel that his main intention and motivation is the production of a dot and a line.

The location of his marking action, the corner of the platform is a salient landmark for him. In an observation to come we will see chimpanzee Nkosi, at age 7 years 2 months 3 days, tapping the four corners of a wooden platform, each in turn, with a single tap of his charcoal-stick.

Corners are attractive landmarks for very young children too, when they make their first generation marking actions. Schiller (1951) also notes the use of the salient landmark of the *corner* in his fascinating study of Alpha, an 18-year-old female chimpanzee. Of extreme interest is that Alpha liked to fold up her drawing-paper, apparently to make new corners at which to aim her pencil. I observed exactly this behaviour in Ben (1 year 11 months 5 days) (35 mm colour-slide recorded observation and hand-written transcription 14 July 1973, Matthews, 1990, p. 153).

If left relatively unrestricted, a human infant may like the chimpanzee aim a marker at noticeable features in the larger physical environment (Ben's use of the corner of the window-frame noted earlier is an example). If restricted to a screen or sheet of paper then the child selects a local landmark there, a corner being a favourite target (though others occur – a pre-existing mark, crease or tear in the paper will do). In the present observation, the corner of the platform is especially noteworthy to Kimoni because it is the corner over which he sees his sister disappear.

This movie ends with Ganga finally removing four boards from the sack and then looking me straight in the eyes.

The first type of charcoal I give the chimpanzees is not the shop-bought, artists' charcoal-sticks written about above but those made from the little logs I mentioned earlier, roughly hewn into a point at one end and burnt into charcoal at this end. These charcoal-sticks happen to be first introduced to the chimpanzees on the day that Ganga's baby, Kimoni, is born.

17 August 2007 Video-recorded observation (Edited DVD) Singapore

(Mark-making, Part 1)

It is raining continuously. Nkosi (7 years 1 month 20 days) hangs by his two forearms from Horizontal Beam 4, Level 3, near Padlock Point. He is suspended in a frontal posture with legs wide open, looking in my direction and with a charcoal-stick sticking out of his mouth. I again get the feeling that he is connecting in his mind the presence of these novel objects with my presence and literally confronting me about this. He contrives to give his slightly swaying pose something of a swagger. With the black stick in his teeth, he looks like a tough gangster chewing a fat cigar.

Meanwhile, his little sister, Rah, now aged 2 years 8 months 17 days, up above on Level 3, also with charcoal-stick in her mouth, is trying to attract his attention. She pretends to stalk him on all fours, her shoulders hunched-up slightly in mock menace. This is communicated pretence rather than realistic deceit. She wants to engage him in play but he does not respond. Suddenly she rears up on two foot-hands but sits down again just as suddenly, the whole up-and-down passage making a vigorous, one–two rhythm. Looking carefully at this up–down phrase in scan-by-scan video-analysis, I note that as she rears up she holds the stick vertically against her mouth momentarily, playing with its precarious balance. Her sudden sitting punctuates her descending trajectory. She sits and chews the blackened end of the charcoal-stick, holding it with two hands vertically against the platform on its other end. She looks around at me, and, holding the stick with the fingers of her right hand curled around it, she thumps her left hand's knuckles in an up-and-down series on its upwardly pointing, black tip, blackening her fingers.

Natural drawing

She stops chewing for a moment and, with the stick now charcoal-point downward and held in her left hand, makes a sudden series of ten stabbing actions with the charcoal tip of the stick against the surface of Word Platform.

This action is an iconic *vertical arc*. This is the first time she has encountered these charcoal-sticks yet she uses them expertly, the vertical-arc action learnt in and carried over from other contexts of striking and bashing. Some of her current downward actions skid on impact and the charcoal tip is dragged toward her body, making long, pull lines.

She lowers her body close to the surface of the platform and sucks the crumbly paste of charcoal and rain-water; her face and lips pressed hard against the wooden surface. In this position she makes a second series of stabbing actions of the stick, the point impacting at a place on the platform very close to where her mouth is pressed. The accuracy of her vertical arc is impressive. She turns the stick gently on its blackened point (here is the winding or screwing action again), so that the charcoal crumbles onto the region of the platform immediately adjacent to her lips and she sucks these particles up too. I realise that aiming the charcoal into a crack in a timber, or into the seam or join between two planks, is an effective strategy to splinter and crumble the charcoal.

Trajectory and moment-of-impact encoded in drawing

17 August 2007 (continued)

Chimpanzee drawing

In another good example of demarcated trajectory-and-impact, Rah (2 years 8 months 10 days) makes a sudden leap to the top of a Corner Post, hitting the point of her charcoal-stick against its flat top to coincide with her moment-of-arrival. Like the corners of the platforms, the roughly circular, sawn-off top of the post serves as an attractive target. Unfortunately however, any mark she makes upon it is hidden from me at my station-point.

As we have noted above, this dynamic-structure, trajectory-and-impact, is produced frequently and intensely by many human children. They gradually realise the potential of mark-making to demarcate visually their moment of impact. I believe Rah is beginning to do this.

Rah clings with her two forehands to the lip of the sawn-off post, the left hand pressing the charcoal against its surface, and then she pulls herself up onto the top of the post, looking briefly at its surface and perhaps at the mark she has made.

She sits, hunkered down on the post for a few moments, holding the charcoal tip in her mouth with both hands and chewing it. She looks down at the surface of the sawn-off post and then jumps up and down upon it, stabbing it once with the charcoal-stick. This action is very reminiscent of her pouncing upon harmless twigs as if they are enemies, except on this occasion she must be leaving a sooty imprint of her impact. On one stroke, she allows the flying arc of the swooping stick to glide closely by the rim of the sawn-off post, at the approximate position of any charcoal mark. Then she echoes this with a variation more consciously controlled, describing, with the handheld stick, another sweeping upward arc, which looks like a perfected and stylised version of the first. She has already modified the primitive vertical arc into a decorous curve.

Here we have a drawing action, derived from a dynamic action of looking, aiming and striking – just as we see in human infancy.

Note that once Rah has made her first strike, the site of the charcoal's impact (perhaps marked by a black spot) now serves as an attractive target for a further strike of the charcoal-stick. Human children also use their own marks as targets for further marking. In childhood, first of all this act seems to have no representational value; the mark just presenting an irresistible target at which to aim the marker.

09 October 1973, 35 mm colour-slide film-recorded child-observation plus hand-written transcription, (Ben) London

Ben, at 2 years 2 months, holding a felt-tip marker in his right hand, makes a powerful downward action with this hand, the point of the marker making a red blob on the paper on the table beneath him. He regards this mark for a moment and then aims his pen again at this target, this time holding the pen-tip against it for a moment before dragging the pen down the page toward its bottom edge.

In the case of this young human, his downward pulling of the marker, which results in the vertical line, is accompanied by a downward-going vocalisation.

In Rah's first series of vertically arcing stabs, when the charcoal hits the surface, it is sometimes held there for a moment at the point of impact and then dragged toward her own body, in a similar fashion as Ben's drawing action and possibly (also like his) making a pull line.

Later on in human development, the same act sometimes acquires an imaginative dimension, in the sense that the visual target represents something which is attacked or hit. A missile hitting a target is a common one, with explosive sounds mapped in one-to-one correspondence with explosive visual effects and moment-of-impact.

02 October 2001 Video-recorded child-observation (Edited DVD) Singapore

Trajectory and impact

Twenty-one days after the 9/11 tragedy, a group of 4- and 5-year-olds in a Singapore kindergarten make drawings about it. One boy raises his felt-tip marker up high and then plunges it against the paper, continuing the impact with a line which soars through an ascending and descending arc on the paper as he synchronises an 'Oooooooooooooooooooooooooo' vocalisation. He repeats this over and over again, representing the moment-of-impact of the hi-jacked plane against the side of the Tower.

17 August 2007 (Part 2) Video-recorded observation (Edited DVD) (continued)

00:24:39 Rah is making high-handed vertical arcs with a charcoal-stick. She is no longer eating the charcoal crumbs. She is surely more interested in the visual and tactile effects she is creating. She presses the charcoal tip determinedly against the surface of the platform and the sawn-off surface of a post.

17 August 2007 (Part 4) Video-recorded observation (Edited DVD) Singapore

00:13:55 Rah is on Swinging Platform in the still heavy rain. She has no charcoal now, though the platform is awash with a diluted paste of the stuff and she experiments with slapping the sopping wet deck and sliding on it, setting it swinging as she does so. She makes push-and-pull strokes upon it too and then presses her lips against its wet and glistening surface, making one of her long, slithery kisses as she slides along on her stomach. This is a development of her sliding techniques but now leaving a recording of its passage in black traces on the slippery surface.

In this observation, Rah seems to be carrying over actions made earlier with the charcoal-sticks, along with the excitement she generated, to a new situation, more like painting, in which her black body becomes the brush whilst the painting surface itself is set wildly rocking by the artist.

Some human artists, when exploring the basis of mark-making, have also used their own bodies (or sometimes someone else's) as a tool for drawing or painting.

Stuart Brisley (1975) hurling his own naked, paint-covered body onto walls and floor, is a good example.

As for the moving support surface, this sort of thing isn't completely unheard of with the human child draughtsperson.

25 November 2009 Child-observation, long-hand written, London

In a coffee-shop, a 4-year-old sits at a table waiting for her mother to bring drinks. She holds an open notebook in both hands, along with a pencil in one of these hands, and she slides the notebook from side-to-side on shiny table-top whilst contriving to write on the paper at the same time.

31 August 2007 Video-recorded observation (raw footage on DVD) Singapore

Simple drawing: Nkosi makes First Generation Marking Actions

It is 13 days later from the last observation about Rara. The weather is dry. Sam the Keeper has again kindly agreed to have some more charcoal-sticks made. This time, they are of various sizes, some quite hefty, of about 30 cm in length by 5 or 6 cm in width and some smaller, more delicate, about 15 cm long and 2 or 3 cm wide. All have sharpened, burnt and blackened points.

Nkosi (7 years 2 months 3 days) takes one up to Level 3, Word Platform and holds it vertically against the wooden surface but by its blunt, sawn-off, non-drawing end. Rah comes to sit close by, near his lap, in an effort to share in his investigation. Nkosi pushes her away with his left hand, but quite gently, as if he is not unsympathetic to his sister's interest. He turns the stick around and investigates and mouths the non-blackened, blunt end. Nkosi gently allows the stick to pivot out of the vertical until it is lying on its side on the platform where he rolls it to-and-from himself. In two weeks time Rah is to do this too but with the addition of a yellow leaf over which she rolls her charcoal-stick like a pastry cook rolls a rolling-pin over a piece of dough (14 September 2007, part 2, editing notes.)

At this moment, I realise that Nkosi has a second stick lying on the platform and he is rolling the two sticks side-by-side, together, on the platform. He raises both sticks into his field-of-view and regards them.

He joggles them about upon their bases, on the wooden platform, between his cupped palms, whilst maintaining their roughly vertical position.

He seems to be interested in the fact that the short, small logs are all about the same length and width and all can be held together, aligned parallel together, as if as a 'set'.

About 5 months later I am to see Ganga display the same kind of interest. We'll make a little digression now to look at this now.

Set theory

> **14 March 2008 Video-recorded observation (raw footage on DVD) Singapore**
> **(Tape 1)**
>
> I wrap some charcoal-sticks in a piece of hessian and have it left in the chimpanzee enclosure. Ganga (14 years 1 month 6 days) finds the package and *without even unwrapping it* takes it to the top of Tall Tree, carrying little Kimoni with her, ventrally.
>
> 00:02:46 Sitting on the sawn-off tree top, Ganga at last withdraws a charcoal-stick from the bag and holds it in her hand. After inspecting it (sniffing both ends, looking at both ends, chewing the charcoal end) she tries to return the stick, point-first and lengthways, alongside its companions into the hessian package!

This is not unlike how human infants investigate a group of objects which constitute a 'collection' or 'set' (Athey 1990/2007; Matthews, 1999, 2003) especially if they are packed inside a clearly defined container.

This is the day when, for safe-keeping, she inserts the charcoal stick vertically into a crack in the sawn-off top of the tree (see pages 254–255).

Meanwhile, back in the present observation . . .

> **31 August 2007 Video-recorded observation (raw-footage on DVD) Singapore**
>
> **Simple drawing (continued): Nkosi makes First Generation Marking Actions**
>
> Rah (2 years 8 months 23 days), unable to see closely her brother's interesting activities, slaps a leaf down by his side, forming a graceful arc in space as she does so, rather like a card-player slamming a card onto the table. It might be by chance alone that she has selected a saliently long leaf (rather than a roundish one), the proportions of which are reminiscent of those of the little logs she desires.

I mentioned earlier (page 109) the glimmering possibility that Rah might make use of leaves as a kind of 'coinage' and perhaps this could be an example here. Is she trying to make an exchange with Nkosi? Rah's long, yellow leaf retains a single likeness to the log – its salient length. In human use a token of exchange, when this last remnant of likeness is shed it becomes the abstract currency of symbolisation. Perhaps money does, after all, grow on trees.

Rah then again tries to nestle into Nkosi's cradling arms, as if to be a part of the action and also to gain his attention and affection.

Nkosi places both sticks on the platform, gets up and strolls away from them as if ignoring them, only to abruptly turn and return to them, as if creeping up on them [in a behaviour reminiscent of Rah's creeping up on her stick playmate – pages 118–119]. Interestingly, he does not take a direct path toward them but chooses the long way round, the first part of his return journey following Edge 4 of Word Platform and so pursuing a vector aimed at missing the sticks. Only on turning the right-angle at Corner 4 does he at last pursue a course which will take him directly to the sticks. When he arrives near them he picks one up and sits with his back to me. He does not want me to see what he is doing.

Using an upward twanging gesture of a hand Rah abruptly vibrates the nearby fire-hose, perhaps to attract Nkosi's attention and initiate a play; perhaps just as an excited demarcation of her trajectory, or perhaps both. She immediately reverses direction and bounds back to the Word Platform giving its edge a double-slap of both front-hands, in a one–two, rhythmic sequence, demarcating her moment-of-arrival and perhaps announcing her return to Nkosi.

With his left hand, Nkosi holds one stick vertically with its burnt end against the platform with its unburnt end pointing upward. With his right hand, and with perfect accuracy, he hammers the upturned charcoal spike of the first stick with the unburnt, blunt end of the other one. He bangs it down eight times maintaining a moderate tempo with the one stick against the other, causing the hammer end of his hammer-stick to collide congruently with the upper end of his 'nail' or, better, 'punch' stick, which is held stationary in his right hand. Stopping, he spreads open his knees, so that he can see the effects of his actions on the wooden surface.

Rara bounds over to his left side to have a look at what her brother is doing. Nkosi pauses for a moment, as if considering. Then he strikes his hammer-stick again against the other stick which is still held in his left hand and still pressed vertically against the platform. He hits the one stick against the other a total of twenty-two times, with a short pause after the tenth strike and a longer rest after the twenty-second. He puts down the hammer-stick and closely studies the charcoal-end of the other stick, the end which has received the blows. I think he is investigating the composition of the charcoal. It is as if he is wondering what the relationship is between this black stuff and the rest of the usual wood? Or, put more simply: *How did the wood get like this?*

After a pause of a couple of seconds he then thumps the stick against the wooden surface upon its unburned end. Again he seems to consider for a few seconds. He turns to look toward me once, with what seems to be a little smile on his lips, as if to include me in his performance and then, with the charcoal-stick in his left hand, he makes the motions of banging the stick experimentally towards seven (?) distinctly different positions on the wooden platform. These are aimed strokes, each with its own separate target. The first three strokes are pretend stabs, made in space and soundless, controlled to halt abruptly a few centimetres above the board, as if stopped at an invisible surface.

After these strokes, he pauses for a second and glances toward me and only then resumes vertical arcing movements. It is as though he has checked to see that I am watching carefully, perhaps even seeking my approval, before striking the surface properly. He uses sharp yet gentle taps but this time fully striking the wooden platform.

'Good boy, Nkosi', I say to him, inserting, as it were, my words within the pause in his action. He then spins on his ischial callosities to his right, to strike the platform three times with his stick, again quite gently. Then he turns his left profile toward me and, choosing another location, stabs the charcoal-stick down three times in an unbroken rhythm, the first two strikes at the same spot but the third one purposely made at a different spot, at Corner 4, a few extra centimetres longitudinally away from him, requiring him to lean forward to make his fullest reach. This third and final stroke strikes almost exactly at Corner 4 of Word Platform, its collision making a sharp, woody sound.

Now he reaches right over toward the sawn-off top of Look-Out Post 2. With the charcoal-stick in his left hand he strikes this flat surface in a series of four vertical arcs. His fourth and last hit slows in tempo and becomes slightly less of a stab and more of a scrape; the stick dragged toward him. He pauses, brings the charcoal-stick down in front of him and seems to trail the stick in a leftward-going arc, against the board, charcoal-end down. He looks down at his action and possibly at its effects.

He turns to look at the sawn-off top of Look-Out Post, against which he struck the charcoal, presumably looking at charcoal powder and marks left there, though, again, stupidly, frustratingly, I cannot see them from my station-point.

About a metre or so below, Rah dodges around to try to get another view of what Nkosi is doing. Exchanging the charcoal-stick to his right hand, Nkosi strikes the sawn-off, flat top of the Look-Out Post twice in succession and then attempts a third stroke but allows the stick to glide down the side of the post in a glancing blow. A barely discernible, vertical smudge appears.

I say into the camera's microphone: 'Oh – he was nearly drawing then!'

The stick hangs loosely in his right hand as he looks at the side of the post. He continues looking there for a couple of seconds. He returns the stick to his lap-space and looks down at it.

Rah looks down at Nkosi's big charcoal-sticks. Bipedal, and with her back to him, she walks backwards, in one of her Silly Walks, into his chest. She nuzzles in close to him and he allows her to toy with one of the two charcoal-sticks rather in the manner which he himself did moments before, wobbling it around within her hands as it rests, roughly vertical, on one of its ends on the platform.

Nkosi holds one stick up to his eyes and looks up at it. He holds it laterally now before his eyes; held in big hands blackened with charcoal. Rah comes up to his right and touches her left fist to the now sooty, stubby end of the stick. She then places this left hand over her brother's right hand; her little hand upon his big one, the sooty fingers of which are coiled around the big stick. She uses both her front-hands to hang upon his hands in this way, looking at him and at what he is doing.

Rah cuddles up to him, her face in his chest and her arms around him. He now teases her by poking her gently with downward pressure of the charcoal-stick. They wrestle affably for a few seconds and then she disappears to the right of the Look-Out Post. She returns immediately however and, in comic slow-motion, walks bipedal and backward into her brother's chest once more; her shoulders swaying from side-to-side in a mock, power display but made backwards, before finally plonking herself again into his lap-space. She makes an alternating tattoo of her front-hands against the floor of the platform, as if emulating or paraphrasing the vertical arcing actions Nkosi just now performed with the charcoal-stick.

Spots before the eyes

Nkosi, supporting himself with his right hand against the platform and with the big stick held in his left hand, palmer-fashion, leans over to his right and strikes the sawn-off top of the post with the stick, twice in succession and then a third time on the side of the post, 4 or 5 cm below its rim. He continues the series of vertical arcs, striking the side of the post seven times in all and making the same number of black spots appear, one at a time. (See Figures 32.11 to 32.12.)

Figure 32.11 Nkosi (7 years 2 months 3 days) makes a black spot on the post with his charcoal stick.

Figure 32.12 He makes several more spots.

Looking closely at the sequence in slow-motion video, the combined forcefulness and accuracy of his marking action becomes apparent. The first dot is made just below the rim of the sawn-off top. His second strike is initially poised over the exact same position as the first spot, but though the stick hovers there for a moment, he changes his mind and chooses a separate location, about 15 cm to the right and lower down, to which he eventually aims his stick. Detailed, video-recorded observations of very young children making vertical arcs with a marker also reveal similar decision-changes in midflight (Matthews, 1997). The main difference between the young child's and the chimpanzee's marking action is that *the chimpanzee's is the more controlled and accurate*. Of course, he is no longer an infant but a juvenile.

The third of Nkosi's jabs is made just 5 cm to the right of the first one; the fourth one made about the same distance from the second one and in a vertical crack in the post, so that the charcoal splinters a little more; the fifth one coincides with fourth one, making an especially dense, black spot. The sixth one is made 2 or 3 cm below the third one and the seventh and last strike is made very close to the same third spot, just underneath it, so close to it that the spots just join.

After he has completed these, he puts his face up close to the highest couple of dots and sucks off some of the charcoal deposit. Rah then jumps onto the top of the post and looks down at the spots. She puts her black finger-tips near the black spots. Then she presses her lips onto the lower spot and licks it. When she moves her face away, we can see the spot is now wetly smudged.

Nkosi now makes off with his little sister Rara on his back, hurtling down into the heart of the Construction. I cannot see if he retains a charcoal-stick.

33 Ida draws

This is the day I ask the Keepers to leave the magenta-coloured, soft cotton-bag containing five plywood boards and five pieces of charcoal in the grass by Hollow Rock. The charcoal sticks are this time shop-bought, artists' charcoal. The bag has got partially wet but this turns out to be an advantage for the research because of the way Ganga makes use of it. This is the day when she rubs this pink remnant in a clear horizontal arc to wipe away the grime from a small area of the underside of a platform (see page 58). As I described above, her action results in an arc-shaped patch, clearly distinguishable against the mottled dirt of the rest of the board, as a small segment of restored wooden surface, brown-painted and shiny.

This action and its resultant shape are exactly those produced by little children in some of their earliest paintings and drawings and which I call the *horizontal arc*. This is a First Generation Marking Action, but on this same day I see skilful drawing far beyond even Third Generation Marking, produced spontaneously by Ida. What is more, she seems to purposely set out to demonstrate, to me, her understanding of drawing. Here's how it happens (see Figures 33.1 to 33.3).

Figures 33.1–33.3 Ida draws. She organises her drawing surface with respect to its steadiness and her viewpoint to it. She single-handedly rotates the charcoal around in her hand so that she can draw with its sharper point. She draws delicate lines and dots. She demarcates the end of one line with a squiggle.

Figure 33.2

Figure 33.3

06 November 2008 Video-recorded observation (raw footage on DVD) Singapore (Tape 1)

Rah (3 years 10 months 29 days) struggles for long minutes to tear open the bag whilst Ida, her mother looks on. Rara is concerned that her mother will take the bag from her but in fact Ida is allowing her daughter to try on her own. It is only after Rah has spent several very frustrated minutes struggling ineffectually with the bag that Ida decides to step in to help.

Ida (34 years 2 months 22 days) holds the open end of the bag to her mouth and bites at the edges of two or three boards, withdrawing them a few centimetres from the bag as, with her two front-hands, she pulls the bag as far open as it will go and peers into it. Rah stands up close to her mother to get a better view of the inside of the bag.

Then Ida does something else which really interests me, which is to stand up on foot-hands and start to walk deliberately towards me, carrying the red bag in her right, front-hand. It is always a special occasion for me when a chimpanzee seeks out my company. Rah hurries after her.

Ida glances up at me as she walks in my direction, negotiating the stone stairway of domed boulders down to the water's edge. Walking now fully bipedal, she holds the bag with two hands, so that the boards inside are horizontally aligned to her stomach, like a waitress with a tray. She looks at me again before sitting down on a fallen log by Hollow Rock at a position only a few metres from me, holding the bag now vertically, with its opening at the top, so that Rah (standing bipedal) can also peer into it.

Finally, Ida makes a determined effort to open the bag. With her teeth gripping a corner of its opened end whilst at the same time her left foot-hand grasps the edges of the boards which poke out, she pulls back with extreme force upon the bag's open end. With these combined, oppositional forces she succeeds in ripping the bag open along its seam.

Sticks of charcoal tumble out into the long grass and Rah notices and almost catches one stick as it falls but then allows the stick to drop, deciding, after all, that her main interest is not the charcoal or the boards but the red bag.

Once again, Ida sits back on her haunches and allows her daughter to try to complete the opening of the bag. Rah inserts her head into its rosy interior, a happy grin on her face, whilst Ida retrieves two or three stacked boards.

It may be that Ida intended Rah to have a share of the boards and charcoal but now that the young female has got her hands on the now empty red bag, she has already lost interest in these other things. Rah drops the wet, red cloth over her upturned face for a moment and then, transferring it to a front-hand, makes off with the remnant, climbing up the tree close by.

Ida glances at me for a moment, then at her disappearing daughter before turning her attention to the boards she holds in her front-hands. The boards, two in her left hand, one in her right, are inclined together roughly like the pages of an opened book, not quite touching at the line where the binding would be in a book. Ida now 'opens up' these wooden pages so that she can look more directly at their flat, smooth surfaces. Then she glances up again to me. 'Well done, Ida', I respond. She smiles softly at me.

00:10:54 Having dropped the boards from her left hand, Ida returns her attention to the one board she still retains in her right hand and, reaching into the grass with her left hand, retrieves a piece of charcoal. In a single, fluid action she brings it, and the board, simultaneously up into her field of view. Her right hand holding the board at its bottom right corner, she positions it almost horizontally in the air in 'portrait' orientation to her and, without hesitation, she draws, on its surface, three or four tentative and light lines, the longest of which is perhaps 4 cm. The lines are approximately parallel with a gap of roughly 2 cm between them.

Then, with stunning dexterity, she rotates the charcoal around through a vertical axis, end to end in her hand, without the help of her other hand, so that what was the non-drawing end of the charcoal becomes the drawing end. Careful scan-by-scan analysis of the video shows that Ida has executed this skilful manoeuvre because she notices that what was originally the non-drawing end of the charcoal has a sharp point formed by a splinter of charcoal. It would appear that she is seeking greater accuracy in her drawing and choosing and using this finer point rather than the blunt, flat, other end of the charcoal. [(Later, trying this manoeuvre myself I find that she is better at it than me.) Of interest is that Paul Schiller (1951) notes that Alpha refused to draw with blunt pencils.]

Ida then returns her line of sight to the board's surface and recommences drawing. She makes three more pull strokes. The little sharpened spike at the end of the charcoal touches the surface precisely, with enough pressure to make a mark, but not too much to break the delicate charcoal point.

'Good girl, Ida.' I say to encourage her.

She finds it uncomfortable to support the board in the air one-handedly from its lower corner so she tries to rest it on top of another board which is still resting on the log on which she sits. The problem now is that her left foot-hand is in the way, squashed between the two boards, making it difficult for her to keep her drawing firm and steady. For a moment she tolerates this and resumes drawing, making three or four little push–pull strokes. But the board on which she draws is too precarious so she contrives to lift her left foot-hand up onto the lower edge of the uppermost board – the one on which she draws. The board is at last flat and steady on the board beneath and almost in congruent alignment with it.

There is no doubt that she is taking the basic steps prior to the act of drawing. Organising the drawing surface and one's orientation to it *by one's self* is the first task for the draughtsperson. This act in itself is indicative of intent to draw.

Ida now presses down on the bottom left corner of the uppermost board (the one which bears her drawing) and uses the knuckles of her right front-hand to press against the lower right corner so as to make it steadier still.

Sometimes, human infants, making their first drawings, have similar difficulties in organising the relationship between marking instrument and drawing surface but they do not always solve these problems as well as Ida. Indeed, for the very young, an adult often arranges the equipment for the child. Part of the scepticism about animal art is the assumption that a human supervisor is required to organise the art materials (Lenain, 1997). 'Tying a brush to the tail of a donkey' is the ultimate derogatory metaphor touted to condemn non-human art.

Yet Ida knows exactly how to prepare herself and her drawing materials. Unlike human infants (I have in mind babies of around 1–2 years of age), she has a completely clear idea what adjustments are required in order to prepare a drawing surface for drawing and has the necessary skills to perform them. She is not of course a baby and does not draw like one. She is drawing carefully, accurately and skilfully as an intelligent adult, non-human primate.

She looks again into the enclosure to see what the others are up to, or perhaps to see if anyone of them has noticed what she is doing, before returning her eye-focus back to her drawing and making five exquisitely fine dots on the board, clustered at around the same locale.

She makes a highly delicate, pushing and pulling action making a little squiggle at the end of one of her previous lines.

With her right hand she then reaches down into the grassy gulley in which boards and charcoal earlier tumbled and, single-handedly and in a single movement, picks up another plywood board along with a second stick of charcoal. She rests this new board congruently upon her other two boards. She passes her new piece of charcoal from right hand to left, where it is held alongside her first stick of charcoal. She holds the new stick of charcoal between first and second finger of this left hand. The other stick, the first stick, is held in the same hand in a waiting position, like she would hold a spare stick-tool in the junction between her thumb and base of index finger. She lowers the new stick, to her new drawing surface. She makes more highly delicate, pushing and pulling actions.

Then she withdraws her hand from the board and just sits on her behind, on the log. She looks toward me and I say: 'Very good, Ida. Good girl.'

Then she stops, lifts the charcoal completely away from the board and again looks away into the enclosure before abruptly looking down at her drawing to make four delicate, little pull lines with the charcoal. The first line is perhaps 4 cm long, the second about 2 cm, the third a barely extended black point whilst the fourth and final mark is a stationary, zero-dimensional point, with the charcoal retained at this position for about a second, perfectly frozen in space and time.

She withdraws her hand from the board and just sits on her behind, on the log. She looks toward me and I say: 'Very good, Ida. Good girl.' She remains looking at me for a few moments as I repeat, 'Good girl'.

She then puts the stick lengthways between her lips. She takes up the board with her latest drawing on it and grips it in her teeth, alongside the stick of charcoal. She starts to knuckle-walk away but after a few steps, has second thoughts and turns back. She walks back and picks up another piece of charcoal, placing it alongside the piece already between her lips so that they rest there parallel to each other. She then manages to replace the board in her teeth at its widthways edge and in this fashion, walks off into the Construction with her charcoal and her drawing.

Two boards are left, almost congruently stacked, on the log on which she sat. Is it just luck that the boards are left inclined conveniently toward my camera's line of sight with the uppermost board and her first drawing, quite clearly configured on the light-coloured plywood? There are six or seven little marks, two of which form clear lines a few centimetres long and one of which has its end demarcated by a little squiggle.

I am beginning to suspect that she knew that I wanted her to make use of the slates and markers. She elected to come down to Hollow Rock to draw, as close as it is possible to be, to me, the human observer, there by the moat. She also seemed to understand that I wanted a clear view of the surface of the board and of her drawing. Does she even leave the finished drawing so that I have a clear view of its surface? We have seen that she and some of the other chimpanzees do arrange clear views of surfaces for themselves and sometimes present clear demonstrations for the gaze of others. Is she in fact doing the drawing for me?

There is, of course, no way of knowing.

What is clear is that Ida understands the first steps in a visual-praxis commonly assumed understood and used only by humans. Her drawing actions are exactly that; they are not the mere by-product of other actions issuing from other intentions; they constitute the genuine, graphic act. After trying the blunt end of the charcoal she turns the stick around and uses its sharp, pointed end which she directs to the board, taking care not to break the fragile tip. She sets out to make lines and dots on a flat surface, which she makes perfectly, whilst looking at what she is doing.

Her very delicate 'in-flight' adjustments of the charcoal are made in the fingers of her left hand with little accompanying support from her other hand which still performs its job of pressing down on the bottom right corner of her drawing board to hold it steady. All of these subtle and skilful in-flight adjustments are made whilst her clear motivation and intention to draw remains uninterrupted. She keeps her mind on the job whilst readjusting her tools.

A line with a squiggle at the end

Demarcated line ending

I mentioned the Jon Amiel film (*Creation*) in which a young orang-utan draws for Charles Darwin. Darwin, (played by Paul Bettany) is writing in a little notebook and the orang-utan comes over to him, takes his pencil and makes clear push–pull markings. In an interview Bettany explains that this was the orang-utan's idea. This is a Second Generation Marking very like that made by little children. Ida demarcates the end of one of her lines with a similar push–pull squiggle. Earlier, we noted that when very young children become interested in the beginnings and ends of lines they often mark them with dots, spots or squiggles. As I also mentioned earlier, demarcated line ending and line beginning is also produced in some elegant paintings by Congo (Morris, 2009, personal communication).

An important difference between Ida's drawing and a comparable one by a human infant is that Ida's line with the little squiggle on top end is produced more carefully, accurately and thoughtfully than those of most moderately skilful human infants in their first drawings.

Ida is a greying, adult female but, as my observations of the infant chimps indicate, even as a baby, Ida would have had more accuracy and motor control than most human infants. Whether, with all her knowledge and years of experience she puts *more* thought into her drawing than would a human infant is an intriguing question to ask! Exactly what those thoughts might be is another one.

Conclusion

Future research should be able to confirm the hypothesis suggested from observational evidence contained in this book that drawing and writing actions are common to both human and non-human primates and issue 'naturally' from the same concerns shared by both species.

When human infants make their first lines they immediately become very interested in them and seem driven to investigate them further. It is not long before they realise the expressive, representational and symbolic possibilities of this activity (long, long before they make recognisable 'pictures'). I have not so far found this to be true of my chimpanzees.

But I could be wrong. This is but a small sample of captive animals observed over a moat and an electric fence for only just over three years. We know very little about what they are capable of and if I could spend longer with such a group who knows what I might see them achieve?

In any case, contrary to most popular opinion and assumptions, mark-making seems to be a significant aspect of chimpanzee life if only in the sense that they deploy it as one of a battery of tests to which they subject novel objects and surfaces. Chimpanzees know about making-marks and they have reasons for making them, just like human infants, to investigate things and to uncover the causal relationships within phenomena. But I think there is more to this than rudimentary survival strategy. Both humans and chimpanzees (and perhaps other members of the primate group, for example, the orang-utan), have feelings and ideas about *shape, location* and *movement* and are driven by a new kind of desire (Vygotsky, 1966, 1986) arising out of *play* to investigate these properties and even use them as the means of expression and, possibly, representation and symbolisation. These early conceptual concerns in chimpanzees seem essentially the same as those of humans, present even at the dawning of neonate cognition and they may form the basis of language (Maher, 1996; Stokoe, 1960). If definitional problems have obscured the nature of very young human children's emergent expressive and representational thought then it should come as no surprise that our understanding of semiotic thought in apes is so limited and hindered by ancient prejudices.

In both human and chimpanzee development, expression, representation and symbolisation grow out a rich background of exploratory, investigative and ludic activities in which patterns of movement in space and time are organised and given meaning.

Although the chimpanzees described in this book are captive beings and the way in which they encounter mark-making materials is obviously different from the way they might happen upon them in the wild, nevertheless I feel that there is sufficient evidence to suggest that essentially similar occurrences could take place in nature with natural materials.

Remember that my chimpanzees have no human contact. The way in which they investigate, use and play with materials (albeit some of which are supplied by a human) is not an artificial contrivance by a human.

On the other hand, I believe that my sympathetic and empathic watching, and my talking to them as intelligent beings did influence them. The fact of this influence would, for some people, invalidate my research findings. This is because they use a research paradigm in which the observer is supposedly – *supposedly* – cut off from the subject as if looking at him or her through a one-way window. Although there is much to be learnt from the checks and balances of the classical scientific approach I do not want to work within this experimental paradigm. Because of my background in working and playing with very young children, in which it is impossible to separate oneself from the observation, I accept my own presence, and analyse my act of watching and my influence too as part of what is going on in the 'ordinary moment' (Forman, Hall, & Berglund, 2001).

Likewise, the criticism of anthropomorphism is a non-issue for me. Chimpanzees *are* like us. There is no getting away from it. One way to understand what they are doing (and certainly not the only way) is to imaginatively identify with them and ask questions like, 'Well, why would Rara do that? What would I being thinking of if I was doing what she is doing? What would my purpose be if I was doing such and such an activity? How would I do that? How would I feel about it?' There is no need to invoke a mysterious 'theory of mind' conjured up by the usual suspects, magic 'modules', unique to the human brain (Leslie, 1987; Leudar & Costall, 2009). With careful and humble observation the chimpanzee's intentions and motivations can be seen in the actions themselves. The art and science of describing and interpreting these actions is in its infancy. I hope there is still time to fully develop it.

Of course, this methodology only works if an enormous amount of honest and painstakingly careful observation is undertaken. My many years of work with young children have helped illuminate the actions of the chimpanzees. Although my approach remains taboo for many scientists, to me, it seems impossible to attain a full understanding without utilising empathic imagination. Without this, knowledge gleaned from experimental studies (about problem-solving, language, numeracy, memory and more) will never fit into a coherent whole. Yes, of course there are dangers in this approach. As Sue Savage-Rumbaugh concedes, yes, of course we will make mistakes and yes, we should retain scepticism.

The issue of anthropomorphism can be turned on its head anyway. It is in dire need of re-casting (Costall, 1998). It seems to me that our present troubles are directly linked to a much grander anthropomorphism which has caused us to sever our connection with our animal heritage. This is a mistake which has led us to the brink of catastrophe. My time in Singapore Zoo has changed me. I now see that all our ideas, thoughts and feelings are *primate* ideas, thoughts and feelings. Likewise, at a deep level of description, if we accept that chimpanzee intelligence is similar to ours then we must also realise that *chimpanzee limitations are our limitations*. We need to recognise this right now and face up to it with intelligence and courage.

But whatever mistakes readers find in this book, the work of an enthusiastic amateur, there remains a strong correlation between chimpanzee development in emergent symbolisation and that which I have observed in human infancy. It is so striking that it cannot be ignored. The drawing activities of the chimpanzees in this study and my recording of them may be sparse and nowhere near as exciting as some captive chimpanzees but the animals take sufficient interest in making marks to warrant our further consideration of what this implies. It is not simply enough to note that the paintings of captive apes look a bit like modern abstract painting and then a little later realise that they don't and give up all further enquiry. What kind of science is that?

As with the drawing and painting of very young children, looking at chimpanzee art in abstract-expressionist terms is no use. One needs to understand early trace-making as an important part of a family of emergent expressive and representational modes. I have tried to show in this book that, as in human childhood development, the chimpanzees show organised and systematic thinking which runs through all their ludic and exploratory and investigative use of media and that, as with very young children, the structure of their actions is echoed in the drawing actions they make. The younger chimpanzees especially (Makhulu, Manzu, Nkosi, Rara and Kimoni) seemed to work through an orderly sequence of structural investigations expressed in terms of dynamic configurations made in space and time. The intentions and motivations behind the making of shapes in space seem to be of the nature of a visual, rhythmic and musical aesthetic. The organisation of these spatiotemporal patterns suggests the presence of enactive-representation and emergent measurement; the topology of the 'knot' being one example.

They understand line-of-sight. They understand something of the optical sample and have techniques to establish field-of-view. They recognise electronic, videographic representations of themselves too. The dawning shock of realisation they seem to experience when they study their portraits in the LCD screen of my video-camera recalls that of very young children when they first encounter their own videographic representation and meet themselves face-to-face in liquid crystal or TV.

Chimpanzees make a little three-dimensional construction and produce the occasional drawing. I believe that these activities are closely related to linguistic thought. It is possible that drawing and writing even precedes, anticipates or at least accompanies the development of language, long, long ago in deep time. Perhaps the act of trace-making stimulated the development of the 'trace' of memory (Derrida, 1976).

Although it might be argued that my chimpanzees, unlike my human children, take only a passing interest in their drawing and constructive activities, we do not yet know what would happen if drawing materials (along with other constructive media) were made available to them for extended periods. My studies of the development of expression, representation and symbolisation in human children shows that if the child is free to return again and again to the materials and media in which his or her thinking is embedded his or her symbolic development is greatly enhanced (Donaldson, 1978). In the wild, chimpanzees are free to return to the same materials over and over again and this may be the main reason why some groups of wild chimpanzees attain high levels of inventiveness, for example, in tool-use (Holmes & Gunton, 2009; Sanz & Morgan, 2007).

Humans have a great deal to learn from the chimpanzee about education. In human children's play and investigation it is precisely the type of amorphous play-material used by the chimpanzee – the piece of string, the wooden block, the card-board box, the blanket, the stick, the marker, the lump of clay (and, I would add, the computer and the video-camera) which, coupled with the interest of 'more experienced learner' (Blenkin, 1990) proffers an abundance of semiotic affordances. We need to know more about the forms of 'scaffolding', or, better, *synergy* which foster and promote cognitive and affective development. We should look at the support given by more experienced or older chimpanzees to their young in their acts of investigation, play and expression. I have a feeling that this aspect of chimpanzee culture, the support given by older more, experienced companions for the young (and not only by the mothers) expressed in acts of giving and sharing, not just of food and things but of time, interest and ideas too, may have been vastly underestimated. This might tell us a great deal about an interactional and ecological approach to learning and development. Such a study might show how closely inter-connected are those processes we term learning, development and evolution, even, perhaps, that they are different aspects of the same process.

Unfortunately, a developmentally referenced approach to children's development and learning is under threat by many governments today because (horror of horrors), not only does it look suspiciously like doing nothing, worse still, it is dangerously successful (Berliner & Biddle, 1995; Kelly, 1990; Simon, 1988).

This book has been about more than just mark-making and drawing activities. I have placed these medium-specific behaviours within the origin and development of symbolic thought as a whole. I suggest that, just as in human childhood, mark-making activities are part of a larger project which involves the whole body and the whole mind. This project comprises the construction of complex action and play-assemblies in space and time which form the precursors for the primates' dawning use of symbols.

As in human childhood, mark-making activities for the chimpanzee form a part of a larger family of emergent expressive, representational and symbolic systems. The way in which these specific activities are arranged in relation to other acts of expression and precursor representation is of immense interest. It shows that all human knowledge is a special form of primate knowledge. Although lifestyle differences confer differences upon the way in which the beginnings of symbolic thought might emerge in chimpanzee life, there nevertheless persist striking similarities between the developmental trajectories of semiotic thought in both humans and chimpanzees.

What is the best way of thinking about this? We know that it would be a mistake to think of the chimpanzees as primitive humans. As regards emergent capacities to symbolise, it is tempting to measure the chimpanzee's development against that of the human. In doing so there is the danger of arriving at the conclusion that the chimpanzee's is an unfinished teleology. Yet, even given endless stretches of evolutionary deep time, there is no guarantee that chimpanzees would eventually end up using symbols in the way in which we do – evolution does not seem to repeat itself. Chimpanzees and humans are two separate species. We didn't evolve from chimpanzees; both they and we share a common ancestor who passed away about 6 million years ago.

Yet, whilst we should guard against the temptation of thinking of chimpanzees as semi-evolved people, the fact that we are so closely related to them means that cautious comparisons might throw light on our own origins and destiny. Chimpanzees are sufficiently close to us for us to ask intriguing questions about the origin of our own symbolic abilities. This is what I have done in this book.

By studying the chimpanzees I have also been re-learning about development in human childhood. Discovering the chimpanzees forced me to start again ('from scratch') a lifetime's work on children. Again, despite differences in the lives of humans and chimpanzees, a significant one being the chimpanzees' ability to move freely through three dimensions of space plus the dimension of time, the teleology of both human infantile development and that of the chimpanzee bear striking similarities.

But here's where it gets tricky. For a start, there is the difference in life-span, the chimpanzee living about 40 years (perhaps 50 in a well-designed zoo with sensitive humans); the human about 80. This makes it problematic to compare what we might think of, for want of a better term, as 'developmental stages'. To cap it all, the chimpanzee's infantile, sensorimotor mark-making is more accurate and controlled than that of the human infant!

However, both the chimpanzee and the human infant seem to pass through a similar sequence of precursor semiotic activities which may be characterised in a number of ways. One way is to think of it as a journey embarked on firstly through topological and dynamic action-assemblies and then which passes through an orderly sequence of spatiotemporal investigation and play to Euclidean and finally projective understandings. In humans this pathway is mapped, step-by-step, onto the drawing surface and I offered slight evidence that chimpanzees may travel some of this journey too, though, so far, no one has observed any *projective* understandings encoded in chimpanzee drawing.

This is but one way of looking at a process which is multilayered like a palimpsest, written over and over in many overlapping ways. The study of visual representation in terms of a straight and narrow road to 'correct' representation has effectively masked the full meaning of its development in children. If then, in our study of human development, definitional issues continue to occlude the true nature and significance of our own children's emergent symbolisation it is hardly surprising that the truth about the origins of symbolisation in our close cousins, the chimpanzees, and perhaps in ourselves before we even became human, remains hid from our eyes.

In my previous work I suggested a way of breaking through what Carolee Fucigna (1983) has called 'definitional problems' in the way in which we discern and describe the development of children's expressive and representational thought. This entails abandoning unquestioned assumptions about art and representation and, rather than measure children's performance against standards and objectives which exist only in the mind of the adult, instead looks carefully at the meaning and structure of the activities which children themselves spontaneously generate through infancy. Starting in this way allows us to see the semantic and organisational characteristic of self-motivated actions children perform naturally and with great interest. These actions have a syntactical structure which persists through development, undergoing changes as the child returns to them again and again. The actions embody emergent semiotic thought. The work of Chris Athey has been most helpful in identifying the modes of embedded thought through which children move and it is this approach I have applied to the chimpanzees.

The action-assemblies I have described exhibit organisational and semantic characteristics akin in the most important respects to those of the actions of play and investigation entwined throughout in human development. Although chimpanzee linguistic capabilities are in dispute, their navigational understandings and the expressive purposes to which their high-speed, three-dimensional activities are put, recall linguistic structure. Perhaps, during evolutionary time linguistic understandings were built upon navigational abilities. Whilst, despite the extraordinary work by the Gardners, Roger Fouts, Sue Savage-Rumbaugh and others the idea that some apes may have language abilities remains controversial it does seem apparent to me that there are secondary levels of meaning in the chimpanzees' organisation and use of body action.

Stephen Mithen suggests that the beginnings of language may have emerged in the form of wholistic 'musical' phrases (as opposed to the idea that language may have started with individual 'words') (Mithen, 2005). This theory seems to make sense in the light of my own observations of the chimpanzees' use of expressive action-assemblies. My chimpanzees' pre-verbal patterns of play seem organised in a musical way. Increased complexity of these play-ensembles coupled with the capability to memorise these as passages which may be repeated and varied as dynamic-structures of interest in themselves make possible ensembles akin to language which may have 'themes' and 'subjects' and which possesses syntax, timing, rhythm and recursion as well as 'asides', 'digressions', 'abbreviations', repetition and even more subtle characteristics, such as prosody, stress, nuance, cadence, humour and even irony. The chimpanzees perform highly patterned action sequences which are very memorable to them. We have seen that these play patterns are by no means rigid or routinised – quite the reverse. They are variable, complex and capable of endless permutation.

Such rich structure lends itself to a polysemic evolution, in which actions of the primate body not initially symbolic in themselves may gradually acquire layers of meaning. That the beginning of language may have been expressed in wholistic phrases shows the sense of Sue Savage-Rumbaugh's and Roger Lewin's initial research question about whether the chimpanzee (or bonobo) has any concept of what a 'word' actually is (Savage-Rumbaugh & Lewin, 1994). Curiously, many researchers of ape language have overlooked this fundamental question.

To be able to make both rhythmic, musical phrases and also break these down into their constituent parts may have been an important step in the evolution of symbolic capacity. That, like human children, chimpanzees do seem able to both construct organised dynamic form and yet be able also to deconstruct it into individual, re-arrangable components would seem to offer a new starting point for primate studies.

An important consequence of thinking of the development of meaning in this way is that when we speak of 'knowledge' what we mean is *primate knowledge*. The things we understand, know and do are not god-like but species-specific – what we know and do is what this particular primate wants and needs to know and do. This knowledge is acquired with the help of fellow primates.

All of this suggests that symbolic systems are learnt just like everything else. There is no moment before which one does not use symbols and after which one does. Representation does not appear out of the blue, Tommy Cooper fashion, 'just like that' (as Alan Costall, 2010, aptly puts it). The development of symbolisation is a continuous process of many small steps starting from actions which are not initially symbolic in themselves. None of the chimpanzees' play and action-assemblies is instinctive but all arise out of cultural, social and interpersonal interaction. In a similar way, language is probably not hard-wired but 'made learnable' for the young by an older companion (Costall, 2009, personal communication).

Future research is best advised to follow the natural development of expression and representation as embodied in the actions the apes make in space and time; the way they extend and amplify these and the developmental variations they make. These are properly the precursors of symbolisation including language, drawing and writing.

The limits of observation

It is true that I didn't have to trek through the forests of Africa for years to obtain these data. Yes, to make this study meant arising before sunrise and making a 20-min walk carrying many kilograms of video and other equipment to the MRT train station. Once aboard the super-efficient Singapore train however it is another 20-min ride to an historic place called Kranji and from there it is a short taxi ride to the zoo. From the zoo entrance I then lug all this stuff along to the chimpanzee enclosure, a final 20-min walk in the now already hot sun or, occasionally, the torrential rain. Yes, of course you need to be a bit tough but it isn't too bad – not as field trips go!

I was never in any physical danger; there were no ravines to fall down, no fierce animals to eat me, no terrorists to kill me or kidnap me. The occasional snake kept out of my way though I was sometimes bitten by the odd mosquito. But, aside from a couple of swathes of diseases which swept the world at this time and provided I washed off any traces of 'zoo-poo' there was little risk of catching a terrible disease faraway from any hospital. Even to get something to eat, I only had to trudge as far as 'Bongo-Burgers' and back.

But Singapore is a tougher jungle (of the concrete kind) than you might think and, as many others will tell you, surviving it is not as easy as it might look. In any case, what I have written in these pages is as accurate as possible a record of the captive chimpanzees I got to know.

Canals on Mars

Perhaps none of the above is true. Perhaps I have made a bad mistake. Perhaps it is nothing but the imaginings of a lonely man in a strange land. After all, it's happened before. Percy Lowell so wanted to see the remains of Martian culture that through his telescope he saw the signs of a complex and intelligent culture struggling to save its dying planet. He too was looking for non-accidental structure and he thought he found it in the form of the *straight line*; the so-called 'canals' of Mars. He made the common error of projecting his own obsessions through his own telescope and, with the assistance of tiny imperfections on his own lens, saw these obsessions inscribed in the dust of a distant planet (Lowell, 1906).

Or perhaps I'm guilty of the exaggerations of the amateur, non-human primate enthusiast, R. L. Garner (1896). He thought he could understand the 'words' of chimpanzee 'language'. He is famous for having a cage constructed in the African jungle so that, from inside it, armed to the teeth with rifle and pistol, he could learn about gorilla culture. Perhaps his enthusiasm got the better of him and he exaggerated his findings. I have sincerely tried to avoid this temptation.

I put these examples in because I am aware of these pitfalls and I want to forestall criticism. Unchecked mental projection or even just plain, careless exaggeration is destructive to science.

But just as dangerous is the sort-of born-again Cartesianism which runs through much contemporary scientific thinking (Costall, 1998). Whatever the nature of the cage I sit in or even the dark glass through which I look, all the ideas in the book come from direct-observations which are as truthful as is humanly possible. The observations are made from a tourists' path in a public zoo. I see only what anyone else can see. I do not use truly expensive technology. My methodology is to simply stand in sun and rain and watch. In this sense, the data collection is made in more or less the same way as that made in my child studies. It is all there for anyone to see. You just need a little humility. I do not invent anything. I do not exaggerate anything either (even when I tell you the effort nearly killed me). Whatever blunders I have made whilst thrashing about in this jungle of words, this is really what I saw – or what I thought I saw – in Singapore Zoo.

References

Amiel, J. (Director). (2009). *Creation* [Feature film]. London: Recorded Picture Company.

Armstrong, D. F., Stokoe, W. C., & Wilcox, S. E. (1995). *Gesture and the nature of language*. Cambridge, UK: Cambridge University Press.

Athey, C. (1977). Humour in children related to Piaget's theory of intellectual development. In A. J. Chapman & H. C. Foot (Eds.), *It's a funny thing, humour*. Oxford: Pergamon.

Athey, C. (1990/2007). *Extending thought in young children: A parent–teacher partnership*. London: Paul Chapman.

Atkinson, D. (2002). *Art in education: Identity and practice*. Dordrecht, The Netherlands: Kluwer Academic Publishers.

Bekoff, M. (1995). Play signals as punctuation: the structure of social play in canids. *Behaviour, 132*, 419–429.

Berliner, D. C., & Biddle, B. J. (1995). *The manufactured crisis: Myths, fraud, and the attack on America's public schools*. Reading, MA: Addison-Wesley.

Biederman, I. (1987). Recognition-by-components: A theory of human image understanding. *Psychological Review, 94*, 115–147.

Blenkin, G. M., & Kelly, A. V. (Eds.). (1988/1996). *Early childhood education: A developmental curriculum*. London: Paul Chapman.

Blenkin, G. M. (1990). Goldsmiths College, University of London [personal communication].

Bower, T. G. R. (1974/1982). *Development in infancy*. San Francisco, CA: Freeman.

Bowlby, J. (1953). *Child care and the growth of love*. London: Penguin.

Boysen, S. T., & Kuhlmeier, V. A. (2002). Representational capacities for pretense with scale models and photographs in chimpanzees (*Pan troglodytes*). In R. W. Mitchell (Ed.), *Pretending and imagination in animals and children* (pp. 210–228). Cambridge, UK: Cambridge University Press.

Brisley, S. (1975). *Moments of decision/indecision* [6 day performance, Warsaw].

Bruce, T. (1987). *Early childhood education*. London: Hodder & Stoughton.

Bruce, T. (1991). *Time to play in early childhood education*. London: Hodder & Stoughton.

Bruner, J. S. (1964). The course of cognitive growth. *American Psychologist, 19*, 1–15.

Bruner, J. S. (1972). The nature and the uses of immaturity. *American Psychologist, 27*, 687–708. Reprinted in Bruner, J. S., Jolly, A., & Silva, K. (Eds.). (1976) *Play: – Its role in development and evolution*. Harmondsworth, UK: Penguin.

Bruner, J. S. (1996). *Toward a theory of instruction*. Cambridge, MA: Harvard University Press.

Budsworth, G. (2004). *The ultimate encyclopedia of knots and ropework*. London: Hermes House.

Cantlon, J., & Brannon, E. (2007). How much does number matter to a monkey (*Macaca mulatta*)? *Journal of Experimental Psychology: Animal Behavior Processes*, *33*, 32–41. Reported in Dingfelder, S. F. (2007, March). Monkey math. *Monitor on Psychology*, *38*(3), 14–15.

Chapman, A. J., & Foot, H. C. (Eds.). (1977). *It's a funny thing, humour*. Oxford, UK: Pergamon.

Child, D. (Director). (2010, 6 January). The secret life of the dog. In *Horizon* [Television series]. London: BBC.

Chomsky, N. (1966). *Cartesian linguistics*. New York, NY: Harper & Row.

Condon, S. (2010). Speech makes babies move. In R. Lewin (Ed.), *Child alive* (pp. 81–90). London: Temple Smith.

Cooter, S. (Producer and Director). (2010, February 10). To infinity and beyond. In *Horizon* [Television series]. London: BBC.

Costall, A. (1985). How meaning covers the traces. In Freeman, N. H. & M. V. Cox (Eds.), *Visual order: The nature and development of pictorial representation* (pp. 17–30). Cambridge: Cambridge University Press.

Costall, A. (1993). Beyond linear perspective: A Cubist manifesto for visual science. *Image and Image Computing*, *11*(6), 334–341.

Costall, A. (1995). The myth of the sensory core: The traditional versus the ecological approach to children's drawings. In C. Lange-Kuettner & G. V. Thomas (Eds.), *Drawing and looking: Theoretical approaches to pictorial representation in children* (pp. 16–26). Hemel Hempstead, UK: Harvester.

Costall, A. (1998). Lloyd Morgan, and the rise and fall of 'animal psychology'. *Society & Animals*, *6*(1), 13–29.

Costall, A. (2001). Introduction to George Luquet's (1927) *Children's drawings ('Le Dessin enfantin')*. London and New York: Free Association Books.

Cousteau, J. Y., Dumas, F., & Dugan, J. (1953). *The silent world*. New York: Harper & Row.

Davila Ross, M., Owren, M. J., & Zimmermann, E. (2009). Reconstructing the evolution of laughter in Great Apes and humans. *Current Biology*, *19*, 1106–1111.

Davis, H., & Memmott, J. (1983). Autocontingencies: Rats count to three to predict safety from shock. *Animal Learning & Behaviour*, *11*, 95–100. Reported in Dingfelder, S. F. (2007, March). Monkey math. *Monitor on Psychology*, *38*(3), 14–15.

de Glanville, S. (2009). *Wildlife finder* (www.bbc.co.uk/wildlifefinder). Reported by Martin Wainwright (2009, September 2009). *The Guardian*, p. 12.

DeLoach, J. S. (2000). Dual representation and young children's use of scale models. *Child Development*, *71*, 329–338.

Derrida, J. (1976). *Of grammatology* (G. C. Spivak, Trans.). Baltimore, MD: Johns Hopkins University Press.

de Waal, F. (1999, November 19). Apes with an oeuvre. *Chronicle of Higher Education*.

de Waal, F. (2000). *Chimpanzee politics: Power and sex among apes*. Baltimore, MD: Johns Hopkins University Press.

de Waal, F. (2005). *Our inner ape*. New York, NY: Riverhead Books.

Diamond, J. (2002). *The rise and fall of the third chimpanzee: How our animal heritage affects the way we live*. London: Vintage.

Donaldson, M. (1978). *Children's minds*. Glasgow: Fontana.

Eisner, E. (2006, April 8–12). *Connoisseurship, criticism and the art of education*. Paper presented on receiving the Palmer O. Johnson Memorial Award at the American Educational Research Association Annual Conference, San Francisco, CA.

Fitch, T. (Contributor). (2009, November 10). Why do we talk. In *Horizon* [Television series]. London: BBC.

Flombaum, J., Junge, J. A., & Hauser, M. D. (2005). Rhesus monkeys (*Macaca* mulatta) spontaneously compute addition operations over large numbers. *Cognition, 97*, 315–225. Reported in Dingfelder, S. F. (2007, March). Monkey math. *Monitor on Psychology, 38*(3), 14–15.

Foer, J., & Nichols, L. (2010, February). The truth about chimps. *National Geographic.* Retrieved from http://ngm.nationalgeographic.com/2010/02/goualougo/goualougo-animation

Forman, G., Hall, H., & Berglund, K. (2001). *The power of ordinary moments.* Child-Care Information Exchange, PO Box 3249, Redmond, WA 98073. Available from www.ChildCareExchange.com.

Fouts, R. (with Tukel Mills, S.). (1997). *Next of kin: My conversations with chimpanzees.* New York, NY: William Morrow.

Freeman, N. H. (1980). *Strategies of representation in young children.* London: Academic Press.

Freeman, N. H., & Cox, M. V. (Eds.). (1985). *Visual order: The nature and development of pictorial representation.* Cambridge, UK: Cambridge University Press.

Fucigna, C. (1983). Research proposal: MA Thesis. Tufts University, MA.

Galdikas, B. (2005). *Great Ape odyssey.* New York, NY: Harry N. Abrams.

Gallup, G. G. Jr. (1970). Chimpanzees: Self recognition. *Science, 167,* 86–87.

Gardner, A., & Gardner, B. (1966). *Project Washoe.* Reno, NV: University of Nevada.

Garner, R. L. (1896). *Gorillas and chimpanzees.* London: Osgood McIlvaine.

Gibson, J. (1979). *The ecological approach to visual perception.* Boston, MA: Houghton Mifflin.

Gibson, J., & Yonas, P. (1967). *The development of graphic activity in the child: A theory and a first experiment.* 'Purple perils', unpublished manuscript, Cornell University, NY.

Gibson, J., & Yonas, P. (1968). A new theory of scribbling and drawing in children. In H. Levin, E. J. Gibson, & J. J. Gibson (Eds.), *The analysis of reading skill.* Washington, DC: U.S. Department of Health, Education and Welfare, Office of Education.

Goodall, J. (2000). *In the shadow of man.* Boston, MA and New York, NY: Mariner Books.

Gray, H. (1978). Learning to take an object from the mother. In A. Lock (Ed.), *Action, gesture and symbol: The emergence of language* (pp. 159–183). London: Academic Press.

Guillaume, P. (1926/1971). *Imitation in children* (E. P. Halperin, Trans.). Chicago, IL: University of Chicago. (Original work *L'imitation chez l'enfant* published 1926).

Gumert, M. (2009, personal communication) E-mail discussion, Nanyang Technological University, Singapore.

Hanna, W., & Barbera, J. (Directors). (from 1940). *Tom & Jerry* [Animated cartoon films]. Century City, LA: MGM Film Studios.

Hayes, C. (1951). *The ape in our house.* New York, NY: Harper.

Herrelko, B. (Contributor). (2010, January 27). The chimpcam project. In *Natural World* [Television series]. London: BBC.

Hess, E. (2008). *The Chimp Who Would Be Human.* Bantam: New York.

Hicks, B. (1991). Grassy knoll. In *Revelations* [Stand-up comedy]. London: Dominion Theatre.

Holmes, M., & Gunton, M. (Producers). (2009, September 29). *Life* [Television programme]. London: BBC.

Jackendoff, R. (1994). *Patterns in the mind: Language and human nature.* New York, NY: Basic Books.

Janson, H. W. (1961). The 'image made by chance' in Renaissance thought. In M. Meiss (Ed.), *Essays in honor of Erwin Panofsky* (pp. 254–266). New York: New York University Press.

Jarvis, P. (2007). Monsters, magic and Mr. Psycho: A biocultural approach to rough and tumble play in the early years of primary school. *Early Years*, *27*, 171–188.

Jessel, J., Arju, T., Gregory, E. and Kenner, C. (2004). Children and their grandparents at home: a mutually supportive context for learning and linguistic development. *English Quarterly*, *36*(4), 16–23.

Kaminski, J. (Contributor). (2010, January 6). The secret life of the dog [Television series episode]. In D. Child (Director), *Horizon*. London: BBC.

Kapoor, A. (2009). Exhibition of the sculpture of Anish Kapoor, Royal Academy of Arts, 26 September to 11 December 2009.

Kellogg, R. (1955). *What children scribble and why*. Palo Alto, CA: National Press.

Kelly, A. V. (1990). *The National Curriculum: A critical review*. London: Paul Chapman.

Keynes, R. (2001). *Annie's box: Charles Darwin, his daughter and human evolution.* London: Fourth Estate.

Khots, N. (1935/2002). *Infant ape and human child. A classic 1935 comparative study of ape emotions and intelligence*. New York, NY: Oxford University Press.

Kitson, R. (2009, October 20). Why Tweddle is a more worthy champion than Button. *The Guardian*, p. 12.

Koestler, A. (1964). *The act of creation*. Harmondsworth, UK: Penguin.

Kohler, W. (1925). *The mentality of apes*. London: Kegan Paul.

Kress, G. (1997). *Before writing: Rethinking the paths to literacy*. London: Routledge.

Lenain, T. (1997). *Monkey painting*. London: Reaktion Books.

Leslie, A. M. (1987). Pretense and representation: The origin of 'Theory of Mind'. *Psychological Review*, *94*, 412–426.

Leudar, I. and Costall, A. (Eds.) (2009). *Against theory of mind.* Basingstoke, UK: Palgrave Macmillan.

Lorenz, K. (1944–1948). *The natural science of the human species: An introduction to comparative behavioral research (The Russian Manuscript, 1944–1948)*. Cambridge, MA: MIT Press.

Lowell, P. (1906). *Mars and its canals*. New York: Macmillan.

Lowenfeld, V. (1955). *Your child and his art: A guide for parents*. New York, NY: Macmillan.

Luke, A. (2007). Pedagogy as a gift. In J. Albright & A. Luke (Eds.), *Pierre Bourdieu and literacy education* (pp. 121–169). London: Taylor & Francis.

Luquet, G. H. (1927/2001). *Children's drawings* ['*Le Dessin enfantin*'] (A. Costall, Trans. and commentary). London and New York: Free Association Books.

MacKarell, P. (1988). *The big jump*. Lecture to post graduate Art Education students, Goldsmiths College, University of London.

Maher, J. (1996). *Seeing language in sign: The work of William C. Stokoe*. Washington, DC: Gallaudet University Press.

Marshall, A. (2009, November 11). Andrea: Queen of the mantas. In *Natural World* [Television series]. London: BBC.

Matsuzawa, T. (2007). Numerical recollection in chimpanzees. Study by Tetsura Matsuzawa, Director of the Primate Research Institute, Kyoto University, Japan. Reported in Randerson, J. (2007, December 4). *The Guardian*.

Matthews, J. (1984). Children drawing: Are young children really scribbling? *Early Child Development and Care*, *18*, 1–39.

Matthews, J. (1990). *Expression, representation and drawing in early childhood.* PhD thesis, University of London, London.

Matthews, J. (1994). Deep structures in children's art: Development and culture. *Visual Arts Research, Fall*, 10–25.

Matthews, J. (1997). Liang ge dong xi bu yi yang (Mandarin-Chinese for 'Two objects not the same'): A study of how Singaporean children between ages 2 and 5 years differentiate, in drawing, between a sphere and an elongated, straight-sided ovoid. *Visual Arts Research, Spring*, 73–96.

Matthews, J. (1999). *The art of childhood and adolescence: The construction of meaning.* London: Falmer.

Matthews, J. (2003). *Drawing and painting: Children and visual representation.* 0–8 Series, Ed. T. Bruce. London: Paul Chapman.

Matthews, J. (2004). The art of infancy. In E. Eisner & M. Day (Eds.), *Handbook of research and policy in art education* (pp. 253–298). Mahwah, NJ: Lawrence Erlbaum Associates.

Matthews, J. (2005). 'Where's Poppy?' A 2-year-old watches herself on television: Representation, e-presentation and time. *Early Childhood Practice: The Journal for Multi-Professional Partnerships, 7*, 42–58.

Matthews, J. (2006a). Very young children's development in moviemaking. *Mind, Culture and Activity, 13*, 129–155.

Matthews, J. (2006b). Presentation, representation and e-presentation in early childhood. *International Journal of Education through Art, 2*, 171–182.

Matthews, J. (2010). Scribble: The development of children's mark-making. In B. Wagoner (Ed.), *Symbolic transformations: The mind in movement through culture and society* (pp. 209–231). London: Taylor & Francis.

Maurer, D., & Riboni, C. (2010). *Early graphic expressions.* 'Wie Bilder "entstehen".' UNESCO Project.: Zurich University of the Arts, Zurich, Switzerland: Verlag Pestalozzianum.

Michotte, A. (1963). *The perception of causality.* Methuen Manual of Modern Psychology. London: Methuen.

Miller, G. (2000). *The mating mind.* New York: Random House.

Mitchell, R. W. (Ed.). (2002a). *Pretending and imagination in animals and children.* Cambridge, UK: Cambridge University Press.

Mitchell, R. W. (Ed.). (2002b). Imaginative animals, pretending children. In R. W. Mitchell (Ed.), *Pretending and imagination in animals and children* (pp. 3–22). Cambridge, UK: Cambridge University Press.

Mithen, S. (2005). *The singing Neanderthals: The origins of music, language, mind and body.* Phoenix, AZ: Orion.

Monty Python. (1969). *Monty Python's Flying Circus* [Television series]. London: BBC.

Morecambe, E. (Writer). (1971). *The Morecambe and Wise Christmas show* [Television programme]. London: BBC.

Morris, D. (1962). *The biology of art: A study of the picture-making behaviour of the Great Apes and its relationship to human art.* New York: Borzoi Books, Alfred A. Knopf.

Morris, D. (1967). *The Naked Ape.* London: Jonathan Cape.

Morris, D. (2008). *Baby: A portrait of the amazing first two years of life.* London: Hamlyn.

Morris, D., with Parker, S. (2009). *Planet ape.* London: Mitchell Beazley.

Morris, D. (2009, December 7). Email with photograph of Congo with large painting showing vertical lines with spots on top ends [personal communication].

Osvath, M. (2009). Spontaneous planning for future stone throwing by a male chimpanzee. *Current Biology*, *19*(5), R190–R191.

Piaget, J. (1951). *Play, dreams and imitation in childhood.* London: Routledge and Kegan Paul.

Piaget, J., & Inhelder, B. (1956). *The child's conception of space*. London: Routledge and Kegan Paul.

Pratt, F. (1983). Intellectual realism in children's copies of cubes and straight lines. In D. Rogers & J. A. Sloboda (Eds.), *The acquisition of symbolic skills* (pp. 53–63). New York, NY: Plenum.

Price, C. (2009, November 10). *Why do we talk?* In *Horizon* [Television series]. London: BBC.

Rauschenberg, R. (2005–2006). *Robert Rauschenberg: Combines* [Exhibition]. New York: Metropolitan Museum of Art.

Reddy, V. (1991). Playing with others' expectation: Teasing and mucking about in the first year. In A. Whiten (Ed.), *Natural theories of mind* (pp. 143–158). London: Blackwell.

Reiss, D. (2010, February 18–22). *Self-awareness and dolphins*. Paper presented at Advancing Science, Serving Society 2010 Annual Meeting, San Diego Convention Center.

Reiss, D., & Marino, L. (2001). Mirror self-recognition in the bottlenose dolphin: A case of cognitive convergence. *PNAS*, *98*, 5937–5942.

Resnick, M. (2002, July 27). *Learning through playful invention and exploration.* Seminar for Science and Technology Education, National Institute of Education, Nanyang Technological University, Singapore.

Reynolds, P. C. (1976). Play, language, and human evolution. In J. Bruner, A. Jolly, & K. Silva (Eds.), *Play: Its role in development and evolution* (pp. 621–635). Baltimore, MD: Penguin Books.

Reynolds, P. C. (2002). Pretending primates: Play and simulation in the evolution of primate societies. In R. W. Mitchell (Ed.), *Pretending and imagination in animals and children* (pp. 196–209). Cambridge, UK: Cambridge University Press.

Sanz, C., & Morgan, D. (2007). Chimpanzee tool technology in the Goualougo Triangle, Republic of Congo. *Journal of Human Evolution*, *52*(4), 420–433.

Sanz, C., Schoning, C., & Morgan, D. (2009). Chimpanzees prey on army ants with specialised tools. *American Journal of Primatology*, *72*, 17–24.

Savage-Rumbaugh, S., & Lewin, R. (1994). *Kanzi: The ape at the brink of human kind.* New York, NY: Wiley.

Savage-Rumbaugh, S., Shanker, S. G., & Taylor, T. J. (1998). *Apes, language and the human mind.* Oxford: Oxford University Press.

Schiller, P. H. (1951). Figural preferences in the drawings of a chimpanzee. *Journal of Comparative and Physiological Psychology*, *44*, 101–111.

Schneemann, C. (2002). *Imaging her erotics: Essays, interviews, projects*. Cambridge, MA: MIT Press.

Schuster, G., Smits, W., & Ullal, J. (2007). *Thinkers of the jungle: The orangutan report, pictures, facts, background.* Pottsdam: h. f. ullman in der Tandom Verlag GmbH.

Simon, B. (1988). *Bending the rules: The Baker 'reform' of education.* London: Lawrence & Wishart.

Smith, N. R. (1983). *Experience and art: Teaching children to paint.* New York, NY: Teachers College Press, Columbia University.

Stamatopoulou, D. (2009). *Symbol formation and the embodied self: A microgenetic case-study examination of the transition to symbolic communication in scribbling activities from 14 to 31 months of age*. Unpublished manuscript, University of Crete.

Stetsenko, A. (1995). The psychological function of children's drawing: A Vygotskian perspective. In C. Lange-Kuttner & G. V. Thomas (Eds.), *Drawing and looking: Theoretical approaches to pictorial representation in children* (pp. 147–158). London and New York, NY: Harvester Wheatsheaf.

Stokoe, W. (1960). *Sign language structure: A outline of the visual communication systems of the American deaf.* Studies in Linguistics No. 8. New York, NY: Department of Anthropology, University of Buffalo.

Sylva, K., Roy, C., & Painter, M. (1980). *Childwatching at playgroup and nursery school.* London: Grant MacIntyre.

Terrace, H. S. (1979). *Nim: A chimpanzee who learned to sign.* New York: Alfred A. Knopf.

Thelen, E., & Smith, L. B. (1994). *A dynamic systems approach to the development of cognition and action.* Cambridge, MA: MIT Press.

Tomasello, M., Kruger, A. C., & Horn Ratner, H. (1993). Cultural learning. *Behavioural and Brain Sciences, 16*, 495–552.

Trevarthen, C. (1995). Mother and baby – seeing artfully eye to eye. In R. Gregory, J. Harris, P. Heard, & D. Rose (Eds.), *The artful eye* (pp. 157–200). Oxford, UK: Oxford University Press.

Trevarthen, C. (2003, June 28). *Feeling chuffed: Mother and baby conversation.* Keynote paper presented at the Pen Green Research and Development Centre for Early Childhood, Corby, Northamptonshire, UK.

Trivedi, S., & Mahnot, S. M. (2002, February). Functional status of pinna muscles in hanuman langur *Semnopithecus entellus*: Speculation on functional loss of pinna muscles in humans. *Primate Report 63*, p. 47.

Vygotsky, L. S. (1966). Play and its role in the mental development of the child. *Soviet Psychology, 12*(6), 62–76.

Vygotsky, L. S. (1986). *Thought and language*. Cambridge, MA: MIT Press.

Wartofsky, M. W. (1980). Visual scenarios: The role of representation in visual perception. In M. A. Hagen (Ed.), *The perception of pictures* (Vol. 2, pp. 131–152). New York, NY and London: Academic Press.

Willats, J. (1997). *Art and representation: New principles in the analysis of pictures.* Princeton, NJ: Princeton University Press.

Willats, J. (2005). *Making sense of children's drawings*. London: Taylor & Francis.

Winner, E. (1986). Where pelicans kiss seals. *Psychology Today, 20*(8), 24–35.

Winnicott, D. (1971). *Playing and reality*. London: Tavistock.

Wolf, D. (1983, April). *The origins of distinct symbolic domains: the waves of early symbolization. The example of event structuring.* Paper presented at the Annual Meeting of the Eastern Psychological Association, Philadelphia, April 1983, Harvard Project Zero, Longfellow Hall, HGSE Cambridge, MA.

Wolf, D., & Fucigna, C. (1983). *Representation before picturing.* Paper presented at the Symposium on Drawing Development, British Psychological Society International Conference on Psychology and the Arts, University of Cardiff, Cardiff, Wales.

Wolf, D., Rygh, J., & Altschuler, J. (1984). Actions and states in play narratives. In I. Bretherton (Ed.), *Symbolic play* (pp. 195–217). London: Academic Press.

Worthington, M., & Carruthers, E. (2006). *Children's mathematics: Making marks, making meaning* (2nd ed.). London: Paul Chapman.

Young, S. (2000). *Young children's spontaneous instrumental music-making in nursery settings.* PhD thesis, Roehampton Institute of Education, University of Surrey, UK.

Author index

Subject index

For Product Safety Concerns and Information please contact our EU representative GPSR@taylorandfrancis.com
Taylor & Francis Verlag GmbH, Kaufingerstraße 24, 80331 München, Germany

www.ingramcontent.com/pod-product-compliance
Ingram Content Group UK Ltd.
Pitfield, Milton Keynes, MK11 3LW, UK
UKHW022007180926
475701UK00003B/243
9781138877610